Lecture Notes in Computer Scie

Commenced Publication in 1973
Founding and Former Series Editors:
Gerhard Goos, Juris Hartmanis, and Jan van Leeuwen

Editorial Board

S. Arumugam   W. F. Smyth (Eds.)

Combinatorial Algorithms

23rd International Workshop, IWOCA 2012
Tamil Nadu, India, July 19-21, 2012
Revised Selected Papers

 Springer

Volume Editors

S. Arumugam
Kalasalingam University
Anand Nagar, Krishanakoil
Tamil Nadu, 626 126, India
E-mail: s.arumugam.klu@gmail.com

W. F. Smyth
McMaster University
Algorithms Research Group
Department of Computing and Software
Hamilton, ON L8S 4K1, Canada
E-mail: smyth@mcmaster.ca

ISSN 0302-9743   e-ISSN 1611-3349
ISBN 978-3-642-35925-5   e-ISBN 978-3-642-35926-2
DOI 10.1007/978-3-642-35926-2
Springer Heidelberg Dordrecht London New York

Library of Congress Control Number: 2012954779

CR Subject Classification (1998): G.2.1, G.2.2, I.1, I.3.5, F.2, E.1, E.4, H.1

LNCS Sublibrary: SL 1 – Theoretical Computer Science and General Issues

Typesetting: Camera-ready by author, data conversion by Scientific Publishing Services, Chennai, India

Printed on acid-free paper

Springer is part of Springer Science+Business Media (www.springer.com)

Preface

This volume contains the papers presented at IWOCA 2012, the 23rd International Workshop on Combinatorial Algorithms.

The 23rd IWOCA was held July 19–21, 2012, at Kalasalingam University (KLU), in rural Tamil Nadu, India, an hour's flight plus an hour's car ride southwest of the state capital, Chennai. The meeting was sponsored and supported financially by the Department of Science and Technology, Government of India, New Delhi; by the Council of Scientific and Industrial Research, New Delhi; and by KLU. It was organized by n-CARDMATH, the National Centre for Advanced Research in Discrete Mathematics of India, whose director, S. Arumugam, was both Co-chair of the Program Committee (with W. F. Smyth) and Chair of the Local Arrangements Committee.

IWOCA descends from the original Australasian Workshop on Combinatorial Algorithms, first held in 1989, then renamed "International" in 2007 in response to consistent interest and support from researchers outside the Australasian region. The workshop's permanent website can be accessed at http://www.iwoca.org/ where links to previous meetings, as well as to IWOCA 2013, can be found.

Using LISTSERVE and other e-mail lists, the IWOCA 2012 call for papers was distributed around the world, resulting in 88 submission. The EasyChair system was used to facilitate management of submissions and refereeing, with three referees from the 45-member Program Committee assigned to each paper. A total of 21 papers (24%) were accepted, subject to revision, for presentation at the workshop, with an additional nine papers accepted for poster presentation. Four invited talks were given:

- Naveen Garg, "Approximation Algorithms for Graphical-TSP"
- Gonzalo Navarro, "Indexing Highly Repetitive Collections"
- Rajeev Raman, "Range Extremum Enqiries"
- Saket Saurabh, "Polynomial Time Preprocessing using Min-Max Theorems in Combinatorial Optimization"

These proceedings contain all 21 presented papers, together with shortened versions of the nine poster papers and extended versions of two of the invited talks.

The workshop also featured a problems session, chaired — in the absence of IWOCA Problems Co-chairs Yuqing Lin and Zsuzsanna Lipták — by Arumugam. The IWOCA problem collection can be found on-line at http://www.iwoca.org/main_problems.php

The 72 registered participants at IWOCA 2012 hold appointments at institutions in 12 different countries on five continents (Africa, Asia, Europe, North America, South America). The nations represented were:

Bangladesh (3), Canada (4), Chile (1), Denmark (1), France (1), India (53), Iran (1), Poland (2), South Africa (1), Taiwan (1), UK (1), USA (3).

Of the 32 papers published here, 19 are principally in the area of graph theory, seven in combinatorics on words, four in applications to algorithms and data structures, and two in miscellaneous combinatorial applications. The papers are grouped by these topic areas in this volume, ordered within each group by the first author's surname.

We thank the authors for their valuable combinatorial contributions and the referees for their thorough, constructive, and enlightening commentaries on the manuscripts.

September 2012   S. Arumugam
                                                                 W.F. Smyth

Organization

Program Committee

Faisal Abu-Khzam	Lebanese American University, Lebanon
Don Adjeroh	West Virginia University, USA
Amihood Amir	Bar-Ilan University and Johns Hopkins University, Israel/USA
Subramanian Arumugam	Kalasalingam University, India
Hideo Bannai	Kyushu University, Japan
Ljiljana Brankovic	University of Newcastle, Australia
Gerth Stølting Brodal	Aarhus University, Denmark
Sunil Chandran	Indian Institute of Science, India
Charles Colbourn	Arizona State University, USA
Maxime Crochemore	King's College London, London, UK and Université Paris-Est, France
Diane Donovan	University of Queensland, Australia
Dalibor Froncek	University of Minnesota Duluth, USA
Roberto Grossi	Università di Pisa, Italy
Sylvie Hamel	University of Montreal, Canada
Jan Holub	Czech Technical University in Prague, Czech Republic
Seok-Hee Hong	University of Sydney, Australia
Costas Iliopoulos	King's College London, UK
Ralf Klasing	LaBRI - CNRS, France
Rao Kosaraju	Johns Hopkins University, USA
Marcin Kubica	Warsaw University, Poland
Thierry Lecroq	University of Rouen, France
Mirka Miller	University of Newcastle, UK
Laurent Mouchard	University of Rouen, France
Ian Munro	University of Waterloo, Canada
Kunsoo Park	Seoul National University, South Korea
Simon Puglisi	Royal Melbourne Institute of Technology, Australia
Jaikumar Radhakrishnan	Tata Institute of Fundamental Research, India
Sohel Rahman	Bangladesh University of Engineering and Technology
Rajeev Raman	University of Leicester, UK
Venkatesh Raman	The Institute of Mathematical Sciences, India
Frank Ruskey	University of Victoria, Canada
Joe Ryan	University of Newcastle, Australia
Joe Sawada	University of Guelph, Canada
Michiel Smid	Carleton University, Canada
William F. Smyth	McMaster University, Canada

Venkatesh Srinivasan   University of Victoria, Canada
Iain Stewart   Durham University, UK
German Tischler   University of Wuerzburg, Germany
Alexander Tiskin   University of Warwick, UK
Lynette Van Zijl   Stellenbosch University, South Africa
Ambat Vijayakumar   Cochin University of Science and Technology,
                           India
Koichi Wada   Nagoya Institute of Technology, Japan
Sue Whitesides   University of Victoria, Canada
Christos Zaroliagis   CTI and University of Patras,
                           The Netherlands/Greece

Additional Reviewers

Ann, Hsing-Yen
Baier, Jan
Basavaraju, Manu
Brlek, Srecko
Diwan, Ajit
Dorbec, Paul
Farràs, Oriol
Feria-Puron, Ramiro
Feroz, Jesun Sahariar
Ferreira, Rui
Foucaud, Florent
Frangioni, Antonio
Froncek, Dalibor
Ghosh, Subir
Golovach, Petr
Goswamy, Partha P.
Goto, Keisuke
Hahn, Gena
Hajiabadi, Mohammad
Holub, Premek
Hsieh, Sun-Yuan
Huang, Chien-Chung
Izumi, Taisuke
Johnson, Matthew
Kamat, Vikram
Kamei, Sayaka
Katayama, Yoshiaki
Kolar, Josef
Kontogiannis, Spyros
Landau, Gad
Lanzi, Leonardo

Lin, Yuqing
Lodaya, Kamal
Marino, Andrea
Matsuzoe, Hiroshi
Miller, Michael
Mishra, Sounaka
Mohar, Bojan
Mondal, Debajyoti
Ono, Hirotaka
Perez-Roses, Hebert
Phanalasy, Oudone
Philip, Geevarghese
Plosila, Juha
Poliak, Martin
Polách, Radomír
Pradhan, Dinabandhu
Prencipe, Giuseppe
Proietti, Guido
Pyatkin, Artem
Rejikumar, K
Saha, Tanay Kumar
Semanicova-Fenovcikova, Andrea
Sillasen, Anita Abildgaard
Singh, Tarkeshwar
Tarantilis, Christos
Thomborson, Clark
Togni, Olivier
Trávníček, Jan
Valicov, Petru
Vetta, Adrian
Zerovnik, Janez

Table of Contents

Bounds on Quasi-Completeness

Malay Bhattacharyya and Sanghamitra Bandyopadhyay

Machine Intelligence Unit, Indian Statistical Institute
203 B. T. Road, Kolkata - 700108, India
{malay_r,sanghami}@isical.ac.in

Abstract. A graph $G = (V, E)$ is γ-quasi-complete ($\gamma \in [0, 1]$) if every vertex in G is connected to at least $\gamma.(|V| - 1)$ other vertices. In this paper, we establish some relationships between the girth and the quasi-completeness of a graph. We also derive an upper bound $\frac{1}{2}\left(1 + \frac{r}{\gamma}\right) + \sqrt{\frac{1}{4}\left(1 + \frac{r}{\gamma}\right)^2 + \frac{2|E|}{\gamma} - \frac{r|V|}{\gamma}}$ for the largest order γ-quasi-complete subgraph in a graph of minimum degree r.

1 Basic Definitions and Preliminaries

Throughout this paper, the term graph is used to denote an unweighted and undirected simple graph (without self-loops or parallel edges) $G = (V, E)$, where V and E are the vertex and edge sets, respectively [5]. The *degree* of a vertex v, denoted as $d(v)$ in a graph, is the number of edges incident to it. A graph is called *r-regular* if every vertex in the graph has *degree* r. Let us denote $\delta(G) = \min_{\forall v \in V} d(v)$, the minimum degree of a graph G. The *order* and *size* defines the cardinality of the vertex set and the edge set of a graph, respectively. A subgraph $G' = (V', E')$ of a graph G ($G' \subseteq G$) is defined such that $V' \subseteq V$ and $E' \subseteq E$. Let us assume that the cardinality of a set S is represented as $|S|$. Now, we have the following basic definitions.

Definition 1 (γ-quasi-complete graph). *A connected graph, $G = (V, E)$, is said to be γ-quasi-complete ($0 < \gamma \leq 1$) if every vertex in the graph has degree value at least $\gamma.(|V| - 1)$, i.e., $\delta(G) \geq \gamma.(|V| - 1)$.*

Definition 2 (γ-quasi-clique). *In a graph $G = (V, E)$, a subset of vertices $V' \subset V$ forms a γ-quasi-clique ($0 < \gamma \leq 1$) if the subgraph induced by V', $G_{V'}$, is a γ-quasi-complete graph.*

Fig. 1(a) shows an example of a quasi-clique in a quasi-complete graph. There are some other versions of the problem for an n vertex subgraph with m edges by the terminology 'γ-clique', where $m \geq \frac{\gamma.n.(n-1)}{2}$ [1] or 'γ-near-clique', where $m \geq \frac{(1-\gamma).n.(n-1)}{2}$ [7], which are however different by definition. Quasi-cliques are generalization of the concept of a clique [6], and therefore, the problem of finding the maximum quasi-clique is computationally harder than the maximum clique problem. Note that, a γ-quasi-clique in any arbitrary graph G is always a

S. Arumugam and B. Smyth (Eds.): IWOCA 2012, LNCS 7643, pp. 1–5, 2012.

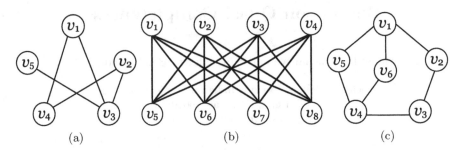

Fig. 1. (a) A $\frac{2}{3}$-quasi-clique induced by the set of vertices $\{v_1, v_2, v_3, v_4\}$ in a $\frac{1}{4}$-quasi-complete graph. (b-c) Some γ-quasi-complete graphs $(\gamma > \frac{1}{|V|-1})$ with *girth* > 3.

γ-clique or a $(1 - \gamma)$-near-clique, but not the vice versa. Therefore, the problem of finding the largest quasi-clique in a graph is computationally different than finding γ-cliques or γ-near-cliques. The decision version and the counting version of the maximum quasi-clique problem are known to be NP-complete and $\#P$-complete, respectively [9].

The quasi-completeness property is not necessarily closed under subgraphs. Note that 1-quasi-complete graph is a complete graph, whereas a 1-quasi-clique is a clique. A *path* in a graph is a sequence of edges such that every pair of subsequent edges share a common vertex. The length of a *path* is the number of edges it includes. These precursory details will be used in the subsequent sections in course of some novel investigations on quasi-cliques.

2 Some Properties

Observation 1. *A connected r-regular graph $G = (V, E)$ is γ-quasi-complete for $\gamma \leq \frac{r}{|V|-1}$.*

An example of this for $r = 2$ is the subgraph induced by the vertices $\{v_1, v_2, v_3, v_4\}$ in Fig. 1. A closed *path* starting and ending with the same vertex in a graph is defined as a *cycle* [5]. Since a connected graph with no *cycle* must have at least one pendant vertex (i.e., a vertex with *degree* 1), we have the following observation.

Observation 2. *Any arbitrary acyclic connected graph $G = (V, E)$, of order at least 2, is γ-quasi-complete for $\gamma \leq \frac{1}{|V|-1}$.*

The minimum length of a *cycle* contained in a graph is defined as its *girth* [5]. Let us define a graph $G = (V, E)$, *(r)-regular* if its *girth* is r and G is *r-regular*. Some interesting relations existing between the *girth* of a graph and its quasi-completeness is presented in the following theorem.

Theorem 1. *Let $G = (V, E)$ be a γ-quasi-complete graph of order at least 3 and $g(G)$ denotes the* girth *of G, then the maximum values γ can attain*

$$
\max \gamma =
\begin{cases}
1, & \text{if } g(G) = 3 \\
\frac{|V|}{2(|V|-1)}, & \text{if } 4 \leq g(G) \leq \left\lceil \frac{2|V|+1}{3} \right\rceil - 1 \text{ and } |V| \text{ is even} \\
\frac{1}{2}, & \text{if } 4 \leq g(G) \leq \left\lceil \frac{2|V|+1}{3} \right\rceil - 1 \text{ and } |V| \text{ is odd} \\
\frac{2}{|V|-1}, & \text{if } g(G) = \left\lceil \frac{2|V|+1}{3} \right\rceil \\
\frac{1}{|V|-1}, & \text{if } \left\lceil \frac{2|V|+1}{3} \right\rceil + 1 \leq g(G) \leq |V| - 1 \\
\frac{2}{|V|-1}, & \text{if } g(G) = |V| \\
\frac{1}{|V|-1}, & \text{if } g(G) = \infty.
\end{cases}
$$

Proof. For the case $g(G) = 3$, the graph G may be, at the largest of its *size*, complete. So, the desired upper bound derives.

When $g(G)$ lies between 4 and $\left\lceil \frac{2|V|+1}{3} \right\rceil - 1$ (both bounds inclusive), the upper bound of $\frac{\lfloor \frac{|V|}{2} \rfloor}{|V|-1}$ is derived for a ④-*regular* graph of even *order*. However for the graphs of odd *order*, this upper bound can be realized separately. Evidently for higher *girth* values, γ will decrease. A ④-*regular* graph of *order* 8 that is $\frac{4}{7}$-quasi-complete is shown in Fig. 1(b).

When $g(G) = \left\lceil \frac{2|V|+1}{3} \right\rceil$, the graph may be in the form of two connected *cycles* (the entire graph is a *cycle* with a single chord), sharing $\left\lceil \frac{|V|+1}{3} \right\rceil$ common vertices and each of them having a length of $\left\lceil \frac{2|V|+1}{3} \right\rceil$. Thus $\delta(G) = 2$, which gives an upper bound of γ to $\frac{2}{|V|-1}$. One such instance is shown in Fig. 1(c). The other possible formations of G will result in a lower value of γ.

For $\left\lceil \frac{2|V|+1}{3} \right\rceil + 1 \leq g(G) \leq |V| - 1$, G will certainly possess a pendant vertex and thus the result.

In the case $g(G) = |V|$, the entire graph is a single cycle and thus 2-*regular*. Hence, Observation 1 derives the required upper bound for γ.

The condition $g(G) = \infty$ arises only when the graph is acyclic, and for that case Observation 2 produces the result.   □

Remark. A *triangle* is a *cycle* of length 3 in a graph. Let $G = (V, E)$ be a triangle-free γ-quasi-complete graph. As G is triangle-free, any two vertices $v_1, v_2 \in V$ must satisfy the condition $d(v_1) + d(v_2) \leq |V|$, and for the γ-quasi-completeness, $d(v_i) \geq \gamma(|V| - 1), \forall v_i \in V$. Combining these two inequalities, we obtain $2\gamma(|V| - 1) \leq |V| \Rightarrow \gamma \leq \frac{|V|}{2(|V|-1)}$. Therefore, if a graph G be γ-quasi-complete for $\gamma > \frac{|V|}{2(|V|-1)}$, then it must contain a *triangle*. The similar result can also be obtained from the second relation produced in Theorem 1.

The diameter of a graph is the maximum *path* length between any pair of vertices. For any arbitrary graph G, the girth $g(G)$ and diameter $\mathcal{D}_G$ is known to satisfy the inequality $g(G) \leq (2\mathcal{D}_G + 1)$ [8]. We combine this relation with the upper bounds of $\mathcal{D}_G$ derived for a γ-quasi-complete graph in [9,10], with respect to the parameter γ and obtain the following result.

Lemma 1. *If $G = (V, E)$ is a γ-quasi-complete graph of order at least 3 and $g(G)$ denotes the girth of G, then*

$$g(G) \begin{cases} = 3, & \text{if } 1 \geq \gamma > \frac{|V|-2}{|V|-1} \\ \leq 5, & \text{if } \frac{|V|-2}{|V|-1} \geq \gamma \geq \frac{1}{2} \\ \leq 6\lfloor \frac{|V|}{\lceil \gamma(|V|-1)\rceil+1}\rfloor - 5, & \text{if } \frac{1}{2} > \gamma > \frac{1}{|V|-1} \text{ and } |V| \bmod (\lceil \gamma(|V|-1)\rceil) = 0 \\ \leq 6\lfloor \frac{|V|}{\lceil \gamma(|V|-1)\rceil+1}\rfloor - 3, & \text{if } \frac{1}{2} > \gamma > \frac{1}{|V|-1} \text{ and } |V| \bmod (\lceil \gamma(|V|-1)\rceil) = 1 \\ \leq 6\lfloor \frac{|V|}{\lceil \gamma(|V|-1)\rceil+1}\rfloor - 1, & \text{if } \frac{1}{2} > \gamma > \frac{1}{|V|-1} \text{ and } |V| \bmod (\lceil \gamma(|V|-1)\rceil) \geq 2 \\ \leq 2|V| - 1, & \text{if } \frac{1}{|V|-1} \geq \gamma > 0. \end{cases}$$

As it is known that the relation $\delta(G) \leq \left\lfloor \frac{2|E|}{|V|} \right\rfloor$ holds good for any arbitrary graph $G = (V, E)$ [5], we note the following.

Observation 3. *For any arbitrary graph $G = (V, E)$, $(\delta(G)|V| - 2|E|) \leq 0$.*

Let us assume that the maximum *order* of a γ-quasi-clique in any arbitrary graph G, i.e., the γ-*quasi-clique* number of the graph, is denoted as $\omega_\gamma(G)$. Then, we can derive an upper bound for $\omega_\gamma(G)$ as follows.

Theorem 2. *For a given graph $G = (V, E)$ with $\delta(G) = r$, we have*

$$\omega_\gamma(G) \leq \frac{1}{2}\left(1 + \frac{r}{\gamma}\right) + \sqrt{\frac{1}{4}\left(1 + \frac{r}{\gamma}\right)^2 + \frac{2|E|}{\gamma} - \frac{r|V|}{\gamma}}. \tag{1}$$

Proof. Let the maximum *order* of the γ-quasi-cliques in G be $\mathcal{Q}$. Then, the total number of edges in G contributed by a γ-quasi-clique of maximum *order* is at least $\frac{\mathcal{Q}}{2}\gamma(\mathcal{Q} - 1) = \frac{\gamma\mathcal{Q}(\mathcal{Q}-1)}{2}$. As the graph G has a minimum *degree* value of r, the vertices outside the γ-quasi-clique contribute at least $\frac{r(|V|-\mathcal{Q})}{2}$ edges in total. Thus, we have

$$|E| \geq \frac{\gamma\mathcal{Q}(\mathcal{Q}-1)}{2} + \frac{r(|V| - \mathcal{Q})}{2}.$$

Simplifying the above relation, we obtain

$$\gamma\mathcal{Q}^2 - (\gamma + r)\mathcal{Q} + r|V| - 2|E| \leq 0.$$

Solving this, the desired upper bound can be obtained. □

Remark. It follows that the function $f(\mathcal{Q}) = \gamma\mathcal{Q}^2 - (\gamma + r)\mathcal{Q} + r|V| - 2|E|$ derived earlier, monotonically increases with $\mathcal{Q}$ where $\mathcal{Q} \geq \frac{\gamma+r}{2\gamma}$, as we have $(r|V| - 2|E|) \leq 0$ (Observation 3). So, the upper bound derived in Eqn. (1) satisfies the monotonicity criterion for $\mathcal{Q} \geq \frac{1}{2}(1 + \frac{r}{\gamma})$.

Corollary 1. *For a given connected graph $G = (V, E)$, we have*

$$\omega_\gamma(G) \leq \frac{(\gamma + 2) + \sqrt{(\gamma + 2)^2 + 8\gamma(|E| - |V|)}}{2\gamma}. \tag{2}$$

Proof. The total number of edges in G contributed by a γ-quasi-clique of maximum *order*, say Q, is at least $\frac{Q}{2}\gamma(Q - 1) = \frac{\gamma Q(Q-1)}{2}$, and as because G is connected, the vertices outside the γ-quasi-clique contribute at least $(|V| - Q)$ edges. So, we have the inequality $|E| \geq \frac{\gamma Q(Q-1)}{2} + (|V| - Q)$, which by solving derives the required result. □

3   Conclusion

Finding almost dense subgraphs like quasi-cliques in a graph is interesting for many real-life applications [3,4]. We present some important properties of quasi-cliques and conclude with some open directions of study. The boundary relations between the *girth* and the γ value of a quasi-complete graph highlighted in Theorem 1 could be tightened. More rigid upper bounds on the *order* of the maximum quasi-cliques are yet to be explored, although there exists a few for maximum cliques [2]. Moreover, a new class of graphs, namely the $(\hat{r})$-*regular* graphs, emerges from our results that can be studied further.

References

1. Abello, J., Resende, M.G.C., Sudarsky, S.: Massive Quasi-Clique Detection. In: Rajsbaum, S. (ed.) LATIN 2002. LNCS, vol. 2286, pp. 598–612. Springer, Heidelberg (2002)
2. Amin, A.T., Hakimi, S.L.: Upper bounds on the order of a clique of a graph. SIAM J. Appl. Math. 22, 569–573 (1972)
3. Bandyopadhyay, S., Bhattacharyya, M.: Mining the Largest Dense Vertexlet in a Weighted Scale-free Graph. Fund. Inform. 96(1-2), 1–25 (2009)
4. Bhattacharyya, M., Bandyopadhyay, S.: Analyzing Topological Properties of Protein-protein Interaction Networks: A Perspective towards Systems Biology. In: Computational Intelligence and Pattern Analysis in Biological Informatics, pp. 349–368. John Wiley & Sons, Inc., Hoboken (2010)
5. Bollobás, B.: Modern Graph Theory. Graduate Texts in Mathematics, vol. 184. Springer, New York (1998)
6. Bomze, I.M., Budinich, M., Pardalos, P.M., Pelillo, M.: The Maximum Clique Problem. In: Du, D.Z., Pardalos, P.M. (eds.) Handbook of Combinatorial Optimization: Supplementary Volume A, pp. 1–74. Kluwer Academic, Dordrecht (1999)
7. Brakerski, Z., Patt-Shamir, B.: Distributed discovery of large near-cliques In: Proceedings of the 23rd International Conference on Distributed Computing, Elche, Spain, pp. 206–220 (2009)
8. Diestel, R.: Graph Theory. Graduate Texts in Mathematics, vol. 173. Springer, Heidelberg (2009)
9. Jiang, D., Pei, J.: Mining Frequent Cross-Graph Quasi-Cliques. ACM Trans. Knowl. Discov. Data 2(4), 16 (2009)
10. Pei, J., Jiang, D., Zhang, A.: On Mining Cross-Graph Quasi-Cliques. In: Proceedings of the 11th ACM SIGKDD, Chicago, Illinois, USA, pp. 228–238 (2005)

Infinite Random Geometric Graphs
from the Hexagonal Metric

Anthony Bonato[1] and Jeannette Janssen[2]

[1] Department of Mathematics
Ryerson University
Toronto, Canada
abonato@ryerson.ca
[2] Department of Mathematics and Statistics
Dalhousie University
Halifax, Canada
janssen@mathstat.dal.ca

Abstract. We consider countably infinite random geometric graphs, whose vertices are points in $\mathbb{R}^n$, and edges are added independently with probability $p \in (0,1)$ if the metric distance between the vertices is below a given threshold. Assume that the vertex set is randomly chosen and dense in $\mathbb{R}^n$. We address the basic question: for what metrics is there a unique isomorphism type for graphs resulting from this random process? It was shown in [7] that a unique isomorphism type occurs for the L_∞-metric for all $n \geq 1$. The hexagonal metric is a convex polyhedral distance function on $\mathbb{R}_2$, which has the property that its unit balls tile the plane, as in the case of the L_∞-metric. We may view the hexagonal metric as an approximation of the Euclidean metric, and it arises in computational geometry. We show that the random process with the hexagonal metric does not lead to a unique isomorphism type.

1 Introduction

Geometric random graph models play an important role in the modelling of real-world networks [21] such as on-line social networks [6], wireless and ad hoc networks [3,17,19], and the web graph [1,16]. In such stochastic models, vertices of the network are represented by points in a suitably chosen metric space, and edges are chosen by a mixture of relative proximity of the vertices and probabilistic rules. In real-world networks, the underlying metric space is a representation of the hidden reality that leads to the formation of edges. In the case of on-line social networks, for example, users are embedded in a high dimensional *social space*, where users that are positioned close together in the space exhibit similar characteristics.

Growth is a pertinent feature of most real-life networks, and most stochastic models take the form of a time process, where graphs increase in size over time. The limit of such a process as time goes to infinity is a countably infinite graph. This study of such limiting graphs is in part motivated by the large-scale nature

S. Arumugam and B. Smyth (Eds.): IWOCA 2012, LNCS 7643, pp. 6–19, 2012.

of real-world complex networks. It is expected that the infinite limit will elucidate the structure that emerges when graphs generated by the process become large. The limiting graphs are also of considerable interest in their own right.

The study of countably infinite graphs is further motivated by two major research directions within graph theory and theoretical computer science. First, there is a well-developed theory of the *infinite random graph*, or the *Rado* graph, written R. The investigation of R lies at the intersection of logic, probability theory, and topology; see the surveys [11,12,15] or Chapter 6 of [5].

Another line of investigation has focused on so-called *graph limits*, developed by Lovász and others [8,9,22]. A framework is given to define the convergence of sequences of graphs of increasing size. Convergence is based on *homomorphism densities*, and the limit object is a symmetric measurable function. Countably infinite graphs that arise as limits of such sequences can be interpreted as random graphs sampled from the limiting object.

In [7] we considered infinite limits of a simple random geometric graph model. In our model, vertices are chosen at random from a metric space, and if the distance between the two vertices is no larger than some fixed threshold, the vertices are adjacent with some fixed probability. More precisely, for a space S with metric d, consider a threshold $\delta \in \mathbb{R}^+$, a countable subset V of S, and a link probability $p \in [0, 1]$. The *Local Area Random Graph* LARG(V, δ, p) has vertices V, and for each pair of vertices u and v with $d(u, v) < \delta$ an edge is added independently with probability p. Note that V may be either finite or infinite. For simplicity, we consider only the case when $\delta = 1$; we write LARG(V, p) in this case. The LARG model generalizes well-known classes of random graphs. For example, special cases of the LARG model include the random geometric graphs (where $p = 1$), and the binomial random graph $G(n, p)$ (where S has finite diameter D, and $\delta \geq D$). We note that the theory of random geometric graphs has been extensively developed (see, for example, [2,14] and the book [23]).

The basic question we consider is whether the graphs generated by this random process retain information about the metric space from which they are derived. In [7] we obtained a positive answer to this question for $\mathbb{R}^n$ with the L_∞-metric (for any $n \geq 1$). In particular, we showed that in this case, if V is countably infinite, dense in $\mathbb{R}^n$ and randomly chosen, then with probability 1, any two graphs generated by the LARG(V, δ, p) (for any fixed δ and p) are isomorphic. Moreover, the isomorphism type is the same for all values of $\delta \in \mathbb{R}^+$ and $p \in (0, 1)$. Thus, we can take the unique infinite graph resulting from this process to represent the geometry of this particular metric space.

The isomorphism result described above for the L_∞-metric lead us to consider following general question.

Geometric Isomorphism Dichotomy (GID): For which metrics on $\mathbb{R}^n$ do we have the property that graphs generated by the random process LARG(V, δ, p), for V countably infinite, dense in $\mathbb{R}^n$ and randomly chosen, and any $\delta \in \mathbb{R}^+$ and $p \in (0, 1)$ are isomorphic with probability 1?

In [7], we showed that two graphs generated by $\text{LARG}(V, \delta, p)$ in $\mathbb{R}^2$ with the L_2-(or Euclidean) metric are non-isomorphic with probability 1, thereby answering the GID in the negative. In the present work, we extend our understanding of the GID to include another metric on $\mathbb{R}^2$, the so-called *hexagonal* (or *honeycomb*) *metric*, written d_{hex}, which is defined by having hexagonal unit balls. This metric may be seen to lie in between the L_∞-metric, where unit balls are squares, and the Euclidean metric, where unit balls are circles. Hexagons have the property that they tile the plane (as squares do), but on the other hand they can be seen as approximate circles. The precise definition of the d_{hex} metric is given in Section 3.

The hexagonal metric arises in the study of Voronoi diagrams and period graphs (see [18]) in computational geometry, which in turn have applications to nanotechnology. The hexagonal metric arises as a special case of *convex polygonal (or polygon-offset) distance functions*, where distance is in terms of a scaling of a convex polygon containing the origin; see [4,13,20]. A precise definition of this metric is given in Section 3. We note that *honeycomb networks* formed by tilings by hexagonal meshes have been studied as, among other things, a model of interconnection networks; see [24].

Our main result is the following theorem.

Theorem 1. *Let V be a randomly chosen, countable, dense set in $\mathbb{R}^2$ with the d_{hex}-metric. Let G and H be two graphs generated by the model $\text{LARG}(V, p)$, where $0 < p < 1$. Then with probability 1, G and H are not isomorphic.*

The theorem answers the *GID* in the negative for $\mathbb{R}^2$ with the hexagonal metric. We devote the present article to a sketch of a proof of Theorem 1. Our techniques are largely geometric and combinatorial (such as Hall's condition), and appear in Section 3. In Section 2, we introduce the hexagonal metric and review some of the concepts developed in [7] which are needed to obtain the non-isomorphism result. We conclude with a conjecture on the GID for a wide family of metrics defined by other polygons.

All graphs considered are simple, undirected, and countable unless otherwise stated. Let $\mathbb{N}$, $\mathbb{N}^+$, $\mathbb{Z}$, and $\mathbb{R}$ denote the non-negative integers, the positive integers, the integers, and real numbers, respectively. Vectors are written in **bold**. For a reference on graph theory the reader is directed to [25], while [10] is a reference on metric spaces.

2   Conditions for Isomorphism

2.1   Hexagonal Metric

We now formally define the hexagonal metric. Consider the vectors

$$\mathbf{a}_1 = \begin{pmatrix} 1 \\ 0 \end{pmatrix}, \mathbf{a}_2 = \begin{pmatrix} \frac{1}{2} \\ \frac{1}{2}\sqrt{3} \end{pmatrix}, \text{ and } \mathbf{a}_3 = \begin{pmatrix} -\frac{1}{2} \\ \frac{1}{2}\sqrt{3} \end{pmatrix}.$$

These are the normal vectors to the sides of a regular hexagon, as shown in Figure 1.

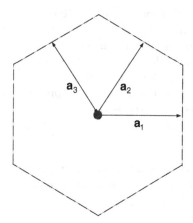

Fig. 1. The vectors $\mathbf{a}_i$

For $\mathbf{x} \in \mathbb{R}^2$ define the *hexagonal norm* of $\mathbf{x}$ as follows:

$$\|\mathbf{x}\|_{\text{hex}} = \max_{i=1,2,3} |\mathbf{a}_i \cdot \mathbf{x}|,$$

where "·" is the dot product of vectors. The *hexagonal metric* in $\mathbb{R}^2$ is derived from the hexagonal norm, and defined by

$$d_{\text{hex}}(\mathbf{x}, \mathbf{y}) = \|\mathbf{x} - \mathbf{y}\|_{\text{hex}}.$$

We may drop the subscript "hex" when it is clear from context. Note that the unit balls with the hexagonal metric are regular hexagons as in Figure 1. We denote the metric space consisting of $\mathbb{R}^2$ with the hexagonal metric by $(\mathbb{R}^2, d_{\text{hex}})$.

2.2 Step-Isometries

For the proof of Theorem 1, we rely on the following geometric theorem. Given metric spaces (S, d_S) and (T, d_T), sets $V \subseteq S$ and $W \subseteq T$, a *step-isometry* from V to W is a surjective map $f : V \to W$ with the property that for every pair of vertices $u, v \in V$,

$$\lfloor d_S(u, v) \rfloor = \lfloor d_T(f(u), f(v)) \rfloor.$$

Every isometry is a step-isometry, but the converse is false, in general. For example, consider $\mathbb{R}$ with the Euclidean metric, and let $f : \mathbb{R} \to \mathbb{R}$ be given by $f(x) = (\lfloor x \rfloor + x)/2$ is a step-isometry, but not an isometry.

A subset V is *dense* in S if for every point $x \in S$, every ball around x contains at least one point from V. We refer to $u \in S$ as points or vertices, depending on the context. A crucial step in the proof of isomorphism results of graphs produced by $\text{LARG}(V, p)$ when V is dense in the underlying metric space (S, d), is that any isomorphism must be a step-isometry. For a rough sketch of this fact, observe that if G is a graph produced by $\text{LARG}(v, d)$, and if u and v are two vertices in V so that $k < \lfloor d(u, v) \rfloor < k+1$, then with probability 1, there exists a path of length k from u to v in G. Since no edge can connect vertices at distance 1 or higher, no path of length less than $k - 1$ can exist between u and v. Thus, the graph distance equals the floor of the distance. Since an isomorphism must preserve graph distance, it therefore must also preserve the distance, up to its integer multiple, and thus, be a step-isometry.

We will use this fact in the form of the following lemma, adapted from [7], whose proof is omitted.

Lemma 1. *Let V be a countable set dense in $\mathbb{R}^2$ with the d_{hex}-metric, randomly chosen, and let G and H be two graphs generated by the model $\text{LARG}(V, p)$, where $0 < p < 1$. If G and H are isomorphic via f, then with probability 1 we have that f is a step-isometry.*

The following theorem is central to our proof of Theorem 1.

Theorem 2. *Let V and W be dense subsets of $\mathbb{R}^2$ with the d_{hex}-metric, with the property that V contains two points $\mathbf{p}_1, \mathbf{p}_2$ so that their distance $d_{hex}(\mathbf{p}_1, \mathbf{p}_2)$ is irrational. Then every step-isometry from V to W is uniquely determined by the images of $\mathbf{p}_1$ and $\mathbf{p}_2$.*

The next section will be devoted to the proof of Theorem 2. Using Theorem 2, we may prove Theorem 1. As the proof of Theorem 1 is analogous to the proof of Theorem 15 in [7], it is omitted.

3   Proof of Theorem 2

To prove Theorem 2, we show first that each finite set of points in the plane introduces a set of lines (which will be recursively defined below). These lines will fall into three parallelism classes, defined by their normals $\mathbf{a}_i$, for $i = 1, 2, 3$. More precisely, for a fixed $i \in \{1, 2, 3\}$, define $\mathcal{F}_i$ to be the family of lines in the plane with normal vector $\mathbf{a}_i$. For $r \in \mathbb{R}$, let $\mathcal{F}_i(r)$ be the family of lines in $\mathcal{F}_i$ which are at integer distance from the line $\mathbf{a}_i \cdot \mathbf{x} = r$. Thus, $\mathcal{F}_i(r)$ contains the lines with equations

$$\mathbf{a}_i \cdot \mathbf{x} = r + z, \text{ for some } z \in \mathbb{Z}.$$

We will show that any step-isometry f between two graphs with dense vertex sets must be "consistent" with these lines (that is, a point in the domain framed by a set of lines must be mapped to a point which is framed by a corresponding set of lines in the range); see Lemma 2. We then show that we can choose these lines to be dense in Lemma 4, and thereby prove that f is in fact uniquely determined by a finite set of points.

We now define this notion of consistency in a more precise fashion. Let $V \subseteq \mathbb{R}^2$ and $f : V \to \mathbb{R}^2$ be an injective map. Let $r \in [0, 1)$ and $i \in \{1, 2, 3\}$, and let σ be a permutation of the index set $\{1, 2, 3\}$. The map f is *consistent* with the family of lines $\mathcal{F}_i(r)$ *with respect to the permutation* σ if there exists $r' \in [0,1)$ so that for all $\mathbf{x} \in V$, for all $z \in \mathbb{Z}$,

$$\mathbf{a}_i \cdot \mathbf{x} < z + r \text{ if and only if } \mathbf{a}_{\sigma(i)} \cdot \mathbf{x}' < z + r',$$

where $\mathbf{x}' = f(\mathbf{x})$.

In the following, we will assume that all sets contain the origin $\mathbf{0}$, and we will assume without loss of generality that any map preserves the origin. Note that we can always replace any set V by an equivalent, translated set $V + \mathbf{b} = \{\mathbf{v} + \mathbf{b} : \mathbf{v} \in V\}$ so that this is indeed the case.

Lemma 2. *Suppose that V and W are dense in $\mathbb{R}^2$, and $f : V \to W$ is a bijection. If f is a step-isometry, then there exists a permutation σ of the index set $\{1,2,3\}$ such that f is consistent with the family of lines $\mathcal{F}_i(0)$ with respect to σ for $i = 1,2,3$.*

Proof. For $1 \leq i \leq 6$, let S_i be the six sections partitioning $\mathbb{R}^2$ formed by the lines $\ell_j : \mathbf{a}_j \cdot \mathbf{x} = 0$, where $j = 1,2,3$. See Figure 2.

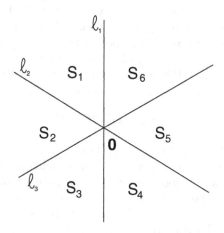

Fig. 2. The sectors S_i and lines $\ell_j : \mathbf{a}_i \cdot \mathbf{x} = 0$

Note that each of the sectors is uniquely defined by its two bounding lines. For example, sector S_1 contains all points $\mathbf{x}$ for which $\mathbf{a}_1 \cdot \mathbf{x} < 0$ and $\mathbf{a}_2 \cdot \mathbf{x} > 0$, and sector S_2 contains all points $\mathbf{x}$ for which $\mathbf{a}_2 \cdot \mathbf{x} < 0$ and $\mathbf{a}_3 \cdot \mathbf{x} > 0$. Thus, any permutation σ of the index set $\{1,2,3\}$ induces a permutation σ^* of the sectors. For example, the permutation $\sigma = (1,2,3)$ induces the permutation $\sigma^* = (1,2,3,4,5,6)$ of the sector indices.

Claim. There exist six points $v_i \in V$, $1 \leq i \leq 6$, and a permutation σ of the index set $\{1,2,3\}$ so that for all i we have that $\mathbf{v}_i \in S_i$ and $f(\mathbf{v}_i) \in S_{\sigma^*(i)}$, where σ^* is the permutation of the sectors induced by σ.

Proof. Let $k = 26$. Choose points $\mathbf{u}_i$, where $i = 0, \ldots, 6k - 1$, such that the following four conditions hold.

1. All points lie in a band at distance between $k - 1$ and k from the origin. More precisely, for all i, $0 \leq i < 6k$, $k - 1 < d(\mathbf{0}, \mathbf{u}_i) < k$.
2. Any two consecutive points lie at less than unit distance from each other. More precisely, for all i, $0 \leq i < 6k$, $d(\mathbf{u}_i, \mathbf{u}_{i+1}) < 1$, where addition in the index is taken modulo $6k$.
3. Any two points that are not consecutive are further than unit distance apart. More precisely, for all i and j, $0 \leq i < 6k$, $2 \leq j \leq 6k - 2$, $d(\mathbf{u}_i, \mathbf{u}_{i+j}) > 1$, where addition in the index is taken modulo $6k$. Note that if these points are adjacent if and only if their distance is less than 1, then the points would form a cycle.
4. For all j, $1 \leq j \leq 6$, the points $\mathbf{u}_{(j-1)k+i}$, $0 \leq i < k$ lie in sector S_j.

For all i, $0 \leq i < 6k$, let $\mathbf{u}'_i = f(\mathbf{u}_i)$, and let $U' = \{\mathbf{u}'_i : 0 \leq i < 6k\}$. Since f is a step-isometry, the points $\mathbf{u}'_i$ must satisfy conditions (1), (2), and (3) for the given $\mathbf{u}_i$. Note that we cannot conclude that (4) holds as well. However, we can deduce that each sector must contain at least $k - 1$ vertices from U'. Namely, let $\mathbf{u}'_s \in U'$ be in sector S_2. Let $\mathbf{u}'_{t_1}$ and $\mathbf{u}'_{t_2}$ be the vertices with lowest index and highest index in $S_2 \cap U'$, respectively. Then $\mathbf{u}'_{t_1-1}$ and $\mathbf{u}'_{t_2+1}$ are in the two sectors adjacent to S_2; without loss of generality $\mathbf{u}'_{t_1-1} \in S_1$ and $\mathbf{u}'_{t_2+1} \in S_3$. By condition (1) and by the geometry of the sectors, $d(\mathbf{u}'_{t_1-1}, \mathbf{u}'_{t_2+1}) \geq k - 1$, and by condition (2), $d(\mathbf{u}'_{t_1-1}, \mathbf{u}'_{t_2+1}) < (t_2 + 1) - (t_1 - 1)$. Thus,

$$(t_2 + 1) - (t_1 - 1) > k - 1,$$

so $t_2 - t_1 + 1 \geq k - 1$. Since the points $\mathbf{u}'_{t_1}, \ldots, \mathbf{u}'_{t_2}$ are in S_2, it follows that S_2 contains at least $k - 1$ points. The same conclusion holds for the other sectors.

A simple counting argument shows that each sector can contain at most $6k - 5(k - 1) = k + 5$ points from U' (as there are $6k$ points $\mathbf{u}_i$). For $j = 1, 2, \ldots, 6$, let $U'_j = \{f(\mathbf{u}) : \mathbf{u} \in U \cap S_j\}$ be the collection of images of points from u that lie in sector S_j. By condition (4), $|U'_j| = k$ for all j. Note that k is chosen large enough so that $tk > (t - 1)(k + 5)$ for any t, $1 \leq t \leq 6$. Thus, by the pigeonhole principle the tk vertices from t of the sets U'_j cannot be contained in less than t sectors. Thus, Hall's condition (see for example, [25]) holds, and we can find a permutation σ^* of the index set $\{1, 2, \ldots, 6\}$ so that $U'_i \cap S_{\sigma^*(i)} \neq \emptyset$. Moreover, because of the cyclic structure of the points in U' implied by conditions (2) and (3), adjacent sets U'_j and U'_{j+1} must be mapped by σ^* to adjacent sectors. This guarantees that σ^* is compatible with a permutation σ of the index set $\{1, 2, 3\}$ of the lines that define the sectors.

For $j = 1, \ldots, 6$, we can then choose $\mathbf{v}_j \in U_j = U \cap S_j$ so that $f(\mathbf{v}_j) \in U'_j \cap S_{\sigma^*(j)}$. This completes the proof of the claim. □

To complete the proof of the lemma, fix the $\mathbf{v}_i$ and σ as in Claim 3. Let

$$A = \{\mathbf{v}_i : 1 \leq i \leq 6\}.$$

Without loss of generality, we set σ to be the identity permutation. For a contradiction, assume there exists $\mathbf{u} \in V$ and $i \in \{1,2,3\}$ so that $\mathbf{a}_i \cdot \mathbf{u} < 0$ but $\mathbf{a}_i \cdot \mathbf{u}' > 0$, where $\mathbf{u}' = f(\mathbf{u})$. Let $L = \{\mathbf{v} \in A : \mathbf{a}_i \cdot \mathbf{v} < 0\}$, and $R = A - L = \{\mathbf{v} \in A : \mathbf{a}_i \cdot \mathbf{v} > 0\}$. Let $L' = \{f(\mathbf{v}) : \mathbf{v} \in L\}$ and $R' = \{f(\mathbf{v}) : \mathbf{v} \in R\}$. By definition, each of the vertices of A lies in a different sector, so we must have that $|L| = |R| = 3$. See Figure 3; we assume in the figure that $i = 1$, so f is not consistent with ℓ_1.

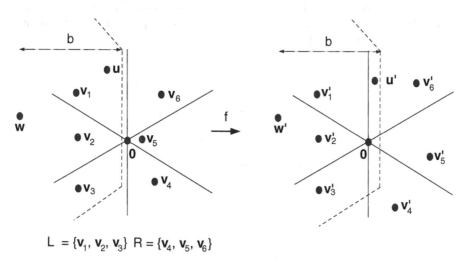

$$L = \{\mathbf{v}_1, \mathbf{v}_2, \mathbf{v}_3\} \quad R = \{\mathbf{v}_4, \mathbf{v}_5, \mathbf{v}_6\}$$

Fig. 3. The original configuration and its image under f. Dotted lines correspond to the boundaries of the (partially shown) balls of radius b centred around $\mathbf{w}$ and $\mathbf{w}'$.

Choose $\mathbf{w}$ so that for some $b \in \mathbb{Z}^+$,

$$d_{\text{hex}}(\mathbf{w}, \mathbf{x}) > b \text{ for all } \mathbf{x} \in R \cup \{\mathbf{0}\}, \text{ and}$$
$$d_{\text{hex}}(\mathbf{w}, \mathbf{x}) < b \text{ for all } \mathbf{x} \in L \cup \{\mathbf{u}\}.$$

More precisely, the ball with radius b centered at $\mathbf{w}$ contains $L \cup \{\mathbf{u}\}$, and is disjoint from $R \cup \{\mathbf{0}\}$.

Since f is a step-isometry, if we set $\mathbf{w}' = f(\mathbf{w})$, then

$$d_{\text{hex}}(\mathbf{w}', \mathbf{x}) > b \text{ for all } \mathbf{x} \in R' \cup \{\mathbf{0}\}, \text{ and}$$
$$d_{\text{hex}}(\mathbf{w}', \mathbf{x}) < b \text{ for all } \mathbf{x} \in L' \cup \{\mathbf{u}'\}.$$

By the definition of A and the assumption that σ is the identity, for each $\mathbf{v} \in A$, if $\mathbf{v}$ is in a certain sector S_j then so is $f(\mathbf{v})$. Thus, $\mathbf{a}_i \cdot \mathbf{v}' < 0$ for all $\mathbf{v}' \in L'$. However, $\mathbf{a}_i \cdot \mathbf{u}' > 0$, so the vertices in $L' \cup \{\mathbf{u}'\}$ lie in four different sectors. Thus, the ball of radius b around $\mathbf{w}'$ must intersect at least four sectors (see Figure 3) and so contains the origin. Since $d_{\text{hex}}(\mathbf{w}', \mathbf{0}) > b$, this ball cannot contain $\mathbf{0}$, which gives a contradiction. □

Let B be a finite subset of $\mathbb{R}^2$. Define a collection of lines $\mathcal{L}(B)$ inductively as follows. To define $\mathcal{L}_0(B)$: for each $\mathbf{w} \in B$, add the three lines through $\mathbf{w}$ in the families $\mathcal{F}_i$, ($i = 1, 2, 3$,) along with their integer parallels. Specifically, these are all the lines with equation $\mathbf{a}_i \cdot \mathbf{x} = \mathbf{a}_i \cdot \mathbf{w} + z$, where $z \in \mathbb{Z}$.

For the inductive step, assume that $\mathcal{L}_j(B)$ has been defined for some $j \geq 0$. To define $\mathcal{L}_{j+1}(B)$, for each point $\mathbf{p} \in \mathbb{R}^2$ which lies on the intersection of two lines in $\mathcal{L}_j(B)$ (which must belong to different, non-parallel families), add the unique line belonging to the third family. See Figure 4. More precisely, if $\mathbf{p}$ is at the intersection of lines ℓ and m, where $l \in \mathcal{F}_{i_\ell}$ and $m \in \mathcal{F}_{i_m}$ (with $i_\ell \neq i_m$), then add to $\mathcal{L}_{j+1}(B)$ the line with equation $\mathbf{a}_j \cdot \mathbf{x} = \mathbf{a}_j \cdot \mathbf{p}$, where j is the unique element in $\{1, 2, 3\} \setminus \{i_\ell, i_m\}$.

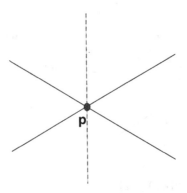

Fig. 4. The solid lines are in $\mathcal{L}_i$ and the dotted line is in $\mathcal{L}_{i+1}$

Finally, define

$$\mathcal{L}(B) = \bigcup_{i \in \mathbb{N}} \mathcal{L}_i(B).$$

We prove the following lemma. A step-isometry $f : V \to W$ *respects the line* ℓ defined by $\mathbf{a} \cdot \mathbf{x} = r$ if for all points $\mathbf{x} \in V$, $\mathbf{a} \cdot \mathbf{x} < r$ implies that $\mathbf{a} \cdot f(\mathbf{x}) = r$ and $\mathbf{a} \cdot \mathbf{x} > r$ implies that $\mathbf{a} \cdot f(\mathbf{x}) > r$. For simplicity, we state this more succinctly by saying that f maps points to the left (right) of ℓ to the left (right) of the line $f(\ell)$.

Lemma 3. *For a fixed finite subset B of $\mathbb{R}^2$, and a step-isometry $f : V \to W$ with V and W dense, f respects all the lines in $\mathcal{L}(B)$.*

Proof. We proceed by a strong induction on k to show that f respects the lines in $\mathcal{L}_k(B)$. The case $i = 0$ follows from Lemma 2. Suppose the statement holds for a fixed $k \geq 0$. For the induction step, suppose that $\ell_1, \ell_2 \in \bigcup_{i=0}^{k} \mathcal{L}_i(B)$. Let ℓ_3 be a fixed line in $\mathcal{L}_{k+1}(B)$ formed by the intersection of ℓ_1 and ℓ_2. In this proof, we will say that a point $\mathbf{p}$ is to the right (left) of a line with equation $\mathbf{a} \cdot \mathbf{x} = t$ if $\mathbf{a} \cdot \mathbf{p} > t$ ($\mathbf{a} \cdot \mathbf{p} > t$). See Figure 5 for a visualization of the proof.

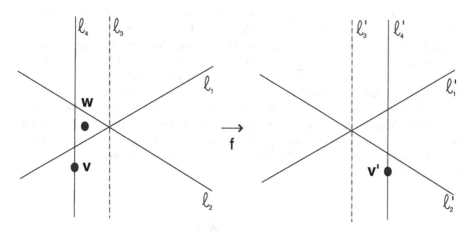

Fig. 5. The point $f(\mathbf{w})$ does not exist

Assume for a contradiction that $\mathbf{v}$ is to the left of ℓ_3, and to the right of the line $f(\ell_3)$ (the case when $\mathbf{v}$ is to the right of ℓ_3 is similar and so is omitted). Let ℓ_4 be the unique line parallel to ℓ_3 and through $\mathbf{v}$.

For $1 \leq i \leq 4$, let ℓ_i' be the line which is the image $f(\ell_i)$. A step-isometry must respect lines in $\mathcal{L}_0(B)$, and so must respect ℓ_4. As V is dense, we may choose $\mathbf{w} \in V$ so that $\mathbf{w}$ is to the right of ℓ_4, left of ℓ_1, and left of ℓ_2. See Figure 5. But then $f(\mathbf{w})$ must be to the right of ℓ_4', to the left of ℓ_1', and to the left of ℓ_2', which is a contradiction.

We need the following lemma.

Lemma 4. *Let $B = \{0, \mathbf{p}\}$, where $\mathbf{a}_i \cdot \mathbf{p} = r \in (0, 1)$. Then the family $\mathcal{L}(B)$ contains all the following lines, where $i \in \{1, 2, 3\}$, $z_1, z_2 \in \mathbb{Z}$:*

$$\mathbf{a}_i \cdot \mathbf{x} = z_1 r + z_2.$$

Moreover, if r is irrational, the set of lines $\mathcal{L}(B)$ is dense in $\mathbb{R}^2$.

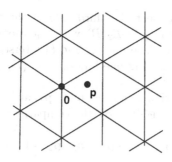

Fig. 6. A triangular lattice

Proof. Consider the triangular lattice formed by all lines in $\mathcal{L}_0(\{\mathbf{0}\})$; that is, all lines with equations $\mathbf{a}_i \cdot \mathbf{x} = z$, where $i = 1, 2, 3$ and $z \in \mathbb{Z}$. See Figure 6. Consider the triangle that contains $\mathbf{p}$. We assume that this triangle is framed by the lines with equations $\mathbf{a}_1 \cdot \mathbf{x} = 1$, $\mathbf{a}_2 \cdot \mathbf{x} = 0$ and $\mathbf{a}_3 \cdot \mathbf{x} = 0$, as shown in Figure 7. (The proof can easily be adapted to cover all other possibilities.) By definition, the line ℓ_1 defined by $\mathbf{a}_1 \cdot \mathbf{x} = r$ is part of $\mathcal{L}_0(B)$.

The line ℓ_1 intersects the two sides of the triangle in $\mathbf{p}_1$ and $\mathbf{p}_2$. The point $\mathbf{p}_1$ lies on the intersection of the lines from the families $\mathcal{F}_1$ and $\mathcal{F}_3$. Thus, $\mathcal{L}_1(\mathbf{w})$ contains the line ℓ_2 in $\mathcal{F}_2$ through $\mathbf{p}_1$ which has equation $\mathbf{a}_2 \cdot \mathbf{x} = \mathbf{a}_2 \cdot \mathbf{p}_2$. Similarly, $\mathcal{L}_1(\mathbf{w})$ contains the line ℓ_3 in $\mathcal{F}_3$ through $\mathbf{p}_2$ which has equation $\mathbf{a}_3 \cdot \mathbf{x} = \mathbf{a}_3 \cdot \mathbf{p}_2$. See Figure 7.

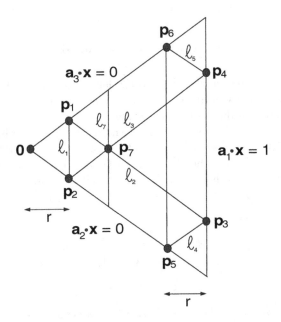

Fig. 7. Generating the line $\mathbf{a}_3 \cdot \mathbf{p}_2 = 1 - r$

The lines ℓ_2 and ℓ_3 intersect the third side of the triangle in $\mathbf{p}_3$ and $\mathbf{p}_4$, generating two lines ℓ_4 and ℓ_5 in $\mathcal{L}_2(B)$ with equations $\mathbf{a}_3 \cdot \mathbf{x} = \mathbf{a}_3 \cdot \mathbf{p}_3$ and $\mathbf{a}_2 \cdot \mathbf{x} = \mathbf{a}_2 \cdot \mathbf{p}_4$, respectively. The lines ℓ_4 and ℓ_5 intersect with the sides of the triangle in $\mathbf{p}_5$ and $\mathbf{p}_6$, generating one line ℓ_6 in $\mathcal{L}_3(B)$ with equation $\mathbf{a}_1 \cdot \mathbf{x} = \mathbf{a}_1 \cdot \mathbf{p}_5 = \mathbf{a}_1 \cdot \mathbf{p}_6$.

From the fact that the triangle formed by $\mathbf{0}$, $\mathbf{p}_1$, and $\mathbf{p}_2$ is isosceles, it follows that $r = \mathbf{a}_1 \cdot \mathbf{p} = \mathbf{a}_2 \cdot \mathbf{p}_1 = \mathbf{a}_3 \cdot \mathbf{p}_2$. By the comparison of similar triangles, we obtain that

$$\mathbf{a}_1 \cdot \mathbf{p}_5 = \mathbf{a}_2 \cdot \mathbf{p}_4 = \mathbf{a}_3 \cdot \mathbf{p}_2 = 1 - r.$$

Now the parallel lines $\mathbf{a}_i \cdot \mathbf{x} = r + z_2, -r + z_2$, $z_2 \in \mathbb{Z}$, may be generated from all similar triangles in the lattice in an analogous fashion.

To complete the proof, consider that the lines ℓ_2 and ℓ_3 intersect in point $\mathbf{p}_7$, which generates a line $\ell_7 \in \mathcal{F}_1$ as indicated in Figure 7. Since the triangle formed by $\mathbf{p}_1, \mathbf{p}_2$, and $\mathbf{p}_7$ is a reflection of the triangle formed by $\mathbf{0}, \mathbf{p}_1$, and $\mathbf{p}_2$, it follows that ℓ_7 has equation $\mathbf{a}_1 \cdot \mathbf{x} = 2r$. This process can be repeated to obtain all the lines $\mathbf{a}_i \cdot \mathbf{x} = z_1 r + z_2$, $z_1, z_2 \in \mathbb{Z}$, $i = 1, 2, 3$.

If r is irrational, then the set $\{z_1 r + z_2 : z_1, z_2 \in \mathbb{Z}\}$ is dense in $\mathbb{R}$ (this is a result from folklore which can be proved by using the pigeonhole principle). That completes the proof of the lemma. □

We now give a proof of the main theorem in this section.

Proof of Theorem 2. Let $f : V \to W$ be a step-isometry. By Lemma 2 there exists a permutation σ of the index set $\{1, 2, 3\}$ such that f is consistent with the family of lines $\mathcal{F}_i(0)$ for $i = 1, 2, 3$ with respect to σ. Without loss of generality, we assume that σ is the identity. Assume that V has two points whose distance is irrational; assume without loss of generality that one of them is the origin. Choose $B = \{\mathbf{0}, \mathbf{p}\} \subseteq V$ so that $d_{hex}(\mathbf{0}, \mathbf{p})$ is irrational. By Lemma 4, the set of lines $\mathcal{L}(B)$ is dense in $\mathbb{R}^2$. By Lemma 3, f respects the lines in $\mathcal{L}(B)$. Thus, any of the points in V is uniquely defined by their position with respect to all lines in $\mathcal{L}(B)$. Moreover, the same is true for W, and the set of lines generated by the images of $\mathbf{0}$ and $\mathbf{p}$. Therefore, once the images of $\mathbf{0}$ and $\mathbf{p}$ are given, each other vertex has a unique image under f. □

4 Conclusion and Further Work

We have shown that the hexagonal metric on a randomly chosen, dense subset of $\mathbb{R}^2$ can lead to non-isomorphic limit graphs in the LARG random process. Our main tool was Theorem 2 which proves that a step-isometry f between randomly chosen dense sets is, with probability 1, determined by the image of two points. Theorem 2 was proven by exploiting that f is consistent with a dense set of lines as proved in Lemmas 2 and 4.

The methods described in this article should extend to other polygonal metrics and higher dimensions. A *polygonal metric* is one where the unit ball is a (convex) point-symmetric polygon. The distance between two points $\mathbf{a}$ and $\mathbf{b}$ given as follows. Translate the polygon until it is centered at $\mathbf{a}$. Let $\mathbf{v}$ be the unique point on the intersection of the ray from $\mathbf{a}$ to $\mathbf{b}$ with the boundary of the polygon. Then the distance is given by the ratio of the (Euclidean) distance between $\mathbf{a}$ and $\mathbf{b}$ to the distance between $\mathbf{a}$ and $\mathbf{v}$. Alternatively, it is the factor by which the polygon centered at $\mathbf{a}$ would have to be enlarged until it touches $\mathbf{b}$.

To be a metric, the polygon needs to be point-symmetric, and thus, has an even number of sides. If the number of sides equals $2n$ for a metric defined in $\mathbb{R}^n$, then the polygon can be transformed into an n-dimensional hypercube by rescaling of the coordinates, and thus the metric is equivalent to the L_∞-metric. We conjecture that this is the only case for which the GID is answered in the affirmative.

Conjecture 1. For all $n \geq 2$, for all convex polygonal distance functions where the polygon has at least six sides, two graphs generated by LARG(V, p), with V randomly chosen and dense in $\mathbb{R}^n$, and $p \in (0, 1)$, are non-isomorphic with probability 1.

We showed earlier that the GID is answered in the negative for the Euclidean metric and $\mathbb{R}^2$. We think that the analogous statement is true for the Euclidean metric in higher dimensions, and that the methods in this paper may suggest a suitable approach to proving this fact.

References

1. Aiello, W., Bonato, A., Cooper, C., Janssen, J., Prałat, P.: A spatial web graph model with local influence regions. Internet Mathematics 5, 175–196 (2009)
2. Balister, P., Bollobás, B., Sarkar, A., Walters, M.: Highly connected random geometric graphs. Discrete Applied Mathematics 157, 309–320 (2009)
3. Barbeau, M., Kranakis, E.: Principles of Ad Hoc Networking. John Wiley and Sons (2007)
4. Barequet, G., Dickerson, M.T., Goodrich, M.T.: Voronoi diagrams for convex polygon-offset distance functions. Discrete & Computational Geometry 25, 271–291 (2001)
5. Bonato, A.: A Course on the Web Graph. American Mathematical Society Graduate Studies Series in Mathematics, Providence, Rhode Island (2008)
6. Bonato, A., Janssen, J., Prałat, P.: The Geometric Protean Model for On-Line Social Networks. In: Kumar, R., Sivakumar, D. (eds.) WAW 2010. LNCS, vol. 6516, pp. 110–121. Springer, Heidelberg (2010)
7. Bonato, A., Janssen, J.: Infinite random geometric graphs. Annals of Combinatorics 15, 597–617 (2011)
8. Borgs, C., Chayes, J., Lovász, L., Sós, V.T., Vesztergombi, K.: Convergent Sequences of Dense Graphs I: Subgraph Frequencies, Metric Properties and Testing. Advances in Math. 219, 1801–1851 (2008)
9. Borgs, C., Chayes, J., Lovász, L., Sós, V.T., Vesztergombi, K.: Convergent Sequences of Dense Graphs II: Multiway Cuts and Statistical Physics, Preprint (2007)
10. Bryant, V.: Metric Spaces: Iteration and Application. Cambridge University Press, Cambridge (1985)
11. Cameron, P.J.: The random graph. In: Graham, R.L., Nešetřil, J. (eds.) Algorithms and Combinatorics, vol. 14, pp. 333–351. Springer, New York (1997)
12. Cameron, P.J.: The random graph revisited. In: Casacuberta, C., Miró-Roig, R.M., Verdera, J., Xambó-Descamps, S. (eds.) European Congress of Mathematics, vol. I, pp. 267–274. Birkhauser, Basel (2001)
13. Chew, L.P., Kedem, K., Sharir, M., Tagansky, B., Welzl, E.: Voronoi diagrams of lines in three dimensions under polyhedral convex distance functions. J. Algorithms 29, 238–255 (1998)
14. Ellis, R., Jia, X., Yan, C.H.: On random points in the unit disk. Random Algorithm and Structures 29, 14–25 (2006)
15. Erdős, P., Rényi, A.: Asymmetric graphs. Acta Mathematica Academiae Scientiarum Hungaricae 14, 295–315 (1963)
16. Flaxman, A., Frieze, A.M., Vera, J.: A geometric preferential attachment model of networks. Internet Mathematics 3, 187–205 (2006)

17. Frieze, A.M., Kleinberg, J., Ravi, R., Debany, W.: Line of sight networks. Combinatorics, Probability and Computing 18, 145–163 (2009)
18. Fu, N., Imai, H., Moriyama, S.: Voronoi diagrams on periodic graphs. In: Proceedings of the International Symposium on Voronoi Diagrams in Science and Engineering (2010)
19. Goel, A., Rai, S., Krishnamachari, B.: Monotone properties of random geometric graphs have sharp thresholds. Annals of Applied Probability 15, 2535–2552 (2005)
20. Icking, C., Klein, R., Le, N., Ma, L.: Convex distance functions in 3-space are different. In: Proceedings of the Ninth Annual Symposium on Computational Geometry (1993)
21. Janssen, J.: Spatial models for virtual networks. In: Proceedings of the 6th Computability in Europe (2010)
22. Lovász, L., Szegedy, B.: Limits of dense graph sequences. J. Comb. Theory B 96, 933–957 (2006)
23. Penrose, M.: Random Geometric Graphs. Oxford University Press, Oxford (2003)
24. Stojmenovic, I.: Honeycomb networks: Topological properties and communication algorithms. IEEE Transactions on Parallel and Distributed Systems 8, 1036–1042 (1997)
25. West, D.B.: Introduction to Graph Theory, 2nd edn. Prentice-Hall (2001)

Saving on Phases: Parameterized Approximation for Total Vertex Cover

Henning Fernau

Fachbereich 4, Abteilung Informatik
Universität Trier, 54286 Trier, Germany
`fernau@uni-trier.de`

Abstract. VERTEX COVER and its variants have always been in the focus of study of Parameterized Algorithmics. This can be also claimed for the emergent area of Parameterized Approximation. While VERTEX COVER is known to be solvable in time $\mathcal{O}^*(c^k)$ with some $c < 2$, this is not the case for variants like CONNECTED VERTEX COVER and others that impose some connectivity requirements on the desired cover. The reason behind is the two-phase approach that is taken for this kind of problems. We show that this barrier can be overcome when we are only interested in approximate solutions. More specifically, we prove that a factor-1.5 approximative solution for TOTAL VERTEX COVER can be found in time $\mathcal{O}^*(1.151^k)$, where k is some bound on the optimum solution.

1 Introduction

1.1 Motivation

Sometimes, under reasonable complexity theoretic (or other) assumptions, no progress can be expected for polynomial-time approximation algorithms. Can we achieve better approximation ratios when allowing fixed-parameter running times? Here, the parameter is basically given by the optimum solution value.

Various techniques have been developed to this end; we discuss four of them in somewhat chronological order (although most of these papers seemed to have been developed in parallel).

- Marx and Razgon [12] derived a parameterized factor-2 approximation for EDGE MULTICUT with a running time of $\mathcal{O}^*(c^{k \log k})$ for some constant c (for some improvements on this kind of constants, we refer to [13]). Under the Unique Games Conjecture of Khot [10], no constant-factor approximation is possible in polynomial time, as shown by S. Chawla, R. Krauthgamer, R. Kumar, Y. Rabani, and D. Sivakumar [5]. The idea is to transform this approximation problem to some other parameterized problem that can be solved using iterated compression.
- VERTEX COVER was first studied by N. Bourgeois, B. Escoffier and V. Th. Paschos [2] from the viewpoint of parameterized approximation. Under the Unique Games Conjecture of Khot [10], no constant-factor approximation better than a factor of two is possible in polynomial time, see [11].

S. Arumugam and B. Smyth (Eds.): IWOCA 2012, LNCS 7643, pp. 20–31, 2012.

- L. Brankovic and H. Fernau [3,4] considered VERTEX COVER and for 3-HITTING SET. Both cases show that an intercalation of so-called worsening steps within a (rather simple) branching can lead to dramatic improvements on the running times of the derived parameterized approximation algorithms compared to, *e.g.*, the approaches of N. Bourgeois, B. Escoffier and V. Th. Paschos [2] and also those of M. Fellows, A. Kulik, F. Rosamond, and H. Shachnai [7] sketched in the next item.
- M. Fellows, A. Kulik, F. Rosamond, and H. Shachnai [7] developed notions of α-fidelity transformations that are based, in their simplest form, on so-called (a, b)-reduction steps.

The first example seems to use techniques that are quite problem-specific, although we would pose it as a challenge to come up with a problem that genuinely uses iterated compression as a solution strategy for parameterized approximation. As explained elsewhere, the second technique can be generalized to a certain extent, allowing for parameterized (and moderately exponential) approximation algorithms for quite some range of problems. It seems to be the case that this approach needs good kernelization algorithms at one stage. The third approach relied so far on the existence of simple branching algorithms for the exact solution. Otherwise, the concept of a worsening step basically corresponds to that of an (a, b)-reduction step in the last approach. In contrast to that last one, these worsening steps are interleaved with the branching, while M. Fellows, A. Kulik, F. Rosamond, and H. Shachnai propose to use them at the very beginning, which enables them to use any (sophisticated) exact parameterized algorithm as a black box in the second phase.

To summarize the task of parameterized approximation, we are always after a compromise between guaranteed approximation quality and running time.

1.2   Our Problem: Total Vertex Cover

Recall that a *vertex cover* is a set of vertices whose removal only leaves isolated vertices in the graph. A *total vertex cover* is a vertex cover that induces a graph without isolated vertices. This notion is formed in analogy to the better known notion of total dominating set, which is a dominating set that induces a graph without isolated vertices. The problem TOTAL VERTEX COVER (TVC) asks to find a minimum total vertex cover or, in its parameterized version, to decide if a vertex cover of size at most k exists. The size of the smallest total vertex cover of a graph is also known as its TVC number. For instance, a cycle C_n has a TVC number of $\lceil \frac{2}{3}n \rceil$, while it possesses a vertex cover set of size $\lceil \frac{1}{2}n \rceil$. As it is known [8] that this problem (under standard parameterization and modulo some complexity-theoretic assumptions) does not admit polynomial-size kernels, the second approach does (most likely) not apply. However, in the terminology of M. Fellows, A. Kulik, F. Rosamond, and H. Shachnai, "Select an edge and put both vertices into the cover" is also here a $(1, 1)$-reduction step. Without going into details, this means that from the $1 + 1 = 2$ vertices put into the approximate cover, at least one vertex belongs to some minimum total vertex cover. We can apply this rule as a preprocessing rule a certain number of times. Observe that this preprocessing never creates single-vertex components within the cover, so that finally we can apply any exact parameterized algorithm. The best current algorithm [9] needs time $\mathcal{O}^*(2.3655^k)$. This leads (leaving out some details at this place) to an $\mathcal{O}^*(2.3655^{(1+(1-\alpha))k})$ time, factor-α approximation for TVC. More concretely, TVC can be factor-$\frac{3}{2}$ approximated in time $\mathcal{O}^*(1.5381^k)$. To explain this running time, consider the following algorithm:

1. Select $k/2$ matching edges.
2. Run the best TVC algorithm with parameter $k/2$ on the remaining graph.

As the reader will see, the first step is just the worsening step used in our algorithm that will allow to factor-$\frac{3}{2}$ approximate TVC in time $\mathcal{O}^*(1.151^k)$, hence largely improving on the time derived when using the sophisticated algorithm [9] as a black box. In fact, that algorithm uses some sort of HITTING SET algorithm (parameterized by the number of edges) in the second phase. This fact alone prohibits any exact algorithm faster than $\mathcal{O}^*(c^k)$ for some $c < 2$ with some two-phase approach. It would be hence quite a challenge to improve on the approximation algorithm presented in this paper using the ideas of fidelity-preserving transformations and some exact parameterized algorithm, as this would (most likely) not help get below $\mathcal{O}^*(\sqrt{2}^k)$ for some factor-1.5 approximation.

1.3 Our Results and the Organization of the Paper

We will mostly focus on presenting several factor-$\frac{3}{2}$ TVC approximation algorithms. In Section 2, we will explain the basic ideas and show that these lead to improved algorithms even with the most simplistic vertex cover branching. It also describes how some annotations can be used along the branching. Section 3 sketches improvements due to further exact and approximate reduction rules and how a different branching strategy (also used in [3] for VERTEX COVER) can be used to finally obtain the claimed running time of $\mathcal{O}^*(1.151^k)$. In Section 4, we discuss how our factor-$\frac{3}{2}$ TVC approximation algorithm scales for any approximation factor between one and two, and we also suggest applying our techniques (and also those from [7]) to obtain results for related problems, concluding with some open problems from the area of parameterized approximation.

2   Putting Worsening Steps Inside of the Branching

As mentioned above, the strategy to design efficient parameterized approximation algorithms for VERTEX COVER and for HITTING SET was to intercalate branching with worsening steps and to analyze the obtained algorithm. By "investing" the worsening steps to work in the direction of better branchings and by developing new (approximate) reduction rules, we were able to improve on algorithms based (for instance) on the approach of [7]. This idea would be hard to follow for the current problem, as the algorithm from [9] is based on some type of Measure&Conquer analysis, which would have to be adapted. Rather, we will use the worsening steps in order to completely avoid the costly second phase of the exact algorithm (at least for approximation factors worse than $\frac{3}{2}$).

This means that we employ the worsening edges to maintain as an invariant of the algorithm that all previously selected cover vertices induce a graph without isolated vertices.

In its simplest version, such an algorithm (aiming first of all at a factor-$\frac{3}{2}$ approximation) might look as follows:

1. Select a high-degree vertex x and put either x or $N(x)$ into the cover.
2. This branching is accompanied with the following *worsening rule*:
   Whenever a vertex u is put into the cover, select some edge $e_u = \{v, w\}$ with $v \in N(u)$ and $u \neq w$; put v and w into the cover, as well.

In either branch, at least one vertex u is put into the cover due to (exact) branching. Selecting an edge e_u in the vicinity ensures to things: (1) Two out of the three vertices put into the cover must be in any minimum solution that contains u. (2) The selected vertices are connected, so that no isolated vertices would show up in the cover. This basic algorithm would already maintain the claimed invariant, and the algorithm would surely have a running time (of $\mathcal{O}^*(\sqrt{2}^k)$), better than any other approach (as sketched above). Item (2) requires that an edge e_u in the vicinity of u can be found. So, what if u has degree zero (so it has no neighbors)? Or if all neighbors of u are of degree one (so, e_u does not exist)? We will overcome these difficulties by using some reduction rules that have been used, in a similar form, already in [9], and also by relaxing the vicinity requirement. To this end, we adapt the idea of marking vertices. To implement this idea, we introduce a *marking function* μ on the vertex set V of the (input) graph G. So, the algorithm will actually deal with *marked graph* instances. We are going to explain the meaning of the marking $\mu : V \to \{-1, 0, 1, 2\}$ in the following:

- $\mu(v) = -1$: unmarked vertex v
  At the very beginning of the algorithm all vertices are unmarked.
- $\mu(v) \geq 0$: marked vertex v
- $\mu(v) = 1$: v belongs to the (partial) total vertex cover
- $\mu(v) = 2$: Like $\mu(v) = 1$, knowing that some $x \in N(v)$ obeys $\mu(x) \geq 1$.
- $\mu(v) = 0$: Some $x \in N(v)$ obeys $\mu(x) \geq 1$.

We will employ the following *marking update rules*:

1. $(\mu(v) = -1 \wedge \exists x \in N(v) : \mu(x) \geq 1) \implies \mu(v) \leftarrow 0$.
2. $(\mu(v) = 1 \wedge \exists x \in N(v) : \mu(x) \geq 1) \implies \mu(v) \leftarrow 2$.
3. If none of the previously listed rules apply, delete all 2-marked vertices.

The correctness of these rules is immediate from the described semantics of the marking. In particular, we can delete 2-marked vertices once all their (possibly previously unmarked) neighbors have been 0-marked or 2-marked. We can assume we are always dealing with a μ-*marked updated graph*. Such a graph has the following properties:

1. There is no unmarked vertex that is known to be in the vertex cover or in the neighborhood of a vertex from the vertex cover.
2. There is no 1-marked vertex that is neighbor of a vertex known to belong to the vertex cover.
3. There is no vertex that is 2-marked.

So, we have exhaustively applied the marking update rules to render a marked graph updated. We always apply the following exact reduction rules:

<u>Degree-0 Rule:</u> If G is a μ-marked updated graph instance and x is a vertex of degree 0, then do the following:

 – If x is 1-marked, then we have a NO-instance.
 – Otherwise, delete x.

The rule is clearly correct: If a vertex is 1-marked, it must still find a partner in its neighborhood, which is impossible for an isolated vertex. So, our μ-marked graph contains no isolated vertices.

Our rule for leaf vertices is much more complicated.

<u>Degree-1 Rule:</u> If G is a μ-marked updated graph instance and x is a vertex of degree 1 with unique neighbor y, then do the following:

1. If x is unmarked or 0-marked, we distinguish two subcases:
   (a) If y has degree 1, then do:
       If $\mu(x) = 0$, then $\mu(x) \leftarrow 2$.
       Else (i.e., x is unmarked) if $\mu(y) = 0$, then $\mu(y) \leftarrow 2$.
       Otherwise, $\mu(x) \leftarrow 1$, $\mu(y) \leftarrow 1$.
   (b) Otherwise, y has degree at least 2.
       If $\mu(y) \leq 0$, then $\mu(y) \leftarrow \mu(y) + 2$.
       Always delete x.
2. Otherwise, x is 1-marked: $\mu(y) \leftarrow 1$.

Lemma 1. *The Degree-1 Rule is correct.*

Proof. Since G is updated, $\mu(z) \leq 1$ for each z. Let x be of degree one, with y being its unique neighbor. We discuss all possible markings separately below.

If x is 1-marked, then (in order to avoid an immediate NO) its unique neighbor y must be 1-marked, as well. The update rules will then change $\mu(x) \leftarrow 2$ and $\mu(y) \leftarrow 2$ and finally delete both x and y.

Now, assume that x is 0-marked.

If y is of degree one, as well, we can use x to cover the edge xy, and x will have a partner in the cover due to the marking. This solution is optimal irrespectively of the current marking of y. In particular, if $\mu(y) = 1$ (which is its maximal value), then x must be in the cover, as well, to satisfy the totality requirement.

If y is of higher degree, we can argue as follows: If y is not in the optimum cover (respecting previous choices), then x and at least one other neighbor x' of y must be in the cover. We can replace such a solution by one that replaces x by

y in the cover. This is reflected by adding two to the marking of y and deleting x, covering the cases $\mu(y) = -1$ and $\mu(y) = 0$. In a sense, the fact that x has already a partner in the cover (as indicated by its marking) does not help.

Finally, assume that x is unmarked.

If y is of degree one, as well, we can use y to cover the edge xy only if $\mu(y) = 0$, as y will have a partner in the cover due to the marking. Otherwise, there is no partner of neither x nor y in the cover (although we might have fixed $\mu(y) = 1$ before): So, x and y must go into the cover together to avoid isolates in the total cover.

If y is of higher degree, we can argue as before.   □

Notice that we did not yet make the handling of the parameter explicit, but as we do not aim at the two-phase approach, we can easily decrement the parameter whenever we fix a vertex by 1-marking it.

We still have to describe the *worsening step* in our scenario (μ-marked graphs) more formally. Observe that we can assume that our μ-marked updated graph has minimum degree at least two. In order to guarantee a sufficient number of worsening steps, we do some book-keeping with the help of the counter *ws-count* that is always initialized by the number of vertices put into the cover by branching.

Worsening Step. If *ws-count* > 0, do:
(a) If there is some vertex u with $\mu(u) = 1$, select some $v \in N(u)$ and some $w \in N(v) - u$.
By the marking update rules, $\mu(v) \leq 0$.
$\mu(v) \leftarrow 1$. $\mu(w) \leftarrow 2$.
(b) If there is no u with $\mu(u) = 1$, select any edge and put both endpoints into the cover.
In both cases, decrement *ws-count*.

Reasoning behind:
To cover the edge vw, at least one vertex (v or w) is needed; we take both. Observe that $\mu(u) = 1$ can be only due to exact branching or reduction rule steps at this point.
The marking update rules will finally delete the whole path uvw (of length two).

We can now present our basic algorithm for finding a factor-$\frac{3}{2}$-approximation to a total cover of a graph G with marking function μ (satisfying $\mu(v) \leq 1$ for all v) and parameter k; this is displayed in Table 1.
"Increment marking" of v means: $\mu(v) \leftarrow \mu(v) + 2$.

Invariant: Before Step 4, $\forall v : \mu(v) \leq 0$.
Namely, this is due to the interplay between the worsening step and the marking update rules. The invariant guarantees that no increment of the marking leads to values larger than two.

We did not make explicit how to deal with the parameter k. As k upperbounds the size of an optimum solution, we can decrease k by one whenever we 1-mark a vertex due to branching or due to the exact reduction rules. However, although the worsening step introduces two 1-marked vertices, we should decrease k only

Table 1. Simple branching with marking

Input: $(G, \mu, k, \textit{ws-count})$
1. Apply marking update rules.
2. If possible: apply low-degree reduction rule and goto 1.
3. If possible: apply a worsening step and goto 1.
4. Possibly, the graph is now empty or the parameter is no longer positive $\rightsquigarrow$ Stop!
5. Let x be a highest-degree vertex (with $\mu(x) \leq 0$).
Branch as follows: either increment the marking of x, or of all of its neighbors, deleting x.

by one, as only one of the two vertices is "for sure". The worsening step is correct, as out of u, v, w, at least two vertices, namely u (due to being 1-marked) and v or w (due to the edge vw) must be in any exact total cover (respecting previous choices). It is important to note that after applying one worsening step, the marking update rules trigger again. They will finally delete all vertices u, v, w that took part in the worsening step. Hence, after applying the marking update rules, a vertex marked 1 was marked due to branching or due to (exact) reduction rules and not due to a worsening step. This is an important second invariant of our algorithm. In particular, in the second recursive branch, the 1-marked (at least two) vertices are worsened one by one. By some additional book-keeping (using the counter $\textit{ws-count}$) we can make sure that, whenever we increment the marking of some y in the branching, k finally drops by at least two. The obvious immediate "danger" for the necessary number of applications of the worsening step comes from possibly neighbored vertices that are 1-marked by branching; these would disappear using the update rules, but we can select one additional edge for worsening purposes anywhere in the graph afterwards.

As the degree of a vertex x that we branch at is at least two, this gives a $\mathcal{O}^*(c^k)$-algorithm for factor-$\frac{3}{2}$ approximation, where $c = \sqrt{\phi} \approx 1.28$, with $\phi \approx 1.62$ being the golden ratio number. Recall again that irrespectively of whether the marking of x or of all of its neighbors is incremented, which would let the parameter drop by some number ℓ, with $\ell = 1$ in the first branch and $\ell \geq 2$ in the second one, by triggering either reduction rules or the worsening rule, the parameter can be decreased by another amount of ℓ. This is already quite nice, but we can further improve on this due to some additional structural properties as explained in the following.

3   Improved Branching

One main source of profit in the VERTEX COVER approximation algorithm from [3] was the following structural lemma:

Lemma 2. *In a triangle-free graph with minimum degree $\delta \geq 2$, any vertex v has a vertex u at distance two.*

In order to make use of it, we must deal with triangles. This can be done by the following rule, which is obviously true for total vertex covers, as well, in the sense of guaranteeing a factor of $\frac{3}{2}$. So, the Triangle Rule is not an exact reduction rule but an approximate one, as we can be (only) sure that two out of the three vertices that we have put into the cover will appear in any optimum solution. This is important for the sequence of rule applications, as the Triangle Rule should be applied after the worsening steps and then trigger again the marking update rules.

Triangle Rule: If G is an μ-marked updated graph instance and x, y, z form a triangle in G, then 1-mark x, y, z in G and decrease the parameter k by two.

To further improve on our branching, we also show how to deal with vertices of degree two; this is again an approximate rule and should be applied after the worsening steps.

Degree-2 Rule: If G is a μ-marked updated triangle-free graph instance with only unmarked and 0-marked vertices, and if y is an unmarked vertex of degree two, with neighbors x and z, then 2-mark x, y, z in G and decrease the parameter k by two.

This rule is correct, as even if y is put into the cover, x or z must be put into the cover to satisfy the totality requirement.

Further notice that the (exact) Folding Rule known from VERTEX COVER, see [6], can be applied if y is 0-marked, as well as its two neighbors x and z; recall that the folding rule allows to decrement the parameter and that the folded vertex should be also 0-marked. This rule is an exact rule and could be applied before the worsening steps.

So, at this point of the algorithm, immediately before branching, the only remaining degree-2 vertices are 0-marked and have at least one unmarked neighbor, as we consider only μ-marked updated triangle-free graphs with unmarked and 0-marked vertices.

Claim: No branching algorithm that works like the one in Table 1, plus integrating the previous two reduction rules, ever picks a vertex of degree two for branching.

Proof. Assume that the claim would be false. Then, the algorithm would pick a 0-marked vertex x for branching. At least one of its neigbors, say y, is unmarked. Clearly, y has degree at most two. But this means that y would have been eliminated by the Degree-2 Rule (or some other rule).   □

We can conclude that the simple branching algorithm would run in time $\mathcal{O}^*(1.211^k)$ for a factor-$\frac{3}{2}$ approximation: in the first branch, one vertex is put into the cover (plus two more due to worsening), and in the second, at least $\ell \geq 3$ (plus 2ℓ more by worsening); this yields a branching vector of $(2, 2\ell)$ which is no worse than $(2, 6)$.

We are now going to explain an even better branching strategy akin to the one employed in [3]. By our arguments above, the graph before doing any branching has minimum degree two and is triangle-free. These are the assumptions of Lemma 2; so there are two vertices x, z with $\text{dist}(x, z) = 2$. Notice again that at this stage of the algorithm, all vertices are unmarked or 0-marked.

We branch on these vertices as follows: Consider a path xyz of length two and apply the following reasoning.

1. If at least one of the end vertices of the path, that is, x or z, is in some minimum cover (respecting the previous choices) then the cover remains unchanged if we add the edge xz and therefore we can treat xyz as a (hypothetical) triangle and add all three vertices to the approximate cover by marking them with one. Clearly, the parameter should drop by two.

2. In the case that neither of the end vertices of the path (x and z) belongs to any minimum vertex cover (respecting the previous choices), then we remove x, z from the graph and add $N(x) \cup N(z)$ to the (approximate) cover by 1-marking these vertices.

   Since we are not after a minimum total vertex cover but rather a factor-$\frac{3}{2}$ approximation, we select $|N(x) \cup N(z)|$ disjoint edges in the subsequent worsening steps and add their end vertices to the approximate cover. This will reduce the parameter by at least $2|N(x) \cup N(z)|$. Notice that if we cannot find enough edges in the worsening steps, then this branch of the search tree will stop, so that this special case does not affect the running time analysis.

Due to our claim, we never invoke this branching on a vertex x with $d(x) = 2$. As we will see, this branching strategy gives a (2,10) branch (or better), leading to a branching number of 1.151.

Theorem 1. TOTAL VERTEX COVER *can be approximated in time* $\mathcal{O}^*(1.151^k)$ *up to a factor of* $\frac{3}{2}$.

Proof. Let us analyze the suggested branching strategy of the algorithm summarized in Table 2.

Clearly, if all vertices u at distance two from some degree-3 or degree-4 vertex x selected for branching satisfy $N(u) \subseteq N(x)$, then $N[N[x]]$ is a small component that can be solved to optimality in constant time.

If $|N(x) \cup N(z)| \geq 5$ for some z with $\text{dist}(x, z) = 2$, then this gives a (2,10) branch (or better), leading to a branching number of 1.151. We refer to this case by $(*)$ in the following. Therefore, the claimed running time holds in particular if $d(x) \geq 4$ due to the previous paragraph, as either $d(x) \geq 5$ or $d(x) = 4$ and some vertex z at distance two from x obeys $N(z) \setminus N(x) \neq \emptyset$. The running time claim can be also confirmed if $|N(x) \cup N(z)| = 4$ by some case analysis that is similar to the one for VERTEX COVER presented in the long version of [3] but distinctively different in several details; more precisely, one of the following cases applies.

(1) There are no vertices at distance three from x; the corresponding small component can be solved in constant time.

(2) There is exactly one vertex w at distance three from x: then, either we have again a small component or w is a cut vertex; then, we can then argue that a *cut vertex branching* is beneficial. (Details on cut vertex branching in more general terms can be found in the long version of [3].)

More precisely, as x has degree three and this is the maximum degree in the graph, the neighbors y, y', y'' of x each have at most two further neighbors. By the Degree-1 Rule, we can assume that all vertices at distance three from x, collected in the set Z, have at least two neigbors. Since w, being a cut vertex,

Table 2. Improved factor-$\frac{3}{2}$ parameterized approximation for TVC

Input: $(G, \mu, k, \textit{ws-count})$
1. Apply marking update rules.
2. If possible: apply exact low-degree reduction rule and goto 1.
3. If possible: apply a worsening step and goto 1.
   {Now, $\forall v : \mu(v) \leq 0$.}
4. If possible: apply approximate reduction rule and goto 1.
   {Now, $\forall v : \mu(v) \leq 0$ and the graph has minimum degree at least two.}
5. Possibly, the graph is now empty or the parameter is no longer positive ⤳ Stop!
6. Let x be a highest-degree vertex. Select a path xyz.
   Branch as follows:
   Either increment marking all of x, y, z (decrease k by two)
   or of all of $N(x) \cup N(z)$ (delete x, z, decrease k by $\textit{ws-count}$, where $\textit{ws-count} \leftarrow |N(x) \cup N(z)|$).

---

has one or two neighbors in Z, so that $2 \leq |Z| \leq 4$. The graph induced by $\{x, y, y', y''\} \cup Z$ has TVC number at least three, and the graph induced by $\{x, y, y', y''\}$ and at least one vertex from Z has TVC number at least two (no triangles), so that we can reduce (due to worsening steps) the parameter k by at least 8 if w goes into the approximate total vertex cover and by $2 \cdot (d(w) + 2) \geq 8$ if w is not put into the cover but $N(w)$, assuming that the small component containing x is solved in constant time, so that the branching continues with the graph component(s) that do not contain x. The branching number is below 1.1. (3) There are at least two vertices w_1, w_2 at distance three from x. Then, we can (again) branch on a pair (x, z) with $z \in Z$. If $N(x) \cup N(z)$ goes into the cover, then first notice that any $z' \neq z$, $z' \in Z$, would now be of degree one (so that the according rule would apply) if $d(z') = 2$ before the branching. As this case can be analyzed as in (∗). So, we can assume that all vertices in Z are of degree three to create a bad case. A similar argument applies to the neighbors of x (now to improve the branching when x and z go into the cover): also they (all) must be of degree three for a bad case. If any $z, z' \in Z$ have a common neighbor, say y', (while y is the "usual" common neighbor of x and z) that is also a neighbor of x, then removing xyz would turn y' into a degree-1 vertex, again yielding a better branching. So, Z-vertices have exactly one neighbor within $N(x)$, and all two other neigbors are at distance 2 and 3 from x. Therefore, $|N(x) \cup N(z)| \geq 5$ for any z at distance two from x, contradicting our case distinction.   □

4   Further Consequences and Questions

Algorithms based on fidelity-preserving transformations [7] scale well on TVC in the sense that, whatever approximation factor we aim at, we can obtain it, together with an according running time upper bound. As we use the same local-ratio type argument and our worsening step is nothing else than a $(1, 1)$-reduction step, we can conclude:

Corollary 1. *For any $\alpha \geq \frac{3}{2}$, there exists an $\mathcal{O}^*(1.3248^{(2-\alpha)k})$ time, factor-α approximation for TVC.*

For better approximation ratios, our technique cannot avoid finally resorting to exact algorithms. Recall that $(1,1)$-reduction steps can be used prior to any exact algorithm. Similarly, we can run our factor-$\frac{3}{2}$ approximation algorithm with some appropriately scaled lower budget and then start the exact algorithm; here, we can profit from the fact that seach-tree algorithm can cope with μ-marked instances. We have to find some maximum β such that $\beta \cdot \frac{3}{2} + (1 - \beta) \leq \alpha$, *i.e.*, $\beta = 2\alpha - 2$. So, we run our optimized factor-$\frac{3}{2}$ approximation algorithm with parameter $\beta \cdot k$ and then the exact algorithm with parameter $(1 - \beta) \cdot k$.

Corollary 2. *For any $\alpha \leq \frac{3}{2}$, there exists an $\mathcal{O}^*(1.3248^{(\alpha-1)k} \cdot 2.3655^{(3-2\alpha)k})$ time, factor-α approximation for TVC.*

A *t-total vertex cover* of a graph G is a vertex cover S that each connected component of the subgraph of G induced by S has at least t vertices. Examples: $t = 1$: Classical VC; $t = 2$: total VC. For the cycle C_n, $n \geq t$, a minimum t-total vertex cover has $\lceil \frac{t}{t+1} n \rceil$ many vertices. Accordingly, we get minimization and parameterized problems termed t-TOTAL VERTEX COVER.

Taking all vertices of a path on t vertices into the cover yields a $(\lfloor \frac{t}{2} \rfloor, \lceil \frac{t}{2} \rceil)$ reduction step for $t \geq 2$. As shown in [8], exact parameterized algorithm running in time $\mathcal{O}^* \left(16.1^{k+\mathcal{O}(\log^2 k)} \right)$ (for any fixed t) are known. The approach of M. Fellows *et al.* [7] would yield:

Corollary 3. *For any $t \geq 2$ and $1 \leq \alpha \leq 2$, t-TVC can be approximated up to a factor of α in time $\mathcal{O}^* \left(16.1^{\gamma_t(\alpha)k+\mathcal{O}(\log^2 k)} \right)$, where $\gamma_t(\alpha) = \frac{t-\alpha\lfloor \frac{t}{2} \rfloor}{\lceil \frac{t}{2} \rceil}$.*

In concrete terms, this is an $\mathcal{O}^* \left(16.1^{\frac{3-\alpha}{2}k+\mathcal{O}(\log^2 k)} \right)$-time, factor-$\alpha$ approximation for 3-TVC, leading to some $\mathcal{O}^*(8.1^{k+\mathcal{O}(\log^2 k)})$ factor-$\frac{3}{2}$ algorithm.

Now, reconsider our factor-$\frac{3}{2}$ algorithm for TVC again. The worsening steps are used in a way to make sure that only paths of cover vertices of length two are created. This is in particular true for the simplistic branching analyzed in Section 2. For the improved branching, additional arguments would be necessary to make it work. Hence, we only state the following somewhat weaker consequence, which is nonetheless much better than what was stated above:

Corollary 4. *3-TVC is approximable in time $\mathcal{O}^*(1.273^k)$ up to a factor of $\frac{3}{2}$.*

We leave out the details of a corresponding algorithm, as we would need to adapt, for instance, the reduction rules to cope with other markings. They should now indicate that a cover vertex already belongs to a cover component with two but not with three (or more) elements, or that a non-cover vertex is neighbor of a cover vertex that (at that point) only stays on its own in the cover.

VERTEX COVER and all variants discussed in this paper allow for polynomial-time approximations up to a factor of two. Actually, the local-ratio technique shows this factor for TVC (and also, for instance, for 4-TVC) by "exhaustively" using the $(1,1)$-reduction step (or $(2,2)$-reduction step for 4-TVC).

Similar problems (showing the same approximation factors) are EDGE DOMINATING SET, CONNECTED VERTEX COVER and also weighted problem variants. The two mentioned problems also admit parameterized algorithms based on a 2-phase approach; see [1]. It is however unclear how to obtain good parameterized approximation algorithms, avoiding the costly second phase.

References

1. Binkele-Raible, D., Fernau, H.: Parameterized measure & conquer for problems with no small kernels. Algorithmica 64, 189–212 (2012)
2. Bourgeois, N., Escoffier, B., Paschos, V.T.: Approximation of max independent set, min vertex cover and related problems by moderately exponential algorithms. Discrete Applied Mathematics 159(17), 1954–1970 (2011)
3. Brankovic, L., Fernau, H.: Combining Two Worlds: Parameterised Approximation for Vertex Cover. In: Cheong, O., Chwa, K.-Y., Park, K. (eds.) ISAAC 2010, Part I. LNCS, vol. 6506, pp. 390–402. Springer, Heidelberg (2010)
4. Brankovic, L., Fernau, H.: Parameterized Approximation Algorithms for HITTING SET. In: Solis-Oba, R., Persiano, G. (eds.) WAOA 2011. LNCS, vol. 7164, pp. 63–76. Springer, Heidelberg (2012)
5. Chawla, S., Krauthgamer, R., Kumar, R., Rabani, Y., Sivakumar, D.: On the hardness of approximating multicut and sparsest-cut. Computational Complexity 15(2), 94–114 (2006)
6. Chen, J., Kanj, I.A., Jia, W.: Vertex cover: further observations and further improvements. Journal of Algorithms 41, 280–301 (2001)
7. Fellows, M.R., Kulik, A., Rosamond, F., Shachnai, H.: Parameterized Approximation via Fidelity Preserving Transformations. In: Czumaj, A., Mehlhorn, K., Pitts, A., Wattenhofer, R. (eds.) ICALP 2012, Part I. LNCS, vol. 7391, pp. 351–362. Springer, Heidelberg (2012)
8. Fernau, H., Fomin, F.V., Philip, G., Saurabh, S.: The Curse of Connectivity: t-Total Vertex (Edge) Cover. In: Thai, M.T., Sahni, S. (eds.) COCOON 2010. LNCS, vol. 6196, pp. 34–43. Springer, Heidelberg (2010)
9. Fernau, H., Manlove, D.F.: Vertex and edge covers with clustering properties: Complexity and algorithms. Journal of Discrete Algorithms 7, 149–167 (2009)
10. Khot, S.: On the power of unique 2-prover 1-round games. In: Reif, J.F. (ed.) Proceedings on 34th Annual ACM Symposium on Theory of Computing, STOC, pp. 767–775. ACM Press (2002)
11. Khot, S., Regev, O.: Vertex cover might be hard to approximate to within $2 - \varepsilon$. Journal of Computer and System Sciences 74, 335–349 (2008)
12. Marx, D., Razgon, I.: Constant ratio fixed-parameter approximation of the edge multicut problem. Information Processing Letters 109(20), 1161–1166 (2009)
13. Raman, V., Ramanujan, M.S., Saurabh, S.: Paths, Flowers and Vertex Cover. In: Demetrescu, C., Halldórsson, M.M. (eds.) ESA 2011. LNCS, vol. 6942, pp. 382–393. Springer, Heidelberg (2011)

On Graph Identification Problems and the Special Case of Identifying Vertices Using Paths[*]

Florent Foucaud[1] and Matjaž Kovše[2,3]

[1] Univ. Bordeaux, LaBRI, UMR5800, F-33400 Talence, France
CNRS, LaBRI, UMR5800, F-33400 Talence, France
`florent.foucaud@gmail.com`
[2] Faculty of Natural Sciences and Mathematics, University of Maribor,
SI-2000 Maribor, Slovenia
[3] Bioinformatics Group, Department of Computer Science and Interdisciplinary
Center for Bioinformatics, Univ. of Leipzig, Härtelstrasse 16-18,
D-04107 Leipzig, Germany
`matjaz.kovse@gmail.com`

Abstract. In this paper, we introduce the identifying path cover problem: an *identifying path cover* of a graph G is a set $\mathcal{P}$ of paths such that each vertex belongs to a path of $\mathcal{P}$, and for each pair u, v of vertices, there is a path of $\mathcal{P}$ which includes exactly one of u, v. This problem is related to a large variety of identification problems. We investigate the identifying path cover problem in some families of graphs. In particular, we derive the optimal size of an identifying path cover for paths, cycles, hypercubes and topologically irreducible trees and give an upper bound for all trees. We give lower and upper bounds on the minimum size of an identifying path cover for general graphs, and discuss their tightness. In particular, we show that any connected graph G has an identifying path cover of size at most $\left\lceil \frac{2(|V(G)|-1)}{3} \right\rceil$. We also study the computational complexity of the associated optimization problem, in particular we show that when the length of the paths is asked to be of a fixed value, the problem is APX-complete.

Keywords: Test cover, Identification, Paths, Approximation.

1 Introduction

This paper aims to study the new optimization problem of identifying the vertices of a graph by means of paths, which we call the *identifying path cover problem*. We first relate this problem to a large number of other problems and review a part of the associated literature, before giving its definition.

---

[*] This research is supported by the ANR Project IDEA - Identifying coDes in Evolving grAphs, ANR-08-EMER-007, 2009-2012.

S. Arumugam and B. Smyth (Eds.): IWOCA 2012, LNCS 7643, pp. 32–45, 2012.

1.1   On Test Covers and the Identification Problem

Identification problems have been addressed many times in the last decades under different denominations and in different contexts. We present two general problems from the literature which have almost the same definition, and which we herein call the minimum test cover problem and the minimum identification problem. Instances of these problems are set systems, i.e. pairs consisting of a set $\mathcal{I}$ of elements ("individuals") and a set $\mathcal{A}$ of subsets of $\mathcal{I}$ ("attributes").

Among these two problems, the *minimum test cover problem*, in short MIN-TC, seems to have been studied first and is probably better known. Given a set system of individuals and attributes, the MIN-TC problem asks for a minimum subset $\mathcal{C}$ of $\mathcal{A}$ such that for each pair I, I' of $\mathcal{I}$, there is an element C of $\mathcal{C}$ such that exactly one of I, I' is *covered* by C, that is, belongs to C (we say that C *separates* I from I'). The MIN-TC problem appears in a large number of papers under different denominations (*minimum test cover problem* [8], *minimum test collection problem* [13], *minimum test set problem* [18]). In fact, a well-celebrated theorem of J. A. Bondy on *induced subsets* [3] can be seen as the first study of this problem.

In this paper and as in a large portion of the literature dealing with special cases of this kind of problems, we are interested in a slight modification of MIN-TC, where not only each pair of individuals has to be separated, but also, each individual has to be covered. We call this problem the *minimum identification problem*, MIN-ID for short (note that it has been studied under the denomination of *discriminating code problem* in [4], but we use our terminology in order to fit to special cases described later). MIN-TC and MIN-ID are very close to each other, since for any solution to one of them, there is a solution to the other one whose size differs by at most 1: any solution to MIN-ID is also one for MIN-TC, and, given a solution $\mathcal{C}$ to MIN-TC which is not a valid solution to MIN-ID, at most one individual I may not be covered by $\mathcal{C}$. It is then sufficient to add an arbitrary attribute A covering I to $\mathcal{C}$ to get a valid solution to MIN-ID.

Both MIN-TC and MIN-ID can be seen as special cases of the well-known *minimum set cover problem* [13,15], MIN-SC for short, where, given a base set $\mathcal{X}$ and a set $\mathcal{S}$ of subsets of $\mathcal{X}$, it is asked to find a minimum subset $\mathcal{C}$ of $\mathcal{S}$ covering all elements of $\mathcal{X}$ [8]. MIN-TC and MIN-ID enjoy the same computational complexity. It is known that both problems are $O(\ln(|\mathcal{I}|))$-approximable (where $\mathcal{I}$ denotes the set of individuals of the input) using a reduction to MIN-SC [18]. On the other hand, both problems are not only NP-hard [4,13] but have also been shown to be NP-hard to approximate within a factor of $o(\ln(|\mathcal{I}|))$ by reduction from MIN-SC [2,8].

A natural restriction of MIN-ID is, given some integer k, the one where the sets of $\mathcal{A}$ all have exactly k elements. We will call this problem MIN-ID-k.

1.2   Related Problems

In this paper, we study a special case of MIN-ID. Just as some particular cases of MIN-SC arising from specific structures have gained a lot of interest (consider for

example all variants of the minimum dominating set problem, or the minimum
vertex cover problem), it is of interest to investigate special cases of the MIN-ID
problem having a particular structure. In this line of research, many specific
cases arising from graph theory are of particular interest since graphs model
networks of all kinds and are found in real world applications. For example, in
the *identifying code problem* [9,12,16], one wants to identify each vertex v using
vertices at distance at most 1 from v. This problem can be seen as MIN-ID where
$\mathcal{I} = V(G)$ and $\mathcal{A}$ is the family of the balls around each vertex. This problem has
been generalized to digraphs [6,11], and to the case where also *sets* of at most ℓ
vertices are to be separated and where vertices can identify at some prescribed
distance $r \geq 1$ [12]. One may also ask to identify the edges of G using edges, i.e.
$\mathcal{I} = E(G)$ and $\mathcal{A}$ is the set of all edge-balls around each edge of G [10]. Rather
than considering full balls, also partial balls may be considered, as in the case
of *watching systems* [1], where $\mathcal{I} = V(G)$, and $\mathcal{A}$ is the family of all stars in G.
Finally, the case where $\mathcal{I} = V(G)$ and $\mathcal{A}$ is the set of all cycles in G has been
considered in [14,20].

1.3   The Identifying Path Cover Problem

In this paper, we study MIN-ID when $\mathcal{I} = V(G)$ and $\mathcal{A}$ is the set of all paths
of G. This problem was first mentioned in a discussion between the first author,
J. L. Sewell and P. J. Slater. We call it *minimum identifying path cover problem*,
MIN-IDPC for short and it studies the following notion:

Definition 1. *Given a graph G, a set $\mathcal{P}$ of paths of G is an* identifying path
cover *if each vertex of G belongs to a path of $\mathcal{P}$ (it is* covered*) and if for each
pair u, v of vertices, there is a path of $\mathcal{P}$ which contains exactly one of u, v (u, v
are* separated*).*

We point out that the covering condition is not implied by the separation condition, since even when all pairs are separated, one vertex of the graph may remain
uncovered. We denote by $p^{\text{ID}}(G)$ the minimum number of paths required in any
identifying path cover of G. Then, MIN-IDPC is the problem, given a graph G,
of determining the value of $p^{\text{ID}}(G)$. An example of an identifying path cover $\mathcal{P}$
of the cube H_3 is given in Fig. 1, where the four thick paths belong to $\mathcal{P}$ (the
full, the densely dotted, the loosely dotted and the dashed-dotted path). Note
that an identifying path cover of G always exists: consider the set of all 0-paths
of G, that is, $\mathcal{P} = V(G)$.

Given an integer $k \geq 1$, we will also discuss the natural variant MIN-IDPC-k
of MIN-IDPC, where one wants to find a minimum *identifying k-path cover of G*,
that is, a set of paths of exactly k vertices forming an identifying path cover of G.
We denote by $p_k^{\text{ID}}(G)$ the size of a minimum identifying k-path cover of G. Unlike
for the general MIN-IDPC problem, not all graphs admit an identifying k-path
cover. We call a graph admitting an identifying k-path cover, *k-path identifiable*.
This is the case if, first of all, each vertex of G lies on a k-path, and if for each pair
u, v of vertices, there is a k-path covering exactly one of u, v. For example, the

Fig. 1. An (optimal) identifying path cover of the hypercube H_3: $p^{ID}(H_3) = 4$

path graphs P_{k-1} and P_{2k-2} are not k-path identifiable. Observe that these two conditions are also sufficient: if both are fulfilled, taking all k-paths of G gives a valid identifying k-path cover of G. Being k-path identifiable is polynomial-time checkable since there are at most $\binom{n}{k} = O(n^k)$ k-paths in G.

1.4 Applications

Problems MIN-TC and MIN-ID have a broad variety of applications, for example in the diagnosis of faults or diseases, biological identification, pattern recognition [8,18]. When the instance of the problem arises from substructures of a graph, the main applications are routing in networks [17] and the location of threats in facilities or networks using sensors [16]: vertices are the "individuals", sensors are the "attributes". Sensors may monitor closed neighbourhoods (identifying codes) or sub-neighbourhoods (watching systems). If sensors are capable of monitoring the vertices lying on a path, we have the situation of an identifying path cover. One can for example imagine sensors in the form of laser detectors, or mobile detecting devices patrolling back and forth along their path.

1.5 Outline of the Paper

We start by giving some preliminary results in Section 2, in the form of bounds from the literature valid for the general MIN-ID problem (which we apply to MIN-IDPC) and some observations valid only for MIN-IDPC. We continue by studying MIN-IDPC in some basic families of graphs in Sections 3 and 4: we give exact values for parameter p^{ID} in paths, cycles, topologically irreducible trees, and an upper bound for trees in general. We use the latter to provide the upper bound $p^{ID}(G) \leq \lceil \frac{2n}{3} \rceil$ for any connected graph. Finally, we show in Section 5 that MIN-IDPC-k is APX-complete for any $k \geq 3$ by means of an L-reduction from the minimum vertex cover problem. We conclude with some open questions in Section 6.

2   Preliminary Observations

The following lower bound was observed in [16] in the context of identifying codes but we refer to [4] for the general statement.

Theorem 2 ([4,16]). *Let $(\mathcal{I}, \mathcal{A})$ be an instance of MIN-ID, and let $\mathcal{C}$ be a solution to it. Then $|\mathcal{C}| \geq \log_2(|\mathcal{I}| + 1)$.*

The following upper bound can be seen as a direct corollary of Bondy's theorem [3]. We refer to [4] for a formal proof in this context.

Theorem 3 ([3,4]). *Let $(\mathcal{I}, \mathcal{A})$ be an instance of MIN-ID, and let $\mathcal{C}$ be an inclusionwise minimal solution to it. Then $|\mathcal{C}| \leq |\mathcal{I}|$.*

Consider an instance of MIN-ID-k. Then, another lower bound holds. This bound was (to our knowledge) first observed in the context of identifying codes in [16], but the proof works in the more general context of MIN-ID-k.

Theorem 4 ([16]). *Let $k \geq 1$ and $(\mathcal{I}, \mathcal{A})$ be an instance of MIN-ID-k. Then for any solution $\mathcal{C}$, $|\mathcal{C}| \geq \frac{2|\mathcal{I}|}{k+1}$.*

Applying theorems 2, 3 and 4 to MIN-IDPC and MIN-IDPC-k, we get:

Theorem 5. *Let G be a graph on n vertices and $k \geq 1$ an integer. Then $\log_2(n+1) \leq p^{ID}(G) \leq n$ and $\max\{\log_2(n+1), \frac{2n}{k+1}\} \leq p_k^{ID}(G) \leq n$.*

It is easily observed that in the complete graph K_n, since we have full freedom to choose the paths in the identifying path cover, $p^{ID}(K_n) = \lceil \log_2(n+1) \rceil$. In fact, much sparser graphs also fulfill this bound, such as the hypercubes: one can easily come up with a solution with $\lceil \log_2(n+1) \rceil$ paths. A similar problem of identification using cycles is addressed in [14,20]; we refer to these papers for the construction. Since removing an edge from a cycle yields a path, their construction is also valid in our case:

Theorem 6 ([14,20]). *Let H_d be the hypercube of dimension d with $n = 2^d$ vertices. Then $p^{ID}(H_d) = \lceil \log_2(n+1) \rceil$.*

One can easily see that the bound $p_k^{ID}(G) \geq \frac{2n}{k+1}$ is tight; given two integers $k \geq 1$ and $p \geq k$, one can construct a graph G with $p_k^{ID}(G) = p$. To do so, one has to take care that for each of the p paths of the solution, there is a vertex that belongs only to this path. All other vertices must belong to distinct sets of exactly two paths of the solution.

Since the set of paths of a graph G is a superset of the set of paths of a subgraph H of G, if H is spanning the vertices of G, any identifying path cover of H will also be one for G. We get the following proposition:

Proposition 7. *Let G be a graph and H a spanning subgraph of G. Then $p^{ID}(G) \leq p^{ID}(H)$.*

The following proposition will be useful. The bound will be shown to be tight for the star (see Thm. 12).

Proposition 8. *If G is a graph having l vertices of degree 1, $p^{ID}(G) \geq \lceil \frac{2l}{3} \rceil$.*

Proof. A vertex v of degree 1 can only be covered by a path P if v is an endpoint of P, and two vertices of degree 1 cannot be covered only by the same path (otherwise they are not separated from each other). If a degree 1 vertex is the endpoint of k paths, then these k paths can cover at most $k+1$ degree 1 vertices. Hence, the minimum is reached when $k = 2$ when two degree 1 vertices are identified with three paths. $\square$

3 Identifying Path Covers of Paths and Cycles

We first investigate identifying path covers in simple graphs such as paths and cycles. The path and the cycle on n vertices are denoted P_n and C_n, respectively. We start with a lower bounds for these graphs.

Proposition 9. *Let G be a connected graph of maximum degree 2 having m edges and l vertices of degree one. Then $p^{ID}(G) \geq \lceil \frac{m+l}{2} \rceil$.*

Proof. Let u,v be two adjacent vertices of G. In any identifying path cover $\mathcal{P}$ of G, there must be a path P that either ends in u and does not contain v, or ends in v and does not contain u (let us say that P *cuts* the edge uv). Moreover, for any vertex of degree 1, there is a path of $\mathcal{P}$ that ends in it. Since one single path can at most cut or cover two edges/degree 1 vertices, the result follows. $\square$

Theorem 10. *For any $n \geq 1$, $p^{ID}(P_n) = \lceil \frac{n+1}{2} \rceil$.*

Proof. The lower bound comes from Prop. 9. For the upper bound, let $V(P_n) = \{v_0, ..., v_{n-1}\}$ and $\mathcal{P} = \{v_i...v_{i+\lceil \frac{n}{2} \rceil} \mid i \in \{0, ..., \lceil \frac{n}{2} \rceil - 1\}\}$ be a set of $\lceil \frac{n}{2} \rceil$ paths. If n is odd, $\mathcal{P}$ is an identifying path cover of cardinality $\lceil \frac{n}{2} \rceil = \lceil \frac{n+1}{2} \rceil$. If n is even, $\mathcal{P}$ separates all pairs of vertices, and covers all vertices but v_{n-1}. Hence, $\mathcal{P} \cup \{v0, ..., v_{n-1}\}$ is an identifying path cover of cardinality $\lceil \frac{n+1}{2} \rceil$. $\square$

Theorem 11. *It holds that $p^{ID}(C_3) = 2$, $p^{ID}(C_4) = 3$ and for any $n \geq 5$, $p^{ID}(C_n) = \lceil \frac{n}{2} \rceil$.*

Proof. For $n \neq 4$, the lower bounds come from Prop. 9, and from Thm. 2 for $n = 4$. We give constructions for the upper bounds. Let $V(C_n) = \{v_0, ..., v_{n-1}\}$. One can check that $\{v_0v_1, v_1v_2\}$ and $\{v_0v_1, v_1v_2, v_2v_3\}$ are valid identifying path covers of C_3 and C_4. For $n \geq 5$, let $\mathcal{P} = \{v_iv_{i+1}v_{i+2} \mid i \text{ even}, i < n - 1\}$. If n is even, $\mathcal{P}$ is a identifying path cover of C_n of cardinality $\lceil \frac{n}{2} \rceil$. Otherwise, the pairs v_0, v_1 and v_{n-2}, v_{n-1} are covered but not separated. Then $\mathcal{P} \cup \{v_{n-1}v_0\}$ is an identifying path cover of C_n of cardinality $\lceil \frac{n}{2} \rceil$. $\square$

4 The Case of Trees with an Application to All Graphs

We start by giving the value of parameter p^{ID} for the star on n vertices, denoted $K_{1,n-1}$. The provided construction and bound will prove useful in what follows.

Theorem 12. *It holds that $p^{ID}(K_{1,n-1}) = \lceil \frac{2(n-1)}{3} \rceil$.*

Proof. The lower bound follows from Prop. 8. Let $v_0, ..., v_{n-1}$ be the leaves of $K_{1,n-1}$ and c its central vertex. Let $\mathcal{P}$ be the set $\mathcal{P} = \{v_icv_{i+1} \mid i \neq 2 \bmod 3\}$ of $2\lfloor \frac{n}{3} \rfloor$ paths. If $n = 0 \bmod 3$, $\mathcal{P}$ is an identifying path cover of $K_{1,n-1}$. If $n = 1 \bmod 3$, $\mathcal{P} \cup \{cv_{n-1}\}$ is, and if $n = 2 \bmod 3$, $\mathcal{P} \cup \{cv_{n-2}, cv_{n-1}\}$ is. $\square$

We call the procedure used in the proof of Thm. 12 "covering three leaves with two paths". A tree is *topologically irreducible* if it has no vertex of degree 2.

Theorem 13. *Let T be a tree with l leaves. Then we have*

(i) if T is topologically irreducible, then $p^{ID}(T) = \lceil \frac{2l}{3} \rceil$,
(ii) if T has t vertices of degree two, then $\lceil \frac{2l}{3} \rceil \leq p^{ID}(T) \leq \lceil \frac{2l}{3} \rceil + \lceil \frac{t}{2} \rceil$.

Proof. The lower bound in both cases follows from Prop. 8.

If T is topologically irreducible, we show how to construct an identifying path cover which size is meeting the lower bound.

First, determine the *center* of T (that is, the set of vertices of minimum largest distance to any other vertex of T). By Jordan's theorem, the center of a tree consists of either a single vertex or a pair of adjacent vertices.

Starting from the center of T, decompose the vertex set of T into *layers* labelled $0, \ldots, h$, where h is the radius of T (the minimum largest distance among pairs of vertices of T). The labels correspond to the distance to the center. For $\ell \in \{0, \ldots, h\}$, let $T_{\leq \ell}$ be the sub-tree of T induced by layers $0, ..., \ell$.

In the case when the center of T consists of a single vertex $T_{\leq 1}$ is isomorphic to a star, and in the case when the center consists of two adjacent vertices, it is isomorphic to a tree with two adjacent vertices of degree at least 3, and all other vertices of degree 1. In both cases, it is straightforward to find a solution of size $\lceil \frac{2l}{3} \rceil$.

For $i \in \{2, \ldots, h\}$ we now describe how to extend the valid solution of $T_{\leq i-1}$ to the solution of $T_{\leq i}$. For any vertex v from layer $i - 1$, we choose an arbitrary neighbour from layer i and extend the corresponding paths (by our construction always exactly one or two paths are ending in a leaf) from the identifying path cover to the neighbour. Now we have two vertices covered by exactly the same set of paths, and still many uncovered vertices, all of them from layer i (at least one more uncovered neighbour for any vertex from layer $i - 1$). Now we contract all covered vertices into a single vertex, obtaining a star. Then we order (arbitrarily) all other remaining vertices and do the procedure "covering three leaves with two paths", as done in the case of the proof of Thm. 12. Doing this we separate all uncovered vertices from layer i. We then expand the obtained star together with the chosen paths into the original tree, where a path between two leaves expands through the unique shortest path in T between these two leaves. Doing this we also separate the two vertices, one from layer $i - 1$ and the other from layer i, which were sharing the same paths. Vertices from layers $0, 1, \ldots, i - 2$ still remain separated and covered.

Doing this procedure until we reach layer h and tree $T_{\leq h} = T$, we obtain at each layer i a solution of size $\left\lceil \frac{2(l_i - 1)}{3} \right\rceil$ for a tree $T_{\leq i}$, where l_i denotes the number of vertices from layer i. At the end we obtain a solution of size $\lceil \frac{2l}{3} \rceil$ for $T_{\leq h} = T$. This concludes the proof of the first part of the theorem.

If T is arbitrary, we first contract all vertices of degree 2 to obtain the topologically irreducible tree T' and find a solution for T' as described above. Then we subdivide edges of T' to obtain tree T, keeping (expanding) the same identifying path cover. Now, observe that we may get pairs of vertices which are not separated from each other. Each such pair contains a vertex of degree 2. Using a similar procedure as in the proof of Thm. 10, we can use $\lceil \frac{t}{2} \rceil$ additional paths to get a solution reaching the upper bound from (ii). □

By Prop. 7, we get an identifying path cover $\mathcal{P}$ of a connected graph G by choosing a spanning tree T of G and constructing $\mathcal{P}$ using Thm. 13 on T. We get the following improvement of Thm. 3 for the case of identifying path covers of connected graphs. Note that by Thm. 12, this bound is tight for stars.

Theorem 14. *For any connected graph G on n vertices, $p^{ID}(G) \leq \left\lceil \frac{2(n-1)}{3} \right\rceil$.*

Unlike for many other variants of identification problems (such as identifying codes, see [9]), Thm. 14 shows that one needs much less sensors than n in order to identify connected graphs, which may prove useful in practice. We remark that the similar upper bound $\frac{2n}{3}$ holds for the size of a watching system (i.e. an "identifying *star* cover") in any connected graph on n vertices [1].

The bound of Thm. 14 can be refined in the following way. Let $\gamma_C(G)$ denote the connected domination number of a graph G (that is, the minimum size of a dominating set of G inducing a connected subgraph) and let $L(G)$ denote the maximum number of leaves in a spanning tree of G. One can observe that for a connected graph G on n vertices, we have $n = \gamma_C(G) + L(G)$. Hence using Prop. 7 and Thm. 13 we get the following upper bound.

Theorem 15. *For any connected graph G on n vertices, it holds that $p^{ID}(G) \leq \left\lceil \frac{2(n-\gamma_C(G))}{3} \right\rceil + \left\lceil \frac{\gamma_C(G)}{2} \right\rceil$.*

5 On the Complexity of MIN-IDPC-k

In this section, we discuss the computational complexity of MIN-IDPC-k. It is shown in [8] that MIN-IDPC-k is approximable within a factor of $O(\ln(k))$ for any $k \geq 1$. In fact, when $k = 1$, we are allowed only paths of length 0 (that is, vertices) and MIN-IDPC-1 is trivial: the only solution consists of the whole set of vertices. When $k = 2$, we want to identify the vertices using paths of two vertices, i.e. edges. This problem is equivalent to MIN-ID-2, where each attribute is common to exactly two individuals. Indeed, an edge can precisely be seen as such an attribute. This case has already been studied in [8], where a strong link between MIN-ID-2 and the maximum P_3-packing problem was established; the authors give a $\frac{7}{6}$-approximation for MIN-ID-2 and show that it is APX-hard by reduction from the maximum 3-dimensional matching problem.

We next prove that MIN-IDPC-k is APX-hard for all $k \geq 3$, i.e. that there exists a constant c (depending on k) for which MIN-IDPC-k is not c-approximable. We use the framework of L-reductions. We recall the definition of an L-reduction between two optimization problems P and Q in Definition 16. It is known that if such a reduction exists and P is APX-hard, then Q is APX-hard as well. For more details, see [19]. Given an optimization problem P and a solution s to an instance x of P, we denote by $cost_P(x, s)$, the value of s, and by $opt_P(x)$, the value of an optimal solution to x.

Definition 16. *Let P and Q be two optimization problems. An* L-*reduction from P to Q is a four-tuple (f, g, α, β) where f and g are polynomial time computable functions and α, β are positive constants with the following properties:*

1. *Function f maps instances of P to instances of Q and for every instance x of P, $opt_Q(f(x)) \leq \alpha \cdot opt_P(x)$.*
2. *For every instance x of P and every solution y of $f(x)$, g maps the pair $(f(x), y)$ to a solution y' of x such that $|opt_P(x) - cost_P(x, g(f(x), y'))| \leq \beta \cdot |opt_Q(f(x)) - cost_Q(f(x), y')|$.*

The problem *minimum vertex cover in cubic graphs*, MIN-VC-3 for short, given a cubic graph, is to find a minimum set of vertices such that each edge is covered by a vertex of the set. MIN-VC-3 is APX-hard [7].

Theorem 17. *Let $k \geq 3$. There is an L-reduction (with parameters $\alpha = 40k^2 - 116k + 47$ and $\beta = 1$) from MIN-VC-3 to MIN-IDPC-k in graphs of maximum degree 4. Hence MIN-IDPC-k is APX-complete, even in this class of graphs.*

Before giving the proof of Thm. 17, we first provide two useful gadgets and exhibit some of their properties. Given $k \geq 3$, we call these gadgets *k-gadget of type A* (see Fig. 3 for an example when $k = 3$ and Fig. 2 when $k = 4$) and *k-gadget of type B* (see Fig. 4). Both gadgets include an *attachment vertex* which will be identified with a vertex of the rest of the graph in our constructions. The k-gadget of type A is the basis for the construction of the k-gadget of type B, which includes $k - 3$ copies of the k-gadget of type A. The k-gadget of type B is described in Fig. 4, but we define the k-gadget of type A more formally. Using these two gadgets, we construct a *vertex k-gadget* and an *edge k-gadget* which will be needed in the reduction (see Fig. 5). The idea of the k-gadget of type A is to attach it at a vertex and make sure that this vertex can be easily covered and identified by a locally optimal solution; the idea of the k-gadget of type B is to force a path from outside the gadget to go through the attached vertex.

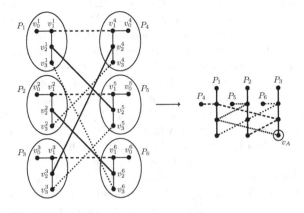

Fig. 2. Construction of the 4-gadget of type A

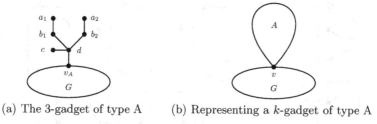

(a) The 3-gadget of type A   (b) Representing a k-gadget of type A

Fig. 3. The k-gadgets of type A

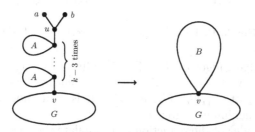

Fig. 4. The k-gadget of type B and its representation

In order to construct the k-gadget of type A, we use a construction of an extremal graph for the lower bound $\frac{2n}{k+1} \leq p_k^{\mathrm{ID}}(G)$ (Theorem 5). This construction is detailed in Definition 18.

Definition 18. Let $k \geq 3$ be an integer. If $k = 3$, the k-gadget of type A with attachment vertex v_A is the graph of Fig. 3. When $k \geq 4$, the k-gadget of type A is constructed as follows (see Fig. 2):

- Let $P_1, \ldots, P_{2(k-1)}$ be $2(k-1)$ vertex-disjoint paths, where for $i \in \{1, \ldots, 2(k-1)\}$, $P_i = \{v_0^i, \ldots, v_{k-1}^i\}$.
- Consider the complete bipartite graph B whose vertices are $P_1, \ldots, P_{2(k-1)}$. It is $(k-1)$-regular.
- Select a $(k-1)$-edge-colouring of B with colours from $\{1, \ldots, k-1\}$ (or equivalently, find a partition of the edges of B into $k-1$ perfect matchings).
- If k is even or $c \notin \{\lfloor \frac{k}{2} \rfloor, \lceil \frac{k}{2} \rceil\}$, for each edge $\{P_i, P_j\}$ (assume $i \leq j$) of B coloured with colour c, identify vertices v_c^i and v_c^j. If k is odd and $c = \lfloor k/2 \rfloor$ (resp. $c = \lceil k/2 \rceil$), identify vertices $v_{\lfloor k/2 \rfloor}^i$ and $v_{\lceil k/2 \rceil}^j$ (resp. $v_{\lceil k/2 \rceil}^i$ and $v_{\lfloor k/2 \rfloor}^j$).
- We let the attachment vertex v_A be vertex v_{k-1}^i for some arbitrary $i \in \{1, \ldots, 2(k-1)\}$.

We let $x_A = 2(k-1)$. Note that when $k \geq 4$, x_A is the number of degree 1 vertices in a k-gadget of type A.

Let G be a cubic graph on n vertices and $m = \frac{3n}{2}$ edges. We construct the graph $f(G)$ by replacing every vertex v by a copy of vertex gadget G_v and each edge

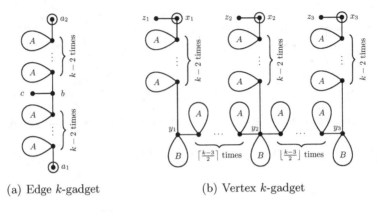

(a) Edge k-gadget   (b) Vertex k-gadget

Fig. 5. Reduction k-gadgets for vertices and edges

e by a copy of edge gadget G_e (see Fig. 5). Given a vertex v incident to edges e_1, e_2, e_3 in G, the vertices x_1, x_2, x_3 of G_v are identified each with either one of the vertices a_1, a_2 of $G_{e_1}, G_{e_2}, G_{e_3}$ in $f(G)$. It is easily noticed that since G is cubic, $f(G)$ has maximum degree 4.

The first main idea of the reduction is to simulate the covering of an edge e by the separation of vertices b and c in G_e by a path going through b. The second main idea is, given a vertex v, to encode the fact that v is part of a vertex cover of G, by having path $y_1 \ldots y_3$ inside the path cover of $f(G)$ (which enables us to "cover" the three edge-gadgets corresponding to the three edges incident to v in G). The proof of the validity of the reduction uses the following Claims 19, 20 and 21 about the gadgets of type A and B.

Claim 19. *Let A be k-gadget of type A ($k \geq 4$). Then, for each pair v, v' of the x_A vertices of degree 1 in A, there is no path of length $k - 1$ between v and v'. Moreover, vertex v_A is at distance at least $k - 1$ of any degree 1-vertex in A.*

Proof. If k is even, then A is bipartite with all degree 1 vertices in the same part. Hence all paths between two degree 1 vertices have even length, but $k - 1$ is odd. If k is odd, by contradiction consider a path P between two degree 1 vertices of A. If for any i, there is no vertex of the form $v^i_{\lfloor k/2 \rfloor}$ or $v^i_{\lceil k/2 \rceil}$ in P, then P cannot be of length $k - 1$ because there is no such path of length more than $k - 3$. Hence P contains some vertex $v^i_{\lfloor k/2 \rfloor}$ or $v^i_{\lceil k/2 \rceil}$. But in either case, P must have at least $k + 1$ vertices, a contradiction. The second part of the statement follows immediately from the choice of vertex v_A in the construction of A.   □

In what follows, we let G be a graph and $\mathcal{P}$, an identifying path cover of G.

Claim 20. *Let A be a k-gadget of type A ($k \geq 3$) attached at vertex v_A in G. Then, there is a set of at least x_A paths of $\mathcal{P}$ having an endpoint in A, and none of these paths can reach a vertex outside of A. Moreover, there is a set of x_A paths in A which can be used to cover and identify all vertices of A.*

Proof. When $k = 3$, we note that we need at least three 3-paths in order to cover the three degree 1 vertices of A. Doing so, we need an additional path to separate either a_1 from b_1 or a_2 from b_2. Finally, the four paths $a_1 \ldots d$, $a_2 \ldots d$, $c \ldots b_1$, $b_2 \ldots v_A$ fulfill the last part of the statement.

If $k \geq 4$, the first part of the statement follows easily from Claim 19: no path can be used to cover two degree 1 vertices in A, and no path starting outside of A can cover a degree 1 vertex of A. Considering the x_A paths of the construction of A proves the second part of the statement.   □

Claim 21. *Let B be a k-gadget of type B attached at some vertex v in G. Then, at least $x_B = x_A(k - 3) + 2$ paths of $\mathcal{P}$ are entirely contained in B. Moreover, if exactly x_B paths of $\mathcal{P}$ are entirely contained in B, then there is an additional path of $\mathcal{P}$ containing vertex v. Finally, there exists such a set of x_B paths.*

Proof. Following Claim 20, we need at least x_A paths in each of the $k - 3$ copies of the k-gadget of type A in B. In order to dominate vertices a and b, we need two additional paths P_a and P_b starting in a and b, which completes the first part of the claim. For the second part, by Claim 20, among these paths, only paths of type P_a or P_b can contain vertex v. If P_a or P_b or both P_a, P_b dominate v, then v is not separated from either a, b or u, proving the second part. Taking the solution from Claim 20 for each copy of the k-gadget of type A together with the paths $a \ldots v$ and $b \ldots v$, we get the last part of the claim.   □

We are now ready to prove Thm. 17.

Proof (Proof of Thm. 17). We first prove that the first part of Def. 16 holds. Let C^* be a minimum vertex cover of G. We construct an identifying k-path cover $\mathcal{P}$ of $f(G)$ as follows. For each copy of a gadget of type A (resp. type B), take the solution of size x_A described in the proof of Claim 20 (resp. of size x_B of Claim 21) into $\mathcal{P}$. Now, for each edge e of G, add an arbitrary path starting in vertex c of G_e. For each vertex v of G, add three arbitrary paths starting in vertices z_1, z_2 and z_3, respectively. Let e_1, e_2, e_3 be the three edges incident to v in G, and b_1, b_2, b_3, the three vertices labelled b in $G_{e_1}, G_{e_2}, G_{e_3}$. If $v \in C^*$, add path $y_1 \ldots y_3$ of G_v, as well as paths $x_2 \ldots b_1$, $x_3 \ldots b_2$ and $x_1 \ldots b_3$ to $\mathcal{P}$. If $v \notin C^*$, add paths $x_1 \ldots y_1$, $x_2 \ldots y_2$ and $x_3 \ldots y_3$ to $\mathcal{P}$.

The reader can check that $\mathcal{P}$ is an identifying k-path cover of $f(G)$, and that:

$$|\mathcal{P}| \leq |C^*| + (2(k - 2)x_A + 1)m + ((3(k - 2) + k - 3)x_A + 3x_B + 6)n \quad (1)$$

Since G is cubic, each vertex of C^* can cover at most three edges and we have $|C^*| \geq \frac{m}{3}$ and hence $m \leq 3|C^*|$ and $n \leq 2|C^*|$. We get: $p_k^{ID}(G) \leq |\mathcal{P}| \leq (16 + (14k - 30)x_A + 6x_B)|C^*|$ and hence (1) of Def. 16 is fulfilled with $\alpha = 16 + (14k - 30)x_A + 6x_B \leq 40k^2 - 116k + 47$.

It remains to prove the second part of Def. 16. Let $\mathcal{P}$ be an identifying k-path cover of $f(G)$. We construct a vertex cover C using $\mathcal{P}$. First of all, by Claim 20, each gadget of type A contains at least x_A paths of $\mathcal{P}$, and by Claim 21, each gadget of type B contains at least $x_B = x_A(k - 3) + 2$ paths of $\mathcal{P}$. Moreover, in each vertex gadget, at least three paths belong to $\mathcal{P}$ in order to cover vertices z_1, z_2, z_3. However, using the structure of the vertex- and edge-gadgets together with Claim 21, if there are exactly that many paths, in each vertex-gadget, we are not able to separate vertices y_1, y_2, y_3 from their respective neighbours, as well as vertices x_1, z_1, x_2, z_2, and x_3, z_3 (similarly, vertices b and c in each edge-gadget). Besides the paths of $\mathcal{P}$ that we already considered, in any vertex-gadget, at least three paths are required in order to cover vertices $x_1, y_1, x_2, y_2, x_3, y_3$. If there are *exactly* three, then they must be $x_1 \ldots y_1$, $x_2 \ldots y_2$, $x_3 \ldots y_3$. We construct C as follows: for each vertex v of G, if there are at least four such additional paths in G_v, we add v to C. Set C is a vertex cover of G: indeed, when $v \in C$, the paths in the copies of the gadgets of type A and B in G_v could be replaced by the "standard" solution given in Claims 20 and 21. Moreover, the (at least) four additional paths of $\mathcal{P}$ in G_v could be replaced by $y_1 \ldots y_3$ and three paths starting at x_1, x_2, x_3 covering the three vertices labelled b in the three edge-gadgets corresponding to the three edges incident to v in G. Hence these edges would be covered. This procedure would give an easy constructable identifying path cover $\mathcal{P}'$ with $\mathcal{P}' \leq \mathcal{P}$, and since all vertices labelled b are covered by a path of $\mathcal{P}'$, C is a vertex cover of G. Furthermore, we have $|C| \leq |\mathcal{P}| - (2(k-2)x_A + 1)m - ((3(k-2) + k - 3)x_A + 3x_B + 6)n$. Applying the construction to a minimum identifying k-path cover, we get $|C^*| \leq |C| \leq p_k^{\mathrm{ID}}(G) - (2(k-2)x_A + 1)m - ((3(k-2) + k - 3)x_A + 3x_B + 6)n$. Together with Equation (1), this implies:

$$|C^*| = p_k^{\mathrm{ID}}(G) - (2(k-2)x_A + 1)m - ((3(k-2) + k - 3)x_A + 3x_B + 6)n \quad (2)$$

From Equations (1) and (2), we get $|C| - |C^*| \leq |\mathcal{P}| - p_k^{\mathrm{ID}}(G)$, which implies $||C^*| - |C|| \leq |p_k^{\mathrm{ID}}(G) - |\mathcal{P}||$; hence (2) of Def. 16 is fulfilled with $\beta = 1$.   □

6   Conclusion and Open Problems

We conclude with some open problems. We gave a procedure to compute the exact value of parameter p^{ID} for topologically irreducible trees, but only gave an upper bound for general trees. It seems not easy to extend the algorithm to the latter case, but it would be interesting to design an(other) algorithm to solve it. Regarding MIN-IDPC-k, we mentioned that not all graphs admit an identifying k-path cover. Identifiable graphs have been studied for some other identification problems [5]; it would be interesting to do so in our context, i.e. studying k-path identifiable graphs. Finally, we have settled the complexity of MIN-IDPC-k by showing that it is APX-complete. However, the question of the complexity of the general MIN-IDPC problem remains open.

Acknowledgements. We would like to thank P. J. Slater and J. L. Sewell for motivating us to study this problem. We also thank O. Schaudt for giving the idea of Theorem 15.

References

1. Auger, D., Charon, I., Hudry, O., Lobstein, A.: Watching systems in graphs: an extension of identifying codes. Discrete Appl. Math. (in press)
2. Berger-Wolf, T.Y., Laifenfeld, M., Trachtenberg, A.: Identifying codes and the set cover problem. In: Proc. 44th Annual Allerton Conference on Communication, Control and Computing (September 2006)
3. Bondy, J.A.: Induced subsets. J. Comb. Theory B 12(2), 201–202 (1972)
4. Charon, I., Cohen, G., Hudry, O., Lobstein, A.: Discriminating codes in bipartite graphs. Adv. Math. Commun. 4(2), 403–420 (2008)
5. Charon, I., Honkala, I., Hudry, O., Lobstein, A.: Structural properties of twin-free graphs. Electron. J. Comb. 14, R16 (2007)
6. Charon, I., Hudry, O., Lobstein, A.: Identifying and locating-dominating codes: NP-completeness results for directed graphs. IEEE T. Inform. Theory 48(8), 2192–2200 (2002)
7. Chlebík, M., Chlebíková, J.: Complexity of approximating bounded variants of optimization problems. Theor. Comput. Sci. 354, 320–338 (2006)
8. De Bontridder, K.M.J., Halldórsson, B.V., Halldórsson, M.M., Hurkens, C.A.J., Lenstra, J.K., Ravi, R., Stougie, L.: Approximation algorithms for the test cover problem. Math. Program. B 98, 477–491 (2003)
9. Foucaud, F., Guerrini, E., Kovše, M., Naserasr, R., Parreau, A., Valicov, P.: Extremal graphs for the identifying code problem. Eur. J. Combin. 32(4), 628–638 (2011)
10. Foucaud, F., Gravier, S., Naserasr, R., Parreau, A., Valicov, P.: Identifying codes in line graphs. To appear in J. Graph Theor.
11. Foucaud, F., Naserasr, R., Parreau, A.: Characterizing extremal digraphs for identifying codes and extremal cases of Bondy's theorem on induced subsets. Graphs Comb. (in press)
12. Frieze, A., Martin, R., Moncel, J., Ruszinkó, M., Smyth, C.: Codes identifying sets of vertices in random networks. Discrete Math. 307(9-10), 1094–1107 (2007)
13. Garey, M.R., Johnson, D.S.: Computers and Intractability: A Guide to the Theory of NP-Completeness. W. H. Freeman (1979)
14. Honkala, I., Karpovsky, M., Litsyn, S.: Cycles identifying vertices and edges in binary hypercubes and 2-dimensional tori. Discrete Appl. Math. 129(2-3), 409–419 (2003)
15. Johnson, D.S.: Approximation algorithms for combinatorial problems. J. Comput. Syst. Sci. 9, 256–278 (1974)
16. Karpovsky, M., Chakrabarty, K., Levitin, L.B.: On a new class of codes for identifying vertices in graphs. IEEE T. Inform. Theory 44, 599–611 (1998)
17. Laifenfeld, M., Trachtenberg, A., Cohen, R., Starobinski, D.: Joint monitoring and routing in wireless sensor networks using robust identifying codes. In: Proc. IEEE Broadnets 2007, pp. 197–206 (September 2007)
18. Moret, B.M.E., Shapiro, H.D.: On minimizing a set of tests. SIAM J. Sci. Stat. Comp. 6(4), 983–1003 (1985)
19. Papadimitriou, C.: Computational Complexity. Addison-Wesley (1994)
20. Rosendahl, P.: On the identification of vertices using cycles. Electron. J. Comb. 10, P1 (2003)

Disjoint Set Forest Digraph Representation for an Efficient Dominator Tree Construction

Wojciech Fraczak[1,2] and Andrew Miller[2]

[1] Université du Québec en Outaouais, Gatineau, Québec, Canada
[2] Benbria Corporation, Ottawa, Ontario, Canada

Abstract. We consider a non-orthodox representation of directed graphs which uses the *"disjoint set forest"* data structure. We show how such a representation can be used in order to efficiently find the dominator tree. Even though the performance of our algorithm does not improve over the already known algorithms for constructing the dominator tree, the approach is new and it gives place to a highly structured and simple to follow proof of correctness.

1 Introduction

In the context of a directed graph (digraph) G with a root (starting) vertex s, vertex x is called a *dominator* of vertex y, if every directed path from s to y contains x. The set $\{x_0, x_1, \ldots, x_n\}$ of all dominators for vertex y is ordered with respect to the order of the first (or last) occurrences of the dominators in a path from s to y (the order remains the same for all such paths). Therefore, we have $x_0 = s$ and $x_n = y$. The vertex x_{n-1}, if it exists, is called the *immediate dominator* of y. The set of pairs (x, y) of vertexes such that x is the immediate dominator of y, defines a tree structure over vertexes, called *"dominator tree"*.

Dominators and dominator trees were initially studied in the context of the flow diagram analysis for computer programs compilation and optimization of computer programs, [4,3]. However, dominators find many applications in other areas of computer science such as networking, parallel computing, or even formal language theory.

In this paper we give a new efficient algorithm for finding the dominator tree for a given digraph with a root (start) vertex. Our algorithm runs in time $O(m \cdot \alpha(m, m))$, where m is the number of edges in the graph and α denotes the inverse of Ackermann's function[1]. This time complexity is achieved by using a non-orthodox digraph representation where vertexes are represented as pairs of sets of incoming and outgoing edges. Thus, a digraph is seen as two partitions of edges, sources and destinations, each represented by the data structure called a *disjoint set forest* [5,6]. Even though the performance of our algorithm does not improve over the already known algorithms for constructing the dominator tree [6,2], the approach is new and it gives place to a highly structured and simple to follow proof of correctness.

---

[1] The inverse of Ackermann's function is a very slowly growing function; for all practical cases, $\alpha(m, n) \leq 4$, see [5].

S. Arumugam and B. Smyth (Eds.): IWOCA 2012, LNCS 7643, pp. 46–59, 2012.
© Springer-Verlag Berlin Heidelberg 2012

2   Preliminaries

We consider directed graphs with a *root* vertex s, such that there exists a directed path from s to every vertex of the graph.

We will deal with dynamic transformations of digraphs, so we need a notion that makes it possible to keep track of changing positions of a directed edge as the graph evolves. Thus we adopt the following, slightly non-orthodox definition of a directed graph (digraph).

Definition 1. *A digraph is a quadruple,* $G = (V, E, \xi, s)$, *where:*

- *V is a finite set of vertexes,*
- *E is a finite set of edges,*
- *ξ is a mapping from E to $V \times V$, and*
- *s is an element from V, called "root" (or "start").*

Intuitively, every edge e of E is associated to its *source* $u \in V$ and its *destination* $v \in V$, defined by mapping ξ, i.e., $\xi(e) = (u, v)$. We refer to source u and destination v of edge e by writing $\xi(e)^-$ and $\xi(e)^+$, respectively. Often, when ξ is known from the context, we will write e^- and e^+ instead of $\xi(e)^-$ and $\xi(e)^+$. The digraph will be subject to dynamic changes which may result in the incremental evolution of its set of vertexes V, its set of edges E, and its mapping ξ. Especially important are dynamic changes of ξ, which may be viewed as migrations of edges.

Notice that our definition allows multi-edges and self-loops, hence many edges may have the same source and destination.

By $E(v)$ we denote the set of all edges outgoing from v, i.e., $E(v) \stackrel{\text{def}}{=} \{e \in E \mid e^- = v\}$. By $E^{-1}(v)$ we denote the set of all edges incoming to v, i.e., $E^{-1}(v) \stackrel{\text{def}}{=} \{e \in E \mid e^+ = v\}$.

A path π in $G = (V, E, \xi, s)$ from a vertex u to v is a finite sequence of edges $\pi = e_1 e_2 \ldots e_n$ such that $e_1^- = u$, $e_n^+ = v$, and $e_i^+ = e_{i+1}^-$ for $i \in [1, n)$. It is assumed that every vertex v is accessible from s, i.e., there exists a path from s to v.

A vertex v is a dominator of a vertex u if and only if v is present on every path from s to u (a vertex is present on a path if it is a source or a destination of one of the edges of the path). The set of all dominators of a vertex u in G is denoted by $D_G(u)$; the order of first (or last) occurrences of the dominators of v is the same in all paths from s to v. Thus, it is natural to represent the set of dominators of v as an ordered list $\delta_G(v)$ of vertexes, $\delta_G(v) = (u_0, u_1, \ldots, u_{n-1}, u_n)$, with $u_0 = s$ and $u_n = v$. The vertex u_{n-1}, if it exists (i.e., when $v \neq s$), is called the *immediate dominator* of v and will be denoted by $P_G(v)$ ("Parent" of v in G).

Proposition 1. *If* $\delta_G(v_n) = (v_0, v_1, \ldots, v_n)$ *then, for* $i \in \{1, \ldots, n\}$, $P_G(v_i) = v_{i-1}$ *and* $\delta_G(v_i) = (v_0, \ldots, v_i)$.

Due to Proposition 1, the dominators of all vertexes of a digraph can be represented in the form of a tree, called the *dominator tree*, defined by P_G as the "parent" relation. The root vertex s is the root of the dominator tree.

Let $G = (V, E, \xi, s)$ be a digraph. The set of dominators of $v \in V$ is related to its "incoming" neighbors in the following way.

Proposition 2. $D_G(v) = \{v\} \cup \bigcap_{e \in E^{-1}(v)} D_G(e^-)$.

We need a nomenclature which generalizes the standard DFS (Depth-First Search, [1]) concepts, such as *DFS tree*, *edge partitioning*, and *discovery order* of vertexes. For that purpose we introduce the notion of a "*digraph annotation*".

Definition 2. *An* annotation *of a digraph* $G = (V, E, \xi, s)$ *is a pair* $\mathcal{A} = (\leq, F)$, *where* $\leq$ *is a total order on* V *and* $F \subseteq E$ *is a set of edges, such that:*

1. F *is a spanning tree of* G *with root* s, *i.e., for any vertex* $v \in V$, *there exists exactly one path from* s *to* v *with edges from* F.
   We write $v \preceq_{\mathcal{A}} u$ *to say that* v *is an ancestor of* u *in* F. *Intuitively,* $v \preceq_{\mathcal{A}} u$ *implies the existence of the unique path from* v *to* u *using edges from* F.
2. *The partial order* $\preceq_{\mathcal{A}}$ *induced by* F *is compatible with* $\leq$, *i.e.,* $v \preceq_{\mathcal{A}} u$ *implies* $v \leq u$, *for all vertexes* $v, u \in V$.
3. *For every edge* $e \in E$, *if* $e^- \leq e^+$ *then* $e^- \preceq_{\mathcal{A}} e^+$.

By $v \npreceq_{\mathcal{A}} u$ we denote that vertexes u and v are not comparable in $\preceq_{\mathcal{A}}$, i.e., neither $v \preceq_{\mathcal{A}} u$ nor $u \preceq_{\mathcal{A}} v$. Whenever an annotation $\mathcal{A}$ is known from the context, we will write $\preceq$ and $\npreceq$ instead of $\preceq_{\mathcal{A}}$ and $\npreceq_{\mathcal{A}}$, respectively. The strict orders induced by $\leq$ and $\preceq$ will be denoted by $<$ and $\prec$, respectively.

The above definition of an "annotation" generalizes the standard DFS as stated in the following lemma.

Lemma 3. *Let* $G = (V, E, \xi, s)$ *be a given digraph. The standard DFS (Depth-First Search) starting in* s, *defines an annotation* $(\leq, F)$ *of* G, *where* F *is the DFS tree and* $u \leq v$ *iff discovery time of* u *is smaller or equal to the discovery time of* v.

Let $\mathcal{A} = (\leq, F)$ be an annotation of $G = (V, E, \xi, s)$. We define the following partitioning of $E = E_T^{\mathcal{A}} \cup E_F^{\mathcal{A}} \cup E_C^{\mathcal{A}} \cup E_B^{\mathcal{A}}$ induced by $\mathcal{A}$:

- $E_T^{\mathcal{A}} \overset{\text{def}}{=} F$ – *tree edges*
- $E_F^{\mathcal{A}} \overset{\text{def}}{=} \{e \in E \mid e^- \prec e^+\} \setminus F$ – *forward edges*
- $E_B^{\mathcal{A}} \overset{\text{def}}{=} \{e \in E \mid e^+ \preceq e^-\}$ – *back edges*
- $E_C^{\mathcal{A}} \overset{\text{def}}{=} \{e \in E \mid e^+ \npreceq e^-\}$ – *cross edges*

Edges from $E_T^{\mathcal{A}}$, $E_F^{\mathcal{A}}$, $E_C^{\mathcal{A}}$, and $E_B^{\mathcal{A}}$ are called *tree edges*, *forward edges*, *cross edges*, and *back edges*, respectively. Intuitively, they correspond to the well-known DFS edge partitioning. Notice, that $e \in E_C^{\mathcal{A}} \cup E_B^{\mathcal{A}}$ if and only if $e^+ \leq e^-$.

3   Deriving Partial Information about Dominators from an Annotation

An annotation $\mathcal{A} = (\leq, F)$ of a digraph $G = (V, E, \xi, s)$ reveals many useful facts about the dominator tree of G, as shown in this section.

All dominators of a vertex v lay on the path of every spanning tree F from the root s to v, i.e., they are ancestors of v in F. Therefore, if $\delta_G(v) = (v_0, v_1, \ldots, v_n)$, then $v_0 \prec_\mathcal{A} v_1 \prec_\mathcal{A} \ldots \prec_\mathcal{A} v_n$, for any annotation $\mathcal{A} = (\leq, F)$ of $G = (V, E, \xi, s)$ and any vertex $v \in V$.

Proper nesting: The pairs of vertexes, $(P_G(v), v)$ are always ordered with respect to $\prec$. The pairs $(P_G(v), v)$ exhibit some nesting properties which are presented graphically in Fig. 1 and stated formally in the following two lemmas.

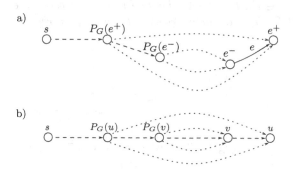

Fig. 1. Illustration for: a) Lemma 4, b) Lemma 5. Solid arrows represent edges, dash arrows represent paths composed only from tree edges, and dot arrows represent any path.

Lemma 4. *Let $e \in E$ be an edge and let $\mathcal{A}$ be an annotation of $G = (V, E, \xi, s)$. If $s \notin \{e^-, e^+\}$ and $e^- \neq P_G(e^+)$ then $P_G(e^+) \preceq_\mathcal{A} P_G(e^-)$.*

Lemma 5. *Let $u, v \in V$ and let $\mathcal{A} = (\leq, F)$ be an annotation of $G = (V, E, \xi, s)$. If $P_G(u) \prec_\mathcal{A} v \prec_\mathcal{A} u$ then $P_G(u) \preceq_\mathcal{A} P_G(v)$.*

Siblings: Let us consider a particular case of a vertex with exactly two incoming edges, one tree edge and one forward edge.

Lemma 6. *Let $\mathcal{A}$ be an annotation of $G = (V, E, \xi, s)$ and $u \in V$ be a vertex such that there are exactly two incoming edges to u, $E^{-1}(u) = \{e, f\}$, with $e \in E_T$ and $f \in E_F$.*

1. *If $e^- = f^-$ then $P_G(u) = e^- = f^-$.*
2. *If $P_G(e^-) \preceq f^- \prec e^-$ then $P_G(u) = P_G(e^-)$. (See Fig. 2)*

Proof. 1.) $P_G(u) = e^- = f^-$, since every path from s to u ends by e or f.

2.) By Proposition 2, $D_G(u) = \{u\} \cup (D_G(e^-) \cap D_G(f^-))$. Because $u \notin D_G(e^-)$ and $e^- \notin D_G(f^-)$, $D_G(u) \setminus \{u\} = (D_G(e^-) \setminus \{e^-\}) \cap D_G(f^-)$. $P_G(e^-)$ is a dominator of f^-; otherwise, there would be a path from s to e^- passing by f^- and avoiding $P_G(e^-)$. By Proposition 1, $D_G(e^-) \setminus \{e^-\} \subseteq D_G(f^-)$. Therefor, $D_G(u) \setminus \{u\} = D_G(e^-) \setminus \{e^-\}$, i.e., $P_G(u) = P_G(e^-)$.   □

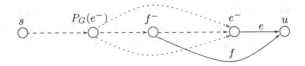

Fig. 2. Illustration for Lemma 6(2)

Cycles:

Lemma 7. *Let $e \in E_B$ be a back edge and u a vertex between e^+ and e^-, i.e., $e^+ \preceq u \preceq e^-$. If $P_G(u) \preceq e^+$ then $P_G(u) = P_G(e^+)$. (See Fig. 3)*

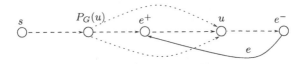

Fig. 3. Illustration for Lemma 7

Proof. By Lemma 5, $P_G(u) \preceq P_G(e^+)$. On the other hand, for every vertex z between $P_G(u)$ and e^+, i.e., $P_G(u) \prec z \prec e^+$, there is a path from s to e^+ ending by e which avoids z. Thus, $P_G(e^+) = P_G(u)$.   □

4   Graph Transformations

In this section we define the notation for local transformations of a digraph, namely: adding, removing, and redirecting edges, and removing and merging vertexes.

Definition 3. *Let $G = (V, E, \xi, s)$ be a digraph, $u, v \in V$, and $e \in E$.*

1. *By $G - e$ we denote the graph obtained by the removal of edge e from G.*
   $G - e \overset{\text{def}}{=} (V, E \setminus \{e\}, \xi_{|E \setminus \{e\}}, s)$, *where $\xi_{|E'}$ for some $E' \subseteq E$, denotes the function ξ restricted to elements of E'.*

2. *By $G + (u, v)$ we denote the graph obtained by the addition of a new edge from u to v in G.*
   $G + (u, v) \overset{\text{def}}{=} (V, E \cup \{f\}, \xi', s)$ *where $f \notin E$, $\xi'_{|E} \overset{\text{def}}{=} \xi$, and $\xi'(f) \overset{\text{def}}{=} (u, v)$.*

3. *By $G[e/(u, v)]$ we denote the graph obtained by the substitution of the source and the destination of edge e in graph G by vertexes u and v, respectively.*
   $G[e/(u, v)] \overset{\text{def}}{=} (V, E, \xi', s)$ *where $\xi'(x) \overset{\text{def}}{=} \begin{cases} (u, v) & \text{if } x = e \\ \xi(x) & \text{otherwise} \end{cases}$.*
   If we want to change only the source or the destination of edge e we will write $G[e/(u, \_)]$ or $G[e/(\_, u)]$ meaning $G[e/(u, e^+)]$ and $G[e/(e^-, u)]$, respectively.

4. By $G - u$ we denote the graph obtained by the removal of vertex u from G together with the removal of all edges connected to u.

$G - u \overset{\text{def}}{=} (V \setminus \{u\}, E', \xi_{|E'}, s)$ where $E' \overset{\text{def}}{=} E \setminus (E(u) \cup E^{-1}(u))$.

The above operations extend naturally to sets. For example, we will write $G - \{e_1, \dots, e_k\}$ instead of $G - e_1 - \dots - e_k$.

5. By $G[u \leftarrow v]$ we denote the graph in which vertex v is merged into vertex u with all self-loops on the merged vertex u removed. Formally:

$G_1 = (V_1, E_1, \xi_1, s_1) = G[E^{-1}(v)/(\_, u)]$,
$G_2 = (V_2, E_2, \xi_2, s_2) = G_1[E_1(v)/(u, \_)]$,
$G_3 = G_2 - (E_2(u) \cap E_2^{-1}(u))$,
$G[u \leftarrow v] \overset{\text{def}}{=} G_3 - v$.

Addition or Removal of Edges. Adding or removing an outgoing edge of a vertex does not change the dominators of the vertex. Moreover, if adding (or removing) an incoming edge to vertex u does not change the dominators of u, then no vertex of the digraph changes its dominators. More formally:

Lemma 8. Let $G = (V, E, \xi, s)$ be a digraph, $u, v \in V$, and $G' = G + (u, v)$.

1. $\delta_G(u) = \delta_{G'}(u)$.
2. If $\delta_G(v) = \delta_{G'}(v)$ then $\delta_G(x) = \delta_{G'}(x)$ for every vertex $x \in V$.

Proof. 1.) Every path π from s to u passing by the new edge e in G' (with $\xi'(e) = (u, v)$) is not simple[2] and thus it passes by all vertexes of a direct path π' from s to u. Thus, the intersection of vertexes of π and π' is the set of vertexes of π', i.e., $\delta_G(u) = \delta_{G'}(u)$. In general, this implies that in order to calculate dominators of a vertex x we can consider only simple paths from s to x.

2.) Consider a vertex $x \in V$. The set of all simple paths in G from s to x can be partitioned into the set of paths passing by v, and the set of paths avoiding v. Every simple path from s to x through v can be split into a simple path from s to v, followed by a simple path from v to x. In G', the set of simple paths from s to x avoiding v and the set of simple paths from v to x, are the same as in G. Since $\delta_G(v) = \delta_{G'}(v)$, all dominators of x in G and G' are the same and they occur in the same order, hence $\delta_G(x) = \delta_{G'}(x)$. □

Migration of Edges. If vertex v is not the immediate dominator of vertex u then an edge from v to u can be replaced by an edge from the immediate dominator of v to u, without changing dominators of any vertex of the graph (see Fig. 4).

Lemma 9. Let e be an edge in a digraph $G = (V, E, \xi, s)$ and $\xi(e) = (v, u)$. If $P_G(u) \neq v$, then $P_{G'}(x) = P_G(x)$ for $G' = G[e/(P_G(v), u)]$ and every $x \in V$.

---

[2] A path $\pi = e_1 e_2 \dots e_n$ is *simple* if it does not contain loops, i.e., for $i, j \in [1, n]$, $e_i^+ = e_j^-$ iff $j = i + 1$.

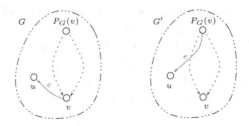

Fig. 4. Illustration for Lemma 9

Proof. In view of Lemma 8, it is sufficient to prove that $\delta_{G'}(u) = \delta_G(u)$, where $G' = G[e/(v', u)]$ and $v' = P_G(v)$.

Let $\Pi_e(u)$, $\Pi'_e(u)$ be the sets of all simple paths from s to u ending by an edge different from e in G and G', respectively. Notice that $\Pi_e(u)$ and $\Pi'_e(u)$ are the same. Let $\Pi(v)$, $\Pi'(v')$ be the sets of all simple paths from s to v in G and from s to v' in G', respectively.

The set of dominators of u in G is the intersection of the set $S_e(u)$ of vertexes occurring on every path in $\Pi_e(u)$, with the set $S(v)$ of vertexes occurring on every path in $\Pi(v)$, i.e., $D_G(u) = S_e(u) \cap (\{u\} \cup S(v))$. The set of dominators of u in G' is: $D_{G'}(u) = S'_e(u) \cap (\{u\} \cup S'(v'))$, where $S'_e(u) = S_e(u)$ and $S'(v')$ denote the set of vertexes occurring on every path in $\Pi'(v')$.

Since no path in $\Pi(v)$ nor in $\Pi'(v')$ uses e, (otherwise, such a path would not be simple), and since no other vertex than $v' = P_G(v)$ on a path from v' to v is a dominator of v, we have $S(v) = S'(v') \cup \{v\}$. Thus $D_G(u) = D_{G'}(u)$. The existence of a simple path from s to u in G avoiding v, which is also a path is G', implies the same order of dominators of u in G and G'.   □

The following lemma says that we can add an edge to a digraph G from v to u without changing any dominators, whenever all dominators of u in G, except u itself, are also dominators of v.

Lemma 10. *Let $u, v \in V$ be vertexes in a digraph $G = (V, E, \xi, s)$. If $D_G(u) \setminus \{u\} \subseteq D_G(v)$, then $P_{G+(v,u)}(x) = P_G(x)$ for every $x \in V$.*

Proof. In view of Lemma 8, it is sufficient to prove that $\delta_{G'}(u) = \delta_G(u)$, where $G' = G + (v, u)$. Consider all simple paths from s to u in G'. Only paths ending with the new edge are not present in G. However, by Lemma 8(1), $\delta_G(v) = \delta_{G'}(v)$, and thus every vertex $x \in D_G(u)$ is present on every path in G' from s to u ending with the new edge, i.e., $D_{G'}(u) = (D_G(v) \cup \{u\}) \cap D_G(u) = D_G(u)$. Since every path in G is also a path in G', the order of dominators does not change, implying $\delta_{G'}(u) = \delta_G(u)$.   □

The next lemma says that if vertexes v, v' have the same immediate dominator, i.e., $P_G(v') = P_G(v)$, and there is a path from s to some vertex u avoiding both v and v', then every edge e from v to u can be replaced by an edge from v' to u without changing any dominators in the graph, as long as every incoming edge of v is duplicated towards v' (see Fig. 5).

Lemma 11. *Let e be an edge in a digraph $G = (V, E, \xi, s)$, $\xi(e) = (v, u)$, and v' a vertex in V such that $P_G(v) = P_G(v')$. If there is a path from s to u avoiding both v and v', then for every $x \in V$ we have $P_{G'}(x) = P_G(x)$, where $G' = G[e/(v', u)] + \{(g^-, v') \mid g \in E^{-1}(v)\}$.*

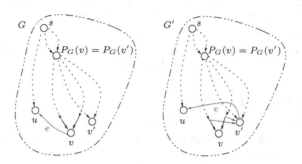

Fig. 5. Illustration for Lemma 11

Proof. We split the construction of G' from G into two steps:

1. $G_1 = G + (v', u) + \{(x^-, v') \mid x \in E^{-1}(v)\}$: adding a new edge f between v' and u and a new edge from every "incoming neighbor" of v towards v';
2. $G_2 = G_1 - e$: removing e. The resulting digraph G_2 is isomorphic to G'. The two digraphs differ only by the name of the new edge (e in G' and f in G_2) from v' to u.

The dominators are preserved by each of the above transformations.

The preservation of the dominators in Step 1 is implied by Lemma 10 by adding the edges one by one in any order.

Consider Step 2, which generates G_2 by removing edge e from G_1. By Lemma 8, it is sufficient to prove that $\delta_{G_2}(u) = \delta_{G_1}(u)$. Because we remove edge e, we have $D_{G_1}(u) \subseteq D_{G_2}(u)$. We prove that $D_{G_1}(u) = D_{G_2}(u)$, by contradiction. Suppose, $z \in D_{G_2}(u)$ and $z \notin D_{G_1}(u)$. There must exist a simple path π from s to u in G_1 avoiding z. The path ends by e, if not it is also present in G_2 or is not simple. I.e., $\pi = \pi'xe$, where $x \in E^{-1}(v)$. For every such path in G_1 there exists a path $\pi'x'f$ in G_2, where f and x' are edges added in Step 1. Thus, $z \in \{v, v'\}$. However, by supposition, there is path from s to u which avoids v or v'. The path is present in G, G_1, and G_2. We conclude that there is no such z. $\qquad\square$

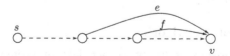

Fig. 6. Illustration for Lemma 13

As a corollary, together with Lemma 8, we state that if $P_G(v)$ is not a dominator of any of the out-neighbors of vertex v (as defined in Lemma 11), then vertex v can be merged into vertex v' without changing the dominators for all remaining vertexes.

Corollary 12. *Let $G = (V, E, \xi, s)$ be a digraph, v, v' two different vertexes in V such that $P_G(v) = P_G(v')$, and $U = \{u \mid \exists e \in E, \xi(e) = (v, u)\}$ the set of all out-neighbors of v.*

If $P_G(v)$ is a dominator for no vertex u from U, i.e., $\forall u \in U \; P_G(v) \notin D_G(u)$, then for every vertex $x \in V \setminus \{v\}$, we have $\delta_G(x) = \delta_{G[v' \leftarrow v]}(x)$.

Elimination of Forward Edges

Lemma 13. *Let $\mathcal{A}$ be an annotation of $G = (V, E, \xi, s)$ and $e, f \in E_F$ be two forward edges sharing the destination, i.e., $e^+ = f^+$. If $e^- \preceq_{\mathcal{A}} f^-$ then $P_G(x) = P_{G-f}(x)$ for every $x \in V$.*

Proof. (Lemma 13, (Fig. 6) Suppose that $D_G(v) \neq D_{G-f}(v)$, i.e., there is a vertex z such that $z \in D_{G-f}(v)$ and $z \notin D_G(v)$. Consider a simple path π in G from s to v which avoids z. It has to end by f, i.e., $\pi = \pi'f$ (if not π is present in $G - f$ or is not simple). For every such path in $G - f$ there is a path in G which instead of f takes the tree edges from f^- to v. Thus, $f^- \preceq z$, i.e., z is not a dominator of u in $G - f$ either.   □

5   Algorithm "Dominator Tree"

The algorithm finds the dominator tree of a given digraph $G = (V, E, \xi, s)$. It is based on a series of transformations of the digraph without modifying the underlying dominator tree. Some transformations will remove edges or modify edges, using Lemmas 9 and 11, and other transformations will remove vertexes.

The idea of the algorithm is the following. During consecutive transformations performed by the algorithm the underlying digraph shrinks and new parts of the dominator tree become available. Upon completion of the algorithm the digraph is reduced to a single vertex and the entire dominator tree is computed.

A vertex u will be removed from the graph as soon as we find its "parent" in the dominator tree, or we find one of its "younger siblings", i.e., a vertex v such that $v < u$ and $P_G(u) = P_G(v)$. For finding the parent of a vertex u, we use Lemma 6(1), and for finding a sibling we use Lemma 6(2) or Lemma 7. All incoming and outgoing edges of the removed vertex u are redirected to the parent or the sibling, respectively, in such a way that the dominators of all remaining vertexes do not change.

5.1   Description of the Algorithm

The algorithm has three parts: firstly, we calculate an initial annotation $\mathcal{A} = (\leq, F)$ for the input graph $G = (V, E, \xi, s)$; secondly, we calculate a representation of the dominator tree; finally, we print out the dominator tree by reporting the immediate dominator of every vertex of the graph.

Part I. Calculate Annotation

Run DFS in order to find an initial annotation $\mathcal{A} = (\leq, F)$ of the input digraph G, see Lemma 3. No edge is marked.

Part II. Find the Dominator Tree

Let $G_i = (V_i, E_i, \xi_i, s_i)$ be a local variable initialized to value of G.

For every vertex u, in decreasing order of $\leq_{\mathcal{A}}$ **do:**

Step 1. **While** there is a non marked edge $e \in E_i$, such that $\xi_i(e) = (u, v)$ and $u < v$, for some $v \in V_i$, (i.e., e is a tree or forward edge) **do:**
  1. Mark e.
  2. **If** all incoming edges of v are marked **then:**
     (a) Examine the tree edge $f \in E_i \cap F$, i.e., $\xi_i(f) = (x, v)$, for some $x \in V_i$.
     (b) **If** $x = u$ **then:**

Output: $P_G(v) = u$	Action: $G_i \Leftarrow G_i[u \leftarrow v]$	(1)

I.e., merge vertex v into u: change the origin of all outgoing edges of v to u and remove the self-loops (see Fig. 7, Action 1). Notice that this operation can add new edges to the list of edges to be considered in this step.
     (c) **Else**, i.e., if $x \neq u$:

Output: $P_G(v) = P_G(x)$	Action: $G_i \Leftarrow G_i[x \leftarrow v]$	(2)

I.e., merge vertex v into x (see Fig. 7, Action 2).
Step 2. **While** there is an edge $e \in E_i$ such that $\xi_i(e) = (v, u)$, $u \leq v$, and $u \preceq_{\mathcal{A}} v$, for some vertex $v \in V_i$, **do:**
  1. **If** e is a self-loop, i.e., $u = v$, **then:**

Output: none	Action: $G_i \Leftarrow G_i - e$	(3)

  2. **Else**, we consider the tree edge $f \in E_i \cap F$ such that $\xi_i(f) = (x, v)$ for some vertex $x \in V_i$:

Output: $P_G(v) = P_G(u)$	Action: $G_i \Leftarrow G_i[x \leftarrow v]$	(4)

I.e., merge vertex v into x (see Fig. 7, Action 4). Notice that the edge e is not removed yet; it will be removed once it becomes a self-loop.

Part III. Using OUTPUT of Part II, Find $P_G(v)$, for Every v

The value for the immediate dominator of a vertex v was reported in Part II in one of two forms: 1) directly, i.e., by $P_G(v) = u$; or 2) indirectly, i.e., by $P_G(v) = P_G(u)$, with $u \preceq_{\mathcal{A}} v$.

In order to directly output the immediate dominator of every vertex v, re-examine all OUTPUTs of Part II in reverse order. In this case every statement $P_G(v) = P_G(u)$ occurs when the value for $P_G(u)$ is already known.

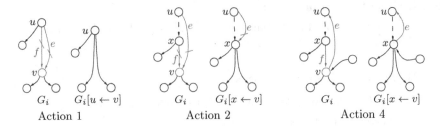

Fig. 7. Examples of Graph Transformations

5.2   Correctness

Lemma 14. *Let G be a digraph, $(\leq, F)$ an annotation of G, and G_i one of the digraph instances generated by Part II of algorithm "**Dominator tree**". At every step the pair $\mathcal{A}_i = (\{(u,v) \in V_i \times V_i \mid u \leq v\}, F \cap E_i)$ (i.e., the initial annotation $(\leq, F)$ of G restricted to elements of G_i) is an annotation of G_i.*

Proof. (Lemma 14) We have to check that all transformations, i.e., Actions 1, 2, 3, and 4, of the digraph $G_i = (V_i, E_i, \xi_i, s)$ are preserving the following properties:

1. $F \cap E_i$ is a spanning tree of G_i.
   Action 3 removes an edge not in F. All other actions, i.e., 1, 2, and 4, remove a vertex v with its incoming tree edge and redirect all incomming tree edges of v towards the parent of v in F. Thus edges in $F \cap E_i$ constitute a spanning tree of G_i after every transformation.
2. Let $e \in E_i$ such that $\xi_i(e) = (x, y)$. If $x \leq y$ then $x \preceq y$, i.e., there is a path from x to y in G_i using edges from $F \cap E_i$.
   Action 3 removes an edge not in F thus does not change any path using edges from $F \cap E_i$.
   Actions 1, 2, and 4, remove a vertex v with its incoming tree edge f and reattach all outgoing edges of v to f^-. Let π be a path using edges from $F \cap E_i$ from vertex x to vertex y (both different from v) before the transformation. The sequence of edges of π restricted to edges from $F \cap E_i \setminus \{f\}$ is also a path in the new graph from x to y.
3. All actions preserve both $\preceq$ and $\leq$ relations on remaining vertexes, thus the initial invariant $\preceq \subseteq \leq$ is preserved.   □

Lemma 15. *Let $\mathcal{A} = (\leq, F)$ be the initial annotation (as calculated in Part I) of the input graph $G = (V, E, \xi, s)$. Let $u \in V$ be the current vertex and $G_i = (V_i, E_i, \xi_i, s)$ the current digraph of the main **For**-loop of Part II.*

1. *For every $v \in V_i$, $P_G(v) \leq u$. Moreover, once the first **While**-loop is terminated, i.e., all outgoing forward edges of u are marked, we have $P_G(v) < u$.*
2. *For every $e \in E_i$, if $e^+ \preceq e^-$ then $e^+ \leq u$.*
   I.e., there is no loop in the processed part of G_i.
3. *Every **OUTPUT** reports correctly $P_{G_i}(v)$, where G_i is the current instance of the graph (before action).*

4. No ACTION changes the dominators of the vertexes remaining in G_i.

Proof. (Lemma 15) For the initial graph $G_i = G$, the two first statements are trivially true because we start the **For**-loop with the largest u, i.e., for all $x \in V$, $x \le u$.

Every transformation preserves the validity of statements 1–4:

Actions 1 and 2:

An edge e is marked only if $u = e^- \le e^+$, i.e., e is a tree edge or a forward edge. Moreover all incoming edges to $v = e^+$ have been marked when visiting vertexes u', with $u \le u' < v$. Every such edge, except the incoming tree edge f of v and the forward edge e, can be removed (Lemma 13) without changing the dominator tree of the digraph. Thus, we can assume that there are exactly two incoming edges to v, the tree edge f and the forward edge e. Lemma 6 applies:

1. If $f^- = u$ then $P_{G_i}(v) = u$, by Lemma 9,
2. If $f^- \neq u$ then $P_{G_i}(v) = P_{G_i}(f^-)$, by Corollary 12.

Let G_i be the graph just after terminating **While**-loop 1, i.e., when all outgoing forward edges of u are marked. We prove that for every vertex v in G_i we have $P_{G_i}(v) \le u$ (knowing already that $P_{G_i}(v) \le u$). By contradiction, assume that the set of vertexes $B = \{v \in V_i \mid u \in D_{G_i}(v)\} \setminus \{u\}$ is not empty. For every $x \in B$ we have: $u \prec x$ and there is at least one unmarked edge g incoming to x. Since g^- has u as a dominator (otherwise u is not a dominator of g^+), we conclude that B contains a loop, which contradicts invariant 2.

Action 3:

By Lemma 8, selfloops can be removed without changing the dominators.

Action 4:

Let e be the back edge $\xi_i(e) = (u, v)$, with $v < u$, $v \prec u$. By Lemma 7, $P_{G_i}(u) = P_{G_i}(v)$. By Corollary 12, Action 4 does not change the dominators in G_i.

After having processed a vertex u, i.e., terminating **While**-loop 2, the properties 1 and 2 are valid for the immediate predecessor of u with respect to $\le$.   □

Theorem 16. *Algorithm "Dominator tree" correctly computes the immediate dominator of any vertex of the input graph.*

Proof. By Lemma 3, the DFS of Part I finds an initial annotation of the input graph. Lemmas 14 and 15 imply the theorem. When the sink vertex is processed, i.e., Part II is terminated, the graph G_i becomes a single sink vertex, which means that all vertexes have been removed and their immediate dominator was reported directly or undirectly. In Part III, the algorithm prints the direct value of the immediate dominator of every vertex (except sink) of the input graph.   □

6   Efficient Implementation of "Dominator Tree"

An efficient implementation of the algorithm can be done by using the data structure called *disjoint set forest* [5,6] which efficiently implements *disjoint set union*.

The disjoint set union problem consists in carrying out a sequence of operations of $\texttt{find}(x)$ and $\texttt{union}(A, B)$ starting with a collection of disjoint sets; $\texttt{find}(x)$ determines the set containing element x, and $\texttt{union}(A, B)$ produces the union of set A and B, destroying A and B.

Connections in a graph $G = (V, E, \xi, s)$ can be represented by two partitions of edges: $\texttt{Src} \overset{\text{def}}{=} E/_{\{(e,f)|e^- = f^-\}}$ and $\texttt{Dst} \overset{\text{def}}{=} E/_{\{(e,f)|e^+ = f^+\}}$, i.e., two edge names are in the same equivalence class of $\texttt{Src}$ or of $\texttt{Dst}$, if they have the same source or the same destination, respectively. For example, an equivalence class A in $\texttt{Src}$ corresponds to a vertex u which is the source of all edges from A. Thus, a vertex u defines an equivalence class $\{e \in E \mid e^- = u\}$ in $\texttt{Src}$, denoted by $\texttt{Src}[u]$, and an equivalence class $\{e \in E \mid e^+ = u\}$ in $\texttt{Dst}$, denoted by $\texttt{Dst}[u]$.

In such a representation, an action $G \Leftarrow G[u \leftarrow v]$ translates into $\texttt{Dst}[u] \Leftarrow \texttt{union}(\texttt{Dst}[u], \texttt{Dst}[v])$ and $\texttt{Src}[u] \Leftarrow \texttt{union}(\texttt{Src}[u], \texttt{Src}[v])$.

In order to make the selection of an edge in **While**-loop of Step 1 and **While**-loop Step 2, in constant time, we attach to every vertex u two lists of edges, $\texttt{out\_list}[u]$ and $\texttt{in\_list}[u]$. List $\texttt{in\_list}$ is initialized as the list of all incoming edges of u. List $\texttt{out\_list}[u]$ is the list of outgoing "forward" edges. List $\texttt{out\_list}[u]$ is initialized as empty; a non-back edge e with $\xi(e) = (u, v)$ is added to $\texttt{out\_list}[u]$ when vertex v is visited. Thus, when u is visited, $\texttt{out\_list}[u]$ contains all outgoing edges of u whose destination has been already visited, i.e., $e \in \texttt{out\_list}[u]$ iff $\xi e = (u, v)$ and $u < v$.

The **While**-loop of the Step 2, Part II, of the algorithm may be rewritten into a **For**-loop over the incoming back edges. Thus, the destination end of every back edge is considered once.

6.1   Complexity

Selecting an edge satisfying the condition of the loop "while" in Step 1 consists in taking an element from list $\texttt{out\_list}[u]$ and hence is done in constant time. Finding the source of the selected edge uses one $\texttt{find}$ operation. Marking an edge in Step 1.(a) and verifying condition 1.(b) is done in constant time. Merging a vertex into another one uses two $\texttt{union}$ operations and a concatenation of $\texttt{out\_lists}$. In Step 2 we consider all incoming edges of u from list $\texttt{in\_list}[u]$. We find e^- using one $\texttt{find}$ operation per edge. If e is a *back* edge, removing all vertexes between e^- and u uses two $\texttt{union}$ operation and one $\texttt{find}$ operation per removed vertex. Otherwise, if e is not a back edge, we add e to $\texttt{out\_list}[e^-]$ which is done in constant time.

In summary, the only non constant time operations $\texttt{union}$ and $\texttt{find}$, which are prefomed $O(m)$ times (where m is number of edges of the initial digraph). In view of [6], this takes $O(m\alpha(m, m))$ time, where α is the inverse Ackermann function.

Acknowledgments. Jurek Czyzowicz and Andrzej Pelc contributed to this work during its initial stage.

References

1. Cormen, T.H., Leiserson, C.E., Rivest, R.L.: Introduction to Algorithms. The MIT Press and McGraw-Hill Book Company (1989)
2. Georgiadis, L., Tarjan, R.E.: Finding dominators revisited: extended abstract. In: Proceedings of the Fifteenth Annual ACM-SIAM Symposium on Discrete Algorithms, SODA 2004, pp. 869–878. SIAM (2004)
3. Lowry, E.S., Medlock, C.W.: Object code optimization. Commun. ACM 12, 13–22 (1969), http://doi.acm.org/10.1145/362835.362838
4. Prosser, R.T.: Applications of boolean matrices to the analysis of flow diagrams. Papers Presented at the Eastern Joint IRE-AIEE-ACM Computer Conference, IRE-AIEE-ACM 1959 (Eastern), December 1-3, pp. 133–138. ACM Press, New York (1959), http://doi.acm.org/10.1145/1460299.1460314
5. Tarjan, R.: Efficiency of a good but not linear set union algorithm. Journal of the Association for Computing Machinery 22(2), 215–225 (1975)
6. Tarjan, R., Leeuwen, J.: Worst-case analysis of set union algorithms. Journal of the Association for Computing Machinery 31(2), 245–281 (1984)

On Some Properties of Doughnut Graphs
(Extended Abstract)

Md. Rezaul Karim[1], Md. Jawaherul Alam[2], and Md. Saidur Rahman[2]

[1] Dept. of Computer Science and Engineering, University of Dhaka,
Dhaka-1000, Bangladesh
rkarim@univdhaka.edu
[2] Dept. of Computer Science and Engineering, Bangladesh University of Engineering
and Technology (BUET), Dhaka-1000, Bangladesh
jawaherul@gmail.com, saidurrahman@cse.buet.ac.bd

Abstract. The class doughnut graphs is a subclass of 5-connected planar graphs. It is known that a doughnut graph admits a straight-line grid drawing with linear area, the outerplanarity of a doughnut graph is 3, and a doughnut graph is k-partitionable. In this paper we show that a doughnut graph exhibits a recursive structure. We also give an efficient algorithm for finding a shortest path between any pair of vertices in a doughnut graph. We also propose a nice application of a doughnut graph based on its properties.

1 Introduction

A five-connected planar graph G is called a *doughnut* graph if G has an embedding Γ such that (a) Γ has two vertex-disjoint faces each of which has exactly p vertices, $p > 3$, and all the other faces of Γ has exactly three vertices; and (b) G has the minimum number of vertices satisfying condition (a). Figure 1(a) illustrates a doughnut graph where F_1 and F_2 are two vertex disjoint faces. Figure 1(b) illustrates a doughnut like embedding of G where F_1 is embedded as the outer face and F_2 is embedded as the inner face. A doughnut graph and their spanning subgraphs admit straight-line grid drawings with linear area [2,3]. The outerplanarity of this class is 3 [3], and it is k-partitionable [5].

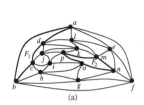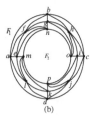

Fig. 1. (a) A doughnut graph G, and (b) a doughnut embedding of G

S. Arumugam and B. Smyth (Eds.): IWOCA 2012, LNCS 7643, pp. 60–64, 2012.
© Springer-Verlag Berlin Heidelberg 2012

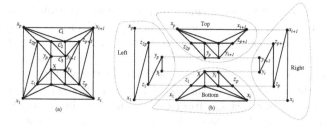

Fig. 2. (a) A straight-line drawing of a p-doughnut graph G where $p = 4$, and (b) illustration for four partition of edges of G

In this paper we present our results on recursive structure, shortest paths and topological properties of a doughnut graph.

2   Recursive Structure of Doughnut Graphs

A class of graphs has a recursive structure if every instance of it can be created by connecting the smaller instances of the same class of graphs. We now show that the doughnut graphs have a recursive structure. Let G be a p-doughnut graph. A doughnut graph G is 5-regular and has exactly $4p$ vertices. Furthermore, G has three vertex-disjoint cycles C_1, C_2 and C_3 with p, $2p$ and p vertices respectively, such that $V(C_1) \cup V(C_2) \cup V(C_3) = V(G)$. Let z_1, z_2, ..., z_{2p} be the vertices of C_2 in counter clockwise order such that z_1 has exactly one neighbor on C_1. Let x_1 be the neighbor of z_1 on C_1, and let x_1, x_2, ..., x_p be the vertices of C_1 in the counter clockwise order. Let y_1, y_2, ..., y_p be the vertices on C_3 in counter clockwise order such that y_1 and y_p are the right neighbor and the left neighbor of z_1, respectively. Let D be a straight-line grid drawing of a p-doughnut graph G with linear area [2], as illustrated in Figure 2(a). We partition the edges of D as follows. The *left partition* consists of the edges - (i) (x_1, x_p), (ii) (z_1, z_{2p}), (iii) (y_1, y_p), (iv)(x_1, z_{2p}) and (v) (z_1, y_p); and the *right partition* consists of the edges - (i) (z_p, z_{p+1}), (ii) the edge between the two neighbors of z_p on C_1 if z_p has two neighbors on C_1 otherwise the edge between the two neighbors of z_{p+1} on C_1, (iii) the edge between the two neighbors of z_p on C_3 if z_p has two neighbors on C_3 otherwise the edge between the two neighbors of z_{p+1} on C_3, (iv) the edge between z_p and its right neighbor on C_1 if z_p has two neighbors on C_1 otherwise the edge between z_{p+1} and its left neighbor on C_1, and (v) the edge between z_p and its right neighbor on C_3 if z_p has two neighbors on C_3 otherwise the edge between z_{p+1} and its left neighbor on C_3. The graph G is divided into two connected components if we delete the edges of the left and the right partitions from G. We call the connected component that contains vertex x_p the *top partition* of edges and we call the connected component that contains vertex x_1 the *bottom partition* of edges.

Figure 2(b) illustrates four partitions of edges (indicated by dotted lines) of a p-doughnut graph G in Figure 2(a) where $p = 4$. We now construct a $(p_1 + p_2)$-

doughnut graph G from a p_1-doughnut graph G_1 and a p_2-doughnut graph G_2. We first construct two graphs G_1' and G_2' from G_1 and G_2, respectively, as follows. We partition the edges of G_1 into left, right, top and bottom partitions. Then we identify the vertex x_{i+1} of the top partition to the vertex y_i of the right partition, the vertex z_{p_1+1} of the top partition to the vertex z_{p_1} of the right partition, and the vertex y_{i+1} of the top partition to the vertex x_i of the right partition. Thus we construct G_1' from G_1. Figure 3(c) illustrates G_1' which is constructed from G_1 in Figure 3(a) where $p_1 = 4$. In case of construction of G_2', after partitioning (left, right, top, bottom) the edges of G_2 we identify the vertex y_{p_2}' of left partition to the vertex x_1' of the bottom partition, vertex z_{2p_2}' of the left partition to the vertex z_1' of the bottom partition, and the vertex x_{p_2}' of left partition to the vertex y_1'. Figure 3(f) illustrates G_2' which is constructed from G_2 in Figure 3(d) where $p_2 = 5$. We finally construct a (p_1+p_2)-doughnut graph

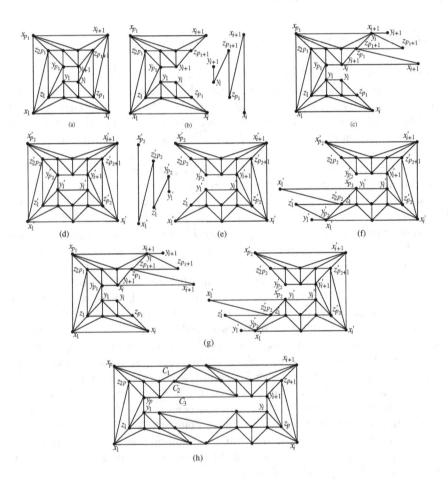

Fig. 3. Illustration for construction of a (p_1+p_2)-doughnut graph G from a p_1-doughnut graph G_1 and a p_2-doughnut graph G_2 where $p_1 = 4$ and $p_2 = 5$

G as follows. We identify the vertices y_{i+1}, z_{p_1+1}, x_{i+1} of G'_1 to the vertices of x'_{p_2}, z'_{2p_2}, y'_{p_2} of G'_2, respectively; and identify the vertices of y_i, z_{p_1}, x_i of G'_1 to the vertices of x'_1, z'_1, y'_1 of G'_2, respectively. Clearly the resulting graph G is a $(p_1 + p_2)$-doughnut graph as illustrated in Fig. 3(h).

We thus have the following theorem.

Theorem 1. *Let G_1 be a p_1-doughnut graph and let G_2 be a p_2-doughnut graph. Then one can construct a $(p_1 + p_2)$-doughnut graph G from G_1 and G_2.*

3 Finding a Shortest Path

A shortest path between any pair of vertices of a doughnut graph can be found efficiently as stated in the following theorem.

Theorem 2. *Let G be a p-doughnut graph. Then a shortest path between any pair of vertices u and v of G can be found in $O(l_s)$ time, where l_s is the length of the shortest path between u and v.*

We have a constructive proof of Theorem 2. The detail is omitted in this extended abstract.

4 Topological Properties of Doughnut Graphs

Let G be a p-doughnut graph. The number of vertices of G is $4p$ where $p(> 3)$ is an integer. A p-doughnut graph is maximal fault tolerant since it is 5-regular. Every p-doughnut graph G has a doughnut embedding Γ where vertices of G lie on three vertex disjoint cycles C_1, C_2 and C_3 such that C_1 is the outer cycle containing p vertices, C_2 is the middle cycle containing $2p$ vertices and C_3 is the inner cycle containing p vertices. Then one can easily see that the diameter of a p-doughnut graph is $\lfloor p/2 \rfloor + 2$. Moreover, a doughnut graph admits a ring embedding since a doughnut graph is Hamilton-connected [5].

Table 1. Topological comparison of doughnut graphs with various Cayley graphs

Topology	number of nodes	diameter	degree	connectivity	Fault tolerance	Hamiltonian
n-cycle	n	$\lfloor n/2 \rfloor$	2	2	maximal	yes
Cube-connected -cycle [6]	$d2^d$	$\lfloor 5d/2 \rfloor - 2$	3	3	maximal	yes
Wrapped around butterfly graph [4]	$d2^d$	$\lfloor 3d/2 \rfloor$	4	4	maximal	yes
d-Dimensional hypercube [1]	2^d	d	d	d	maximal	yes
p-doughnut graphs [2]	$4p$	$\lfloor p/2 \rfloor + 2$	5	5	maximal	yes

5   Conclusion

We have shown that doughnut graphs exhibit recursive structure. We have proposed an efficient algorithm to find shortest path between any pair of vertices which exploit the structure of the graph. We have also found that doughnut graph has smaller diameter, higher degree and connectivity, maximal fault tolerance and ring embedding. There are several parameters like connectivity, degree, diameter, symmetry and fault tolerance which are considered for building interconnection networks [7]. Table 1 presents the topological comparison of various Cayley graphs, which are widely used as interconnection networks, with doughnut graphs. The table shows that topological properties of doughnut graphs are very much similar to interconnection networks. We may have an efficient routing scheme using shortest path finding algorithm. Thus doughnut graphs may find nice applications as interconnection networks.

Acknowledgement. This work is supported by Bangladesh Academy of Sciences.

References

1. Bhuyan, L., Agarwal, D.P.: Generalized hypercube and hyperbus structure for a computer network. IEEE Trans. Comput. 33, 555–566 (1984)
2. Karim, M.R., Rahman, M.S.: Straight-line grid drawings of planar graphs with linear area. In: Proceedings of Asia-Pacific Symposium on Visualisation 2007, pp. 109–112. IEEE (2007)
3. Karim, M.R., Rahman, M.S.: On a class of planar graphs with straight-line grid drawings on linear area. Journal of Graph Algorithms and Applications 13(2), 153–177 (2009)
4. Leighton, F.T.: Introduction to parallel algorithms and architectures: Arrays-trees-hypercubes. Morgan Kaufmann Publishers (1992)
5. Nahiduzaman, K.M., Karim, M.R., Rahman, M.S.: A linear-time algorithm for k-partitioning doughnut graphs. INFOCOMP 8(1), 8–13 (2009)
6. Preparata, F.P., Vuillemin, J.: The cube-connected-cycles: A versatile network for parallel computation. Communications of the ACM 24, 300–309 (1981)
7. Xu, J. (ed.): Topological Structure and Analysis of Interconnection Networks. Kluwer Academic Publishers (2001)

On the Steiner Radial Number of Graphs

K.M. Kathiresan[1], S. Arockiaraj[2], R. Gurusamy[2], and K. Amutha[3]

[1] Center for Research and Post Graduate Studies in Mathematics
Ayya Nadar Janaki Ammal College
Sivakasi - 626 124,Tamil Nadu, India
[2] Department of Mathematics
Mepco Schlenk Engineering College
Mepco Engineering College (PO)-626005
Sivakasi, Tamil Nadu, India
[3] Department of Mathematics
Sri Parasakthi College, Courtallam, Tamil Nadu, India
{kathir2esan,sarockiaraj_77}@yahoo.com,
{sahama2010,amuthakaruppasamy}@gmail.com

Abstract. The Steiner n-radial graph of a graph G on p vertices, denoted by $SR_n(G)$, has the vertex set as in G and $n(2 \leq n \leq p)$ vertices are mutually adjacent in $SR_n(G)$ if and only if they are n-radial in G. While G is disconnected, any n vertices are mutually adjacent in $SR_n(G)$ if not all of them are in the same component. When $n = 2$, $SR_n(G)$ coincides with the radial graph $R(G)$. For a pair of graphs G and H on p vertices, the least positive integer n such that $SR_n(G) \cong H$, is called the Steiner completion number of G over H. When $H = K_p$, the Steiner completion number of G over H is called the Steiner radial number of G. In this paper, we determine 3-radial graph of some classes of graphs, Steiner radial number for some standard graphs and the Steiner radial number for any tree. For any pair of positive integers n and p with $2 \leq n \leq p$, we prove the existence of a graph on p vertices whose Steiner radial number is n.

Keywords: n-radius, n-diameter, Steiner n-radial graph, Steiner completion number, Steiner radial number.

1 Introduction

Throughout this paper, we consider finite undirected graphs without multiple edges and loops. Let G be a graph on p vertices and S a set of vertices of G. In [2], the Steiner distance of S in G, denoted by $d_G(S)$, is defined as the minimum number of edges in a connected subgraph of G that contains S. Such a subgraph is necessarily a tree and is called a Steiner tree for S in G. The Steiner n-eccentricity $e_n(v)$ of a vertex v in a graph G is defined as $e_n(v) = \max\{d_G(S) : S \subseteq V(G)$ with $v \in S$ and $|S| = n\}$. The n-radius $rad_n(G)$ of G is defined as the smallest Steiner n-eccentricity among the vertices of G, and the n-diameter $diam_n(G)$ of G is the largest Steiner n-eccentricity. The concept of Steiner distance was further developed in [3,6,5].

S. Arumugam and B. Smyth (Eds.): IWOCA 2012, LNCS 7643, pp. 65–72, 2012.

In [4], KM. Kathiresan and G. Marimuthu introduced the concept of radial graphs. Two vertices of a graph G are said to be radial to each other if the distance between them is equal to the radius of the graph. The radial graph of a graph G, denoted by $R(G)$, has the vertex set as in G and two vertices are adjacent in $R(G)$ if and only if they are radial in G. If G is disconnected, then two vertices are adjacent in $R(G)$ if they belong to different components of G.

Any n vertices of a graph G are said to be n-radial to each other if the Steiner distance between them is equal to the n-radius of the graph G. The Steiner n-radial graph of a graph G, denoted by $SR_n(G)$, has the vertex set as in G and $n(2 \leq n \leq p)$ vertices are mutually adjacent in $SR_n(G)$ if and only if they are n-radial in G. If G is disconnected, any n vertices are mutually adjacent in $SR_n(G)$ if not all of them are in the same component. For the edge set of $SR_n(G)$, draw K_n corresponding to each set of n-radial vertices. By taking $n = 2, SR_n(G)$ coincides with $R(G)$. Consider the graph G given in Figure 1.

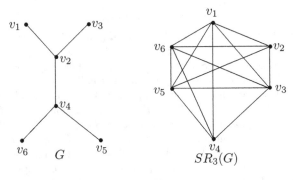

Fig. 1.

If we let $n = 3$, we get that $rad_3(G) = 3$ and that $S_1 = \{v_1, v_2, v_5\}$, $S_2 = \{v_1, v_2, v_6\}$, $S_3 = \{v_1, v_3, v_4\}, S_4 = \{v_1, v_4, v_5\}$, $S_5 = \{v_1, v_4, v_6\}$, $S_6 = \{v_2, v_3, v_5\}$, $S_7 = \{v_2, v_3, v_6\}, S_8 = \{v_2, v_5, v_6\}$, $S_9 = \{v_3, v_4, v_5\}$ and $S_{10} = \{v_3, v_4, v_6\}$ are the sets of 3-radial vertices of G. Hence the Steiner 3-radial graph of G is as shown in Figure 1. A graph G is called a Steiner 3-radial graph if $SR_3(H) \cong G$ for some graph H. If G does not contain K_3 as a subgraph, then G is not a Steiner 3-radial graph. The converse of this statement is not true. For example, the graph G given in Figure 2 is not a Steiner 3-radial graph.

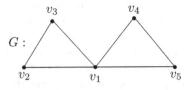

Fig. 2.

Let G and H be two graphs on p vertices. If there exists a positive integer n such that $SR_n(G) \cong H$, then H is called a Steiner completion of G. The positive integer n is said to be Steiner completion of G over H if n is the least positive integer such that $SR_n(G) \cong H$. For example, the Steiner completion number of bistar B_{p_1,p_2} over $K_{p_1+p_2+2} - e$ is $p_1 + p_2$. If there is no n such that $SR_n(G) \cong H$, then the Steiner completion number of G over H is ∞. The Steiner completion number of G over H is not necessarily equal to the Steiner completion number of H over G. For the graphs G and H given in Figure 3, the Steiner completion number of G over H is 3 but the Steiner completion number of H over G is ∞.

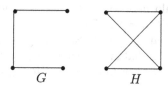

Fig. 3.

When $H = K_p$, the Steiner completion number of G over H is called the Steiner radial number of G. That is, the Steiner radial number $r_S(G)$ of a graph G is the least positive integer n such that the Steiner n-radial of G is complete. In this paper, we determine the Steiner radial number for some classes of graphs and obtain the Steiner radial number for any tree. Also we prove that for every pair of integers n and p with $2 \le n \le p$, there exists a graph on p vertices whose Steiner radial number is n. For graph theoretic terminology we follow [1].

2   Steiner 3-Radial Graphs of Some Classes of Graphs

Proposition 1. *Let P_p be any path on $p \ge 3$ vertices. Then $SR_3(P_p) = K_2 + \overline{K_{p-2}}$ where K_{p-2} is the complete graph on $p-2$ internal vertices of P_p.*

Proof. Let $P_p : v_1 v_2 \cdots v_{p-1} v_p$ by any path on $p \ge 3$ vertices. Then the Steiner 3-eccentricity of each vertex of P_p is $p-1$ and hence $rad_3(P_p) = p-1$. Now for every vertex $v_i, 2 \le i \le p-1$, we have $d(\{v_i, v_1, v_p\}) = p-1$. Hence $\{v_i, v_1, v_p\}$ where $2 \le i \le p-1$ forms a K_3. So assume $p \ge 4$. Also for every pair of vertices v_i and $v_j, i \ne j$ and $2 \le i, j \le p-1$, there exists no $v_k, 1 \le k \le p, k \ne i$ and $k \ne j$ such that $d(\{v_i, v_j, v_k\}) = p-1$ and hence there is no edge between v_i and $v_j, i \ne j$ and $2 \le i, j \le p-1$. Therefore $SR_3(P_p) = K_2 + \overline{K_{p-2}}$ where K_{p-2} is the complete graph on $p-2$ internal vertices of P_p.

Proposition 2. *Let C_p be any cycle of length $p \ge 5$. Then*

$$SR_3(C_p) = \begin{cases} C_p(k) & \text{if } p = 3k \\ C_p(k, k+1) & \text{if } p = 3k+2 \\ C_p(k-1, k, k+1) & \text{if } p = 3k+1. \end{cases}$$

Where the circulant graph $C_p(n_1, n_2, \ldots, n_l)$ is obtained from a cycle on p vertices by joining each vertex v_i, $1 \le i \le p$ with the vertices v_{i-n_j} and v_{i+n_j}, the subscripts being taken modulo p, for $1 \le j \le l$. When $p = 3$, $SR_3(C_p) = K_3$ and when $p = 4$, $SR_3(C_p) = K_4$.

Proof. Clearly $SR_3(C_p) = K_3$ or K_4 when $p = 3, 4$ respectively.

Let $C_p : v_0, v_1, v_2, \ldots, v_{p-1}, v_0$ be any cycle of length $p \ge 5$.

Case 1. $p = 3k$.

In this case v_{i-k} and v_{i+k} are Steiner 3-eccentric vertices of v_i where the subscripts are taken modulo p. Also Steiner 3-eccentricity of a vertex v_i is $2k$ for every i and hence $rad_3(C_p) = 2k$. Thus v_{i-k} and v_{i+k} are the only vertices which are at a 3-radius distance with v_i. So that $\{v_i, v_{i-k}, v_{i+k}\}$ forms a K_3 in the corresponding Steiner 3-radial graph. Hence v_i is adjacent to v_{i-k} and v_{i+k} only. This is true for every i. Therefore we get $SR_3(C_p) = C_p(k)$.

Case 2. $p = 3k + 2$.

In this case we have three sets of Steiner 3-eccentric vertices of v_i namely $S_1 = \{v_{i+(k+1)}, v_{i-k}\}$, $S_2 = \{v_{i+k}, v_{i-(k+1)}\}$ and $S_3 = \{v_{i+(k+1)}, v_{i-(k+1)}\}$, where the subscripts are taken modulo p. Also their distance with v_i is $2k+1$. Thus $v_{i+(k+1)}$ and v_{i-k} are vertices which are at a 3-radius distance with v_i. So S_1 forms a K_3 in the corresponding $SR_3(G)$. Similarly each set of Steiner 3-eccentric vertices forms a K_3. Hence v_i is adjacent to $v_{i+k}, v_{i-k}, v_{i+(k+1)}$ and $v_{i-(k+1)}$. Therefore $SR_3(C_p) = C_p(k, k + 1)$.

Case 3. $p = 3k + 1$.

In this case we have six sets of Steiner 3-eccentric vertices of v_i namely, $\{v_{i+(k+1)}, v_{i-k}\}, \{v_{i+(k+1)}, v_{i-(k+1)}\}, \{v_{i+(k+1)}, v_{i-(k-1)}\}, \{v_{i+k}, v_{i-(k+1)}\}, \{v_{i+k}, v_{i-k}\}$ and $\{v_{i+(k-1)}, v_{i-(k+1)}\}$, where the subscripts are taken modulo p. Also their distance with v_i is $2k$. If we proceed as in the proof of Case 2, we get v_i is adjacent to $v_{i+k}, v_{i-k}, v_{i+(k+1)}, v_{i-(k+1)}, v_{i+(k-1)}$ and $v_{i-(k-1)}$ and hence $SR_3(C_p) = C_p(k - 1, k, k + 1)$.

Proposition 3. *Let G be any cycle C_p of length $p \ge 5$. Then $SR_3(\overline{G}) = C_p(1, 2)$ where $\overline{G}$ is the complement of G. When $p = 3, 4$, $SR_3(\overline{G}) = K_3, K_4$ respectively.*

Proof. If $G \cong C_3, C_4$ then $rad_3(\overline{G}) = \infty$. Hence we get $SR_3(\overline{G}) = K_3, K_4$ respectively. Let G be any cycle C_p of length $p \ge 5$ having the vertices $v_0, v_1, \cdots, v_{p-1}$. Then any vertex v_i in $\overline{G}$ is adjacent to all the vertices of $\overline{G}$ except v_{i-1} and v_{i+1}. Here $\{v_{i-1}, v_{i+1}\}, \{v_{i+1}, v_{i+2}\}$ and $\{v_{i-1}, v_{i-2}\}$ are the sets of Steiner 3-eccentric vertices of v_i. Also their distance with v_i is 3. That is in $\overline{G}$ $d(\{v_i, v_{i-1}, v_{i+1}\}) = d(\{v_i, v_{i+1}, v_{i+2}\}) = d(\{v_i, v_{i-1}, v_{i-2}\}) = 3$. So $e_3(v_i) = 3$ for all i and hence $rad_3(\overline{G}) = 3$. Thus $\{v_i, v_{i-1}, v_{i+1}\}, \{v_i, v_{i+1}, v_{i+2}\}$ and $\{v_i, v_{i-1}, v_{i-2}\}$ each forms a K_3 in the corresponding $SR_3(\overline{G})$. Similarly, each set of Steiner 3-eccentric vertices forms a K_3. Therefore v_i is adjacent to $v_{i-1}, v_{i+1}, v_{i+2}$ and v_{i-2}. Hence we get $SR_3(\overline{G}) = C_p(1, 2)$ for $p \ge 5$.

Proposition 4. *For $p_1 \leq p_2$,*

$$SR_3(K_{p_1,p_2}) = \begin{cases} K_{1+p_2} & \text{if } p_1 = 1 \\ K_{2+p_2} & \text{if } p_1 = 2 \\ K_{p_1} \cup K_{p_2} & \text{if } p_1 \geq 3. \end{cases}$$

Proof. Let $X = \{x_1, x_2, \ldots, x_{p_1}\}$ and $Y = \{y_1, y_2, \ldots, y_{p_2}\}$ be the bipartition of K_{p_1,p_2}.

Case 1. $p_1 = 1$.

Then $d(\{x_1, y_i, y_j\}) = 2$. We have $e_3(x_1) = 2$ and $rad_3(G) = 2$. So in the corresponding Steiner 3-radial graph, $\{x_1, y_i, y_j\}$ forms a K_3. Since y_i and y_j are arbitrary vertices of G, we have $SR_3(K_{1,p_2}) = K_{1+p_2}$.

Case 2. $p_1 = 2$.

Here $|X| = 2$ and $|Y| \geq 2$. Then $e_3(x_1) = e_3(x_2) = 2$ and hence $rad_3(K_{p_1,p_2}) = 2$. Thus for every pair of vertices $v_i, v_j \in V(K_{p_1,p_2})$, there exists a vertex v_k such that $d(\{v_i, v_j, v_k\}) = 2$. Therefore any two vertices are adjacent in $SR_3(K_{p_1,p_2})$. Hence we get $SR_3(K_{p_1,p_2}) = K_{2+p_2}$.

Case 3. $p_1 \geq 3$.

Here $|X| \geq 3$ and $|Y| \geq 3$. Let x_i be any vertex of X. Then $e_3(x_i) = 3$ and every two vertices in X different from x_i are the Steiner 3-eccentric vertices of x_i. Similarly for $y_i \in Y, e_3(y_i) = 3$ and every two vertices in Y different from y_i are the Steiner 3-eccentric vertices of y_i. Therefore $rad_3(K_{p_1,p_2}) = 3$ and hence $SR_3(K_{p_1,p_2}) = K_{p_1} \cup K_{p_2}$.

Theorem 5. *If G is a disconnected graph of order $p \geq 3$, then $SR_3(G) \cong K_p$.*

Proof. Let G be a disconnected graph with two components say G_1 and G_2. Then every vertex in G_1 is Steiner 3-eccentric with a vertex in G_2. Thus $e_3(v_i) = \infty$ for all i and hence $rad_3(G) = \infty$. Therefore for every two vertices v_i and v_j, there exists a vertex v_k in G such that $d(\{v_i, v_j, v_k\}) = \infty$. Hence we get $SR_3(G) = K_p$.

Theorem 6. *For any integer $n \geq 2$, there exists a graph G such that $rad_3(G) = n$.*

Proof. Let $n \geq 2$ be any integer. Construct a graph G by adding the vertices u, v, x and y with a path $P_{n-1} : v_1 v_2 \cdots v_{n-1}$ and join u, v to v_1 and x, y to v_{n-1} as shown below.

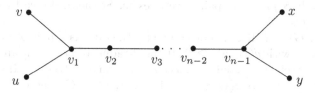

Fig. 4.

Then $e_3(v_i) = n$ for every $v_i \in V(P_{n-1})$. Also $e_3(u) = n + 1$. Similarly $e_3(v) = e_3(x) = e_3(y) = n + 1$. Then $rad_3(G) = \min\{e_3(v_i) : v_i \in V(P_{n-1}), e_3(u), e_3(v), e_3(x), e_3(y)\} = n$.

Problem 7. *Characterize all Steiner 3-radial graphs.*

3   Steiner Radial Number

Observation 8 *It follows from the definition that for any connected graph G on p vertices, $2 \leq r_S(G) \leq p$.*

Proposition 9. *If $r_S(G) = n$, then K_p is the only Steiner m-radial graph for G for $m \geq n$.*

Proof. For a graph G, let $r_S(G) = n$ and let r be the n-radius of G. Then there exists a vertex v in $V(G)$ such that $e_n(v) = r$. Let N be a n-element set containing v with Steiner distance r. Consider the set $N \cup \{x\}$, where $x \in V(G) - N$. Since Steiner n-eccentricity of any n-element set containing v is almost r, the set $N \cup \{x\}$ is of Steiner distance either r or $r + 1$ for any $x \in V(G) - N$. Otherwise a n-element subset of $N \cup \{x\}$ with v is of Steiner distance more than r.

By the same argument, $e_n(u) \leq e_{n+1}(u)$ for all $u \in V(G)$. The result follows when $(n + 1)$-radius of G is r. If $(n + 1)$-radius of G is $r + 1$, then the vertex v in $V(G)$ has the minimum Steiner $(n + 1)$-eccentricity $r + 1$. Let v_i and v_j be any two vertices of G. Since v_i and v_j are adjacent in Steiner n-radial of G, there exists an n-element set S with Steiner distance r. If $r + 1 = p - 1$, then any set of $n + 1$ elements containing v_i and v_j has the Steiner distance $r + 1$. Suppose $r + 1 < p - 1$. If v does not belong to the Steiner tree containing S, then $S \cup \{v\}$ has Steiner distance $r + 1$. If $v \in S$, then also we adjoint a vertex w in S which does not belong to the Steiner tree containing S such that the Steiner distance of $S \cup \{w\}$ is $r + 1$. Hence any two vertices v_i and v_j are adjacent in Steiner $(n + 1)$-radial of G. Hence the result follows.

Theorem 10. $r_S(G) = 2$ *if and only if G is either complete or totally disconnected.*

Proof. When G is complete (respectively a totally disconnected graph), 2-radius is 1 (respectively ∞) and any pair of vertices has Steiner distance 1 (respectively ∞). Hence $r_S(G) = 2$.

Suppose $r_S(G) = 2$. If G is not complete, then it has a pair of non-adjacent vertices u and v with $d(\{u, v\}) \geq 2$. If the 2-radius of G is 1, u and v are not adjacent in the Steiner 2-radial of G, a contradiction to $r_S(G) = 2$. If the 2-radius of G is ≥ 2, then we have $d(\{x, y\}) = 1$ for all $x, y \in V(G)$ where $(x, y) \in E(G)$, hence x and y are not adjacent in the Steiner 2-radial graph of G, so the edge-set must be empty.

Proposition 11. *For any star graph with* p *vertices,*

$$r_S(K_{1,p-1}) = \begin{cases} 2 & \text{for } p = 2 \\ 3 & \text{for } p \geq 3. \end{cases}$$

Proof. The case $p = 2$ follows directly from Theorem 10 as $K_{1,1} = K_2$. When $p = 3$, 2-radius of $K_{1,2}$ is 1 and Steiner 2-radial of $K_{1,2}$ is not complete. Also 3-radius of $K_{1,2}$ is 2 and Steiner 3-radial of $K_{1,2}$ is K_3. For $p \geq 4$, let v_1 be the vertex of degree $p - 1$ and $v_2, v_3, \ldots, v_p$ be the pendant vertices of $K_{1,p-1}$. By Theorem 10, $r_S(K_{1,p-1})$ can not be 2. The 3-radius of $K_{1,p-1}$ is 2, since $e_3(v_1) = 2$ and $e_3(v_i) = 3, 2 \leq i \leq p$. In Steiner 3-radial of G, v_1 is adjacent to each vertex $v_i, 2 \leq i \leq p$, since the set $\{v_1, v_i, v_j(j \neq 1, i)\}$ has the Steiner distance 2. Also v_i is adjacent to v_j for $2 \leq i, j \leq p$ and $i \neq j$, since the set $\{v_1, v_i, v_j\}$ has the Steiner distance 2.

Proposition 12. *For any complete bipartite graph* K_{p_1,p_2} *with* $p_1 \leq p_2$ *and* $p_1 \neq 1, r_S(K_{p_1,p_2}) = p_1 + 1.$

Proof. Let $\{u_1, u_2, \ldots, u_{p_1}\}$ and $\{v_1, v_2, \ldots, v_{p_2}\}$ be the two partitions of K_{p_1,p_2}. When $n \leq p_1, e_n(u_i) = n, 1 \leq i \leq p_1$ and $e_n(v_i) = n, 1 \leq i \leq p_2$. Hence $rad_n(K_{p_1,p_2}) = n$. In Steiner n-radial of G, u_i is not adjacent to v_j, since the n-element sets containing u_i and v_j have only the Steiner distance $n - 1$. Therefore, $r_S(K_{p_1,p_2}) > p_1$. When $n > p_1, e_n(u_i) = n - 1, 1 \leq i \leq p_1$ and $e_n(v_i) \geq n - 1, 1 \leq i \leq p_2$. Hence $rad_n(K_{p_1,p_2}) = n - 1$.

In Steiner $(p_1 + 1)$-radial of G, u_i is adjacent to u_j for $1 \leq i, j \leq p_1$, u_i is adjacent to v_j for all $1 \leq i \leq p_1, 1 \leq j \leq p_2$ and v_i is adjacent to v_j for all $1 \leq i, j \leq p_2$, since each of the sets $\{u_1, u_2, \ldots, u_{p_1}, v_j\}, \{u_1, u_2, \ldots, u_{p_1}, v_j\}$ and $\{v_i, v_j, u_2, u_3, \ldots, u_{p_1}\}$ have the Steiner distance p_1 respectively. Hence Steiner $(p_1 + 1)$-radial of K_{p_1,p_2} is $K_{p_1+p_2}$.

Theorem 13. *For every tree* T *with* $m(\neq p-1)$ *pendant vertices* $r_S(T) = m+2$.

Proof. Let T be a tree with m pendant vertices $x_1, x_2, \ldots, x_m$ and the remaining vertices be $v_1, v_2, \ldots, v_{p-m}$. Then $e_n(v_i) = p-1$ for $n = m+1$ and hence $(m+1)$-radius is $p - 1$. If $v_i v_j$ is a non-pendant edge in T, then the set $\{v_i, v_j\} \cup X$, where $X \subseteq \{x_1, x_2, \ldots, x_m\}$ with $|X| = m - 1$, has Steiner distance $< p - 1$. Therefore, v_i is not adjacent with v_j in Steiner $(m+1)$-radial of G. Since $(m+2)$-radius is $p - 1$ and any set $\{v_i, v_j, x_1, x_2, \ldots, x_m\}$ has Steiner distance $p - 1$ for $1 \leq i, j \leq p - m$, Steiner $(m + 2)$-radial of G is K_p.

Corollary 14 *For every positive integer* $k \geq 2$, *there exists a graph having Steiner radial number* k.

Proposition 15. *For any wheel,* $r_S(W_p) = \begin{cases} 2 & \text{for } p = 4 \\ 3 & \text{for } p \geq 5. \end{cases}$

Proof. When $p = 4$, the result follows from Theorem 10. So assume $p \geq 5$. Let v_1 be vertex of degree $p - 1$ in W_p and $v_2, v_3, \ldots, v_p$ be the vertices on the cycle of W_p. Since W_p is not complete by Theorem 10, $r_S(W_p) > 2$. Since $e_3(v_1) = 2$ and $e_3(v_i) = 3, 2 \leq i \leq p, rad_3(G) = 2$. For $2 \leq i, j \leq p$ and $i \neq j$, the set $\{v_1, v_i, v_j\}$ has the Steiner distance 2 and hence the Steiner 3-radial of W_p is complete.

Theorem 16. *For any pair of integers n and p with* $2 \leq n \leq p$, *there exists a graph on p vertices whose Steiner radial number is n.*

Proof. When $p = 2$, the result is obvious. When $p = 3$, the only connected graph on 3 vertices are P_3 and K_3 in which $r_S(P_3) = 3$ and $r_S(K_3) = 2$. When $p = 4, r_S(K_4) = 2, r_S(C_4) = 3$ and $r_S(P_4) = 4$. When $p \geq 5, r_S(W_p) = 3$ by Proposition 15. Also $r_S(K_p) = 2$ and $r_S(T) = m + 2$ where m is the number of pendant vertices in T and $2 \leq m \leq p - 2$.

Acknowledgments. The authors thank the anonymous referees for careful reading and detailed comments which helped in restructuring the paper in the present form.

References

1. Buckley, F., Harary, F.: Distance in graphs. Addison-Wesley, Reading (1990)
2. Chartrand, G., Oellermann, O.R., Tian, S., Zou, H.B.: Steiner distance in graphs. Casopis Pro Pestovani Matematiky 114(4), 399–410 (1989)
3. Day, D.P., Oellermann, O.R., Swart, H.C.: Steiner distance-hereditary graphs. SIAM J. Discrete Math. 7, 437–442 (1994)
4. Kathiresan, K.M., Marimuthu, G.: A study on radial graphs. Ars Combin. 96, 353–360 (2010)
5. Oellermann, O.R., Tian, S.: Steiner centers in graphs. J. Graph Theory 14(5), 585–597 (1990)
6. Raines, M., Zhang, P.: The Steiner distance dimension of graphs. Australasian J. Combin. 20, 133–143 (1999)

Further Results on the Mycielskian of Graphs

T. Kavaskar

Department of Mathematics, Bharathidasan University, Tiruchirappalli-620024, India
t_kavaskar@yahoo.com

Keywords: Mycielskian of a graph, acyclic chromatic number, dominator chromatic number, independent domination number.

1 Introduction

All graphs considered here are finite, undirected, connected and non-trivial. In the mid 20th century there was a question regarding, triangle-free graphs with arbitrarily large chromatic number. In answer to this question, Mycielski [7] developed an interesting graph transformation as follows: For a graph $G = (V, E)$, the Mycielskian of G is the graph $\mu(G)$ with vertex set consisting of the disjoint union $V \cup V' \cup \{u\}$, where $V' = \{x' : x \in V\}$ and edge set $E \cup \{x'y : xy \in E\} \cup \{x'u : x' \in V'\}$. We call x' the twin of x in $\mu(G)$ and vice versa and u, the root of $\mu(G)$. We can define the *iterative Mycielskian* of a graph G as follows: $\mu^m(G) = \mu(\mu^{m-1}(G))$, for $m \geq 1$. Here $\mu^0(G) = G$. It is well known [7] that if G is triangle free, then so is $\mu(G)$ and that the chromatic number $\chi(\mu(G)) = \chi(G) + 1$. There had been several papers on Mycielskian of graphs. Few of the references are [2], [3], [5], [7], [8]. Several graph parameters, especially in domination theory, on Mycielskian of graphs have been discussed in [2], [8].

We say that a set $S \subseteq V(G)$ is a *dominating set* of G if every vertex $x \in V(G) \backslash S$ has at least one neighbor in S. The *domination number* of G is defined by $\gamma(G) = \min\{|S| : S$ is a dominating set of $G\}$. A set $S \subseteq V(G)$ is said to be an *independent dominating set* of G [4] if S is a dominating set of G and the induced subgraph $\langle S \rangle$ has no edge in G. The *independent domination number* of G is defined by $i(G) = \min\{|S| : S$ is an independent dominating set of $G\}$. A set $S \subseteq V(G)$ is said to be *acyclic domination set* of G if it is a dominating set of G and $\langle S \rangle$ is a forest. The *acyclic dominating number* of G is defined by $\gamma_a(G) = \min\{|S| : S$ is an acyclic domination set of $G\}$.

A proper k-coloring, $V_1, V_2, \ldots, V_k$, of G is said to be *acyclic k-coloring* if for $1 \leq i < j \leq k$, $\langle V_i \cup V_j \rangle$ is a forest. The *acyclic chromatic number* of G is defined by: $\chi_a(G) = \min\{k : G$ has an acyclic k-coloring$\}$. A proper k-coloring of G is said to be *dominator k-coloring* [6] if for every vertex $v \in V(G)$, $N[v]$ contains at least one color class. The *dominator chromatic number* of G is defined by: $\chi_d(G) = \min\{k : G$ has a dominator k-coloring$\}$.

In this paper, we determine the independent domination number, acyclic chromatic number and dominator chromatic number of the Mycielskian and the iterated Mycielskian of a graph with respect to their parent graphs.

S. Arumugam and B. Smyth (Eds.): IWOCA 2012, LNCS 7643, pp. 73–75, 2012.
© Springer-Verlag Berlin Heidelberg 2012

2 Main Results

We start with the following well-known result on the domination number of the Mycielskian.

Theorem 21 ([2]). *For any graph* G, $\gamma(\mu(G)) = \gamma(G) + 1$.

We now show that similar results hold good for the independent domination number of $\mu(G)$.

Theorem 22. *For any graph* G, $i(\mu(G)) = i(G) + 1$.

Proof. Let S be an independent dominating set of G with $|S| = i(G)$. Then $S \cup \{u\}$ is an independent dominating set of $\mu(G)$ and hence $i(\mu(G)) \leq i(G) + 1$. Suppose $i(\mu(G)) \leq i(G)$. Let T be an independent dominating set of $\mu(G)$ with $|T| = i(\mu(G)) \leq i(G)$. If $u \in T$, then $T \backslash \{u\}$ is an independent dominating set of G and hence $i(\mu(G)) - 1 = |T \backslash \{u\}| \geq i(G)$, a contradiction. Thus $u \notin T$, which implies that $T \cap V' \neq \emptyset$. Similarly we get a contradiction for this case also. Thus $i(\mu(G)) \geq i(G) + 1$ and hence $i(\mu(G)) = i(G) + 1$.

Iteratively applying Theorem 22, we get

Corollary 23. *For any graph* G, $i(\mu^m(G)) = i(G) + m$.

Next we discuss the acyclic domination number of $\mu(G)$. Let S be an acyclic dominating set of G with $|S| = \gamma_a(G)$, then $S \cup \{u\}$ is an acyclic dominating set of $\mu(G)$. Hence we have the following result.

Proposition 24. *For any graph* G, $\gamma_a(\mu(G)) \leq \gamma_a(G) + 1$.

We have an example for $(\gamma_a(G) + 1) - \gamma_a(\mu(G))$ is arbitrarily large. For example, consider a graph G is obtained by joining k pendent vertices to each vertex of the cycle C_n for any $k \geq 2$ and $n \geq 3$. One can easily check that $\gamma_a(G) = n + k - 1$ and $\gamma_a(\mu(G)) = n + 1$.

We now recall that the well-known result on the chromatic number of the Mycielskian.

Theorem 25 ([7]). *For any graph* G, $\chi(\mu(G)) = \chi(G) + 1$.

Next we show that similar results hold good for the acyclic chromatic number and dominator chromatic number of $\mu(G)$.

Theorem 26. *For any graph* G, $\chi_a(\mu(G)) = \chi_a(G) + 1$.

Proof. Let $V_1, V_2, \ldots, V_k$ be a acyclic k-coloring, where $k = \chi_a(G)$. Now set $U_i = V_i \cup V_i'$, for i, $1 \leq i \leq k$ and $U_{k+1} = \{u\}$. Then $U_i's$ forms a acycile coloring of $\mu(G)$ using $k + 1$ colors and hence $\chi_a(\mu(G)) \leq \chi_a(G) + 1$. Next to prove $\chi_a(\mu(G)) \geq \chi_a(G) + 1$. Suppose $\chi_a(\mu(G)) < \chi_a(G) + 1$. Let $k = \chi_a(\mu(G))$. Then $\mu(G)$ has a acyclic k-colors. We have to show that G has a acyclic $(k - 1)$-coloring. Proof of this is similar to the Theorem 25 given in [7]. Thus $\chi_a(\mu(G)) = \chi_a(G) + 1$.

Iteratively applying Theorem 26, we get

Corollary 27. *For any graph G, $\chi_a(\mu^m(G)) = \chi_a(G) + m$.*

Finally we determine the dominator chromatic number of $\mu(G)$.

Theorem 28. *For any graph G, $\chi_d(\mu(G)) = \chi_d(G) + 1$.*

Iteratively applying Theorem 28, we get

Corollary 29. *For any graph G, $\chi_d(\mu^m(G)) = \chi_d(G) + m$.*

Acknowledgement. This research was supported by Dr.D.S. Kothari Post Doctoral Fellowship, University Grants Commission, Government of India grant F.4-2/2006(BSR)/13-511- 2011(BSR) dated 26-08-2011 at the Department of Mathematics, Bharathidasan University, Tiruchirappalli-620024, India.

References

1. Coekayne, E.J., Hedetniemi, S.T., Miller, D.J.: Properties of hereditary hypergraphs and middle graphs. Canad. Math. Bull. 21(4), 461–468 (1978)
2. Fisher, D.C., McKenna, P.A., Boyer, E.D.: Hamiltonicity, diameter, domination, packing, and biclique partitions of Mycielski's graphs. Discrete Applied Mathematics 84(1-3), 93–105 (1998)
3. Chang, G.J., Huang, L., Zhu, X.: Circular chromatic numbers of Mycielski's graphs. Discrete Mathematics 205(1-3), 23–37 (1999)
4. Halldorsson, M.M.: Approximating the minimum maximal independence number. Information Processing Letters 46(4), 169–172 (1993)
5. Larsen, M., Propp, J., Ullman, D.: The fractional chromatic number of Mycielski's graphs. J. Graph Theory 19, 411–416 (1995)
6. Chellali, M., Maffray, F.: Dominator Colorings in Some Classes of Graphs. Graphs and Combinatorics 28, 97–107 (2012)
7. Mycielski, J.: Sur le coloriage des graphs. Colloq. Math. 3, 161–162 (1955)
8. Lin, W., Wu, J., Lam, P.C.B., Gu, G.: Several parameters of generalized Mycielskians. Discrete Applied Mathematics 154, 1173–1182 (2006)

Approaches and Mathematical Models for Robust Solutions to Optimization Problems with Stochastic Problem Data Instances

Niraj Ramesh Dayama* and Ketki Kulkarni

Indian Institute of Technology Bombay,
Powai-400 076, India

1 Introduction

Practical applications of scheduling, routing and other generic constrained optimization problems often involve an uncertainty in the values of the data presented in the problem data instances. On the contrary, most of the established algorithms for typical classes of well-studied problems in the field of constrained optimization assume that deterministic precise values of data would be known. Hence, any solution developed for a specific optimization problem with a given problem data instance would become non-optimal and/or infeasible when applied to another data instance with even slight perturbation. We argue the fallacy of using solutions developed based on the mean values of data for real life problems having stochastic data. This paper presents frameworks for solving typical constrained optimization problems where the exact values of the problem data instances are not known; only the mean, variance and probable distribution of every governing parameter is available. The solutions, thus generated, should continue to be feasible, close-to-optimal and relatively "good" for a range of values around the initially assumed precise deterministic values. Thus, if the input parameters of a particular optimization problem are stochastic with a known distribution, then the techniques being presented yield a robust solution that is applicable to almost any instance in the entire range of values possible. We develop results for the traveling salesman problem and sequence dependent job scheduling problem with time constraints.

TSP with time windows (TSPTW) corresponds to the problem of job scheduling with sequence dependent processing time. This is a practical industrial application where the sequence dependent setup time between a pair of jobs cannot be exactly determined but will typically be a probabilistic distribution around a mean value. TSPTW with time windows is an excellent test criterion to decide the efficacy of the techniques and algorithms presented here because, for sufficiently restrictive time windows, an optimal solution for a given problem data instance may actually become infeasible with even small change in the parameters from the problem instance.

---

* Corresponding author.

S. Arumugam and B. Smyth (Eds.): IWOCA 2012, LNCS 7643, pp. 76–80, 2012.

The efficacy of the proposed algorithm is discussed using two criteria, which are quantified as follows:

1. Robust solution : A solution $\widehat{S}$ is said to be robust if it is infeasible for less than a specified fraction of the permissible problem data instances. Robustness has been quantitatively defined in the next paragraph.

2. Quality of solution for data instance: For a problem data instance described by (A, B, E), let the optimal route be S and the optimal objective value of cumulative traversal cost (for S) be c. Also, for another proposed solution $\widehat{S}$, let the cumulative traversal cost be $\widehat{c}$. Then the quality of $\widehat{S}$ for the data instance (A, B, E) is $c/\widehat{c}$. In general, we would like to have the quality as high as possible. Also, $c/\widehat{c} \leq 1$.

3. Overall quality of solution: The overall quality of a proposed solution $\widehat{S}$ is the norm of quality of that solution for all observed data instances where that solution is feasible. For this paper, we take the norm as the maximum (worst case) value.

4. Overall robustness of a solution: The overall robustness of a solution is the ratio of the number of problem instances where it is feasible to the number of total observed problem instances.

2   Problem Definition and Formulation

The problem is defined on a directed complete graph $G = (V, E)$, where, the node set $V = 0, 1, \ldots, n, n+1$ is a set of $(n+2)$ nodes. Here, the nodes $1, 2, \ldots, n$ indicate individual cities to be visited. The nodes 0 and $n+1$ are notional depot nodes (indicating start or end of routes). There exists a non-negative real cost value $E_{i,j}$ associated with every edge $\{i, j\} \in E$. Traversal of edge $\{i, j\} \in E$ will imply that the salesman traversing that edge $\{i, j\}$ spends time E_{ij}. Every node in $i \in V$ has a time window $[A_i, B_i]$ associated with it. The problem now is to find a route of minimum cumulative traversal cost (collective cost of edges traversed); the route starts from depot 0, terminates at the depot $n+1$ and visits every node in $i \in V$ exactly once, within its own time window $[A_i, B_i]$. The values of A_i, B_i are precisely known; values of E_{ij} are stochastic with known distribution.

We assume that a solution procedure based on MIP formulation for deterministic TSPTW is available. However, as compared to a standard formulation for TSPTW, our MIP formulation is expected to have one additional feature - besides specifying the sequence dependent traversal time between any pair of successive nodes, an additional wait time C_i can be imposed at any node node i. Other details of this MIP formulation (while a non-trivial exercize) is beyond the scope of this document. We combine this MIP formulation with simulation process to form a hybrid optimization-simulation (HOS) algorithm .

Outline of HOS Algorithm: We start by constructing the mean edge cost matrix $\bar{E}$ from the mean values of individual E_{ij}. Then, a standard MIP formulation is used to find the optimal route $\bar{R}$ for $(A, B, \bar{E})$. $\bar{R}$, considered as a seed solution, is given as input to simulation. Around $25 \sim 50$ iterations with ~ 30 replications each are performed for each suggested solution. If the route violates any time window restriction in any city, then the violated city and the maximum value of time violation is noted. A listing of time violations is resent to the MIP algorithm as C_i values. The MIP algorithm modifies the route so as to wait C_i at every preceding city for the time as listed in the time violations for the next city in the route. Based on this, the new optimal route as determined by the MIP algorithm is now passed back to the simulation model. It is expected that the new route reports fewer number of violations during the iterations of the simulation model. This new route would probably be sub-optimal for several of the permissible problem data instances; but should be feasible for many more of them. Hence, overall robustness of solution improves, while possibly getting a lower overall quality of the solution. If overall robustness is better than desired, the algorithm terminates; otherwise further iterations are possible. This HOS algorithm can be summarized as follows:

1. Get seed optimum solution from MIP solver.
2. Perform multiple runs of stochastic simulation model for the seed solution.
3. From analysis of violated nodes, choose suitable truncation point. If no violations found or if termination criterion is met, STOP.
4. Using data of violated nodes, get new sequence from MIP and go to Step 2

The simulation model has been implemented using Anylogic 6.5.0. The model takes the route or sequence of cities to be visited as the input. Multiple runs are performed using the Parameter Variation experiment in Anylogic. If a sequence violates any time window restriction, the model is terminated and violated time instance is noted along with the city that violated it. The data of violated nodes and corresponding times is collected for 300 samples and analyzed to generate feedback for iteration of the DP solver.

Explanation of Simulation Model: We assume that the edge costs E_{ij} have normal distribution $(N(\mu, \sigma))$. Let $\widehat{S}(E)$ be a solution for data instance with Edge cost matrix E. The efficacy of $\widehat{S}(E)$ solution for any data instance with edge costs in the range $E \pm \sigma$ is of interest. Let P be the set of feasible solutions for the constraint set defined by (A,B). A solution $\widehat{S}(E) \in P$ is a robust solution for the data instance $(A, B, E \sim N(\mu, \sigma))$ with significance level $k\sigma$ if and only if it belongs to the set P', where $P' = \left\{ \widehat{S}(E) \in P | \widehat{S}(E + \lambda\sigma) \in P \forall 0 \leq \lambda \leq k \right\}$

Computational Procedure: Consider that a real life situation involves finding a route or schedule where the number of nodes to be traversed (say n), the time windows for every node (say A, B) are known precisely. Values for edge costs edge costs $E = E_{ij}$ are not known precisely but their distribution is known. For this situation, a proposed route or schedule $\widehat{S}$ has been identified. Now, m more permissible problem data instances based on $E^1, E^2, \ldots, E^m$ are developed within the specified ranges and with the specified distribution. Then, the simulation model is executed to find how $\widehat{S}$ performs for these problem data instances $E^1, E^2, \ldots, E^m$. For every problem data instance, we check whether $\widehat{S}$

is feasible for $E^k \dots k \in 1, 2, \dots m$. If $\widehat{S}$ is feasible for some E^k, then determine the objective value (cumulative traversal cost) of $\widehat{S}$ for E^k. If not feasible, find the values C_i $\forall i \in V$ such that $C_i >$max(violation at city i in E^k) and call the MIP to regenerate the route. This procedure repeats till the truncation point.

Let the cumulative traversal costs of these m problem data instances be $c_1, c_2, \dots, c_m$. If $\widehat{S}$ is feasible for E^k, then we set c_k equal to the cumulative traversal cost of $\widehat{S}$ for E^k. If $\widehat{S}$ is infeasible for E^k, we set c_k equal to a very large integer.

The utopian objective was to **find a route $\widehat{S}$ that minimizes the maximum value of c_k.** In practice, the variance and range for E might too large to get any solution $\widehat{S}$ that is feasible over the entire range. So, the more practical objective is to determine a solution $\widehat{S}$ that remains feasible over all specified (very large) number of possible values of E, while minimizing the worst value of c_k on those feasible situations. In this interpretation of the objective, we are neglecting the (very few) infeasibility cases while determining the worst c_k but we would restrict the maximum possible number of such infeasibility cases. Thus, the algorithm yields **a route $\widehat{S}$ that minimizes the maximum value of c_k for all feasible cases, while ensuring that infeasibility is not permitted for more than a pre-specified number of cases.** This maximum number of infeasible cases can be specified as a fraction of m.

3   Computational Experiments

Developing Problem Data Instance

1. Select a particular problem size (n). Here, $n = 5, 10, 15, 20$.
2. For given n, randomly generate n points on a 500×500 grid. Only integer values of coordinates are considered.
3. Choose time window values A, B so that only around n different independent feasible routes can traverse all n nodes within the time restrictions. Having too few or too many different feasible routes makes the time windows too loose or too tight and violates the rationale behind the discussion.
4. Repeat the steps 1-3 to generate about 50+ instances of every problem size.

Details of Computations: Consider a problem data instance of size $n = 5$. We get the values of (A,B,E) as follows:

$A = \begin{pmatrix} 0 & 0 & 0 & 0 & 0 \end{pmatrix}$, $B = \begin{pmatrix} 100 & 20 & 8 & 40 & 50 \end{pmatrix}$

$$\text{The Edge cost matrix E} = \begin{pmatrix} 0 & 1 & 2 & 3 & 4 & 5 & 6 \\ 1 & 0 & 5 & 9 & 5 & 3 & 3 \\ 1 & 1 & 0 & 3 & 3 & 3 & 3 \\ 4 & 4 & 1 & 0 & 1 & 1 & 1 \\ 1 & 1 & 1 & 1 & 0 & 1 & 1 \\ 1 & 1 & 1 & 1 & 1 & 0 & 1 \\ 1 & 1 & 1 & 1 & 1 & 1 & 0 \end{pmatrix}$$

This yields the initial seed solution $\widehat{S} = \begin{pmatrix} 3 & 4 & 2 & 1 & 5 \end{pmatrix}$. Also consider another candidate solution $\widehat{S}' = \begin{pmatrix} 2 & 3 & 4 & 1 & 5 \end{pmatrix}$. Now, we question the efficacy of $\widehat{S}$ and

$\widehat{S}'$ for permissible problem data instances obtained by perturbation of E. Let a permissible problem data instance be (A, B, E'), where $E' = E_{ij} \pm \lambda\sigma$. During the simulation model runs, we get several different values for E'. For example, consider:

$$\begin{pmatrix} 0 & 1 & 2 & 3.5 & 4 & 5 & 6 \\ 1 & 0 & 5 & 9 & 5 & 2.14 & 3 \\ 1 & 0.75 & 0 & 3 & 3 & 3 & 3 \\ 4 & 4 & 1 & 0 & 0.75 & 1 & 1 \\ 1 & 1 & 0.68 & 1 & 0 & 1 & 1 \\ 1 & 1 & 1 & 1 & 1 & 0 & 1 \\ 1 & 1 & 1 & 1 & 1 & 1 & 0 \end{pmatrix}$$

Quality of Solutions: We find that for this particular sequence $\widehat{S}$, the cumulative traversal cost (objective value) for this E' is 11.4. However, it is noted that the optimal route for E' is $\widehat{S}'$ and the cumulative traversal cost (objective value) for $\widehat{S}'$ on E' is 8.89.

So, the quality of $\widehat{S}$ for (A,B,E') is $\frac{8.89}{11.4} = 0.78$ and the quality of $\widehat{S}'$ for (A,B,E') is (1.0). Atleast in this specific case, $\widehat{S}'$ has a much better quality than $\widehat{S}$.

In a similar fashion, the values of the qualities for many different values of E' can be determined by the simulation model and hence the overall quality of the solution can be calculated.

Robustness of Solutions: After multiple (300) runs of the simulation model for different permissible values of (A, B, E') , 300 samples are obtained. Of these 300, 4 samples show violations for a particular city node (city 4, for our data instance) for the solution sequence $\widehat{S}$. On the contrary, for the solution sequence $\widehat{S}'$, we get 258 violations out of 300. So, $\widehat{S}'$ had a better solution quality for one specific problem data instance but poor overall robustness as compared to $\widehat{S}$.

We notice that $\widehat{S}$ was feasible for the given E' value. If, it continues to be feasible for all other E' values, then the solution $\widehat{S}$ is totally feasible. Otherwise, its overall robustness can be calculated and was found to be very good (296 feasible out of 300 is near to 6σ). However, quality of solution for $\widehat{S}$ is found to be quite poor.

A suitable truncation point is selected to cover 99.99%(6σ) of time violations. This information is given to the DP solver, which suggests a new sequence. This may or may not take multiple iterations, depending on the size of the problem and uncertainty in the underlying system. For the chosen data instance, satisfactorily robust solution $\widehat{S}$ with near 6σ robustness is achieved after two iterations.

Conclusion: We have designed a new framework in which stochastic inputs for optimization problems can be defined and solved. We have defined the terminologies for such problems and thereafter presented algorithms which can handle such situations.

Faster Replacement Paths Algorithm for Undirected, Positive Integer Weighted Graphs with Small Diameter

Jay Mahadeokar and Sanjeev Saxena

Dept. of Computer Science and Engineering,
Indian Institute of Technology,
Kanpur, India-208 016

Abstract. We consider the replacement path problem for undirected graphs in case of edge failures. Given a 2-edge connected graph $G(V, E)$, where $n = |V|$ and $m = |E|$, for each edge e on the shortest $s - t$ path of G, we are to report the shortest $s - t$ path in $G \setminus e$. If d is the diameter of the graph, the proposed algorithm takes $O(m + d^2)$ time.

For graphs where $d = O(\sqrt{m})$, typically dense graphs, or graphs with small diameter we have a linear time solution.

1 Introduction

The replacement paths problem in case of edge failure is:

given a 2-edge connected graph $G(V, E)$, for each edge e on the shortest $s - t$ path of G, report the shortest $s - t$ path in $G \setminus e$.

There exists a trivial solution where replacement path for every edge e is computed by running shortest path algorithm independently on $G \setminus e$. Let d be the diameter of the graph. As there can be at most d edges on the shortest path, this takes $O(d(m + n \log n))$ time.

Malik et.al.[6] describe an $O(m + n \log n)$ time solution for the problem on undirected graphs with positive weights. This algorithm was rediscovered by Hershberger and Suri [3] (see e.g. [7,4] for brief history). Their algorithm consists of two main parts:

- Finding the shortest path trees rooted at s and t.
- Reporting the replacement paths for each edge on $s-t$ path using the shortest path trees.

In this paper we describe a new algorithm for the second part. The proposed algorithm takes $O(m + d^2)$ time. If $d = O(\sqrt{m})$ this results in a linear time algorithm for the second part. If the graph has integer weights, then we use the linear time algorithm of Thorup [11] for the first part. Or, if the graph is planar, then we can use the $O(n)$ time shortest path algorithm described by Henzinger et al.[5] instead, to get a linear time algorithm for the problem.

S. Arumugam and B. Smyth (Eds.): IWOCA 2012, LNCS 7643, pp. 81–85, 2012.

Nardelli et al [8] have described an $O(m\alpha(m,n))$ time algorithm for solving the most vital edge problem, which also solves the replacement paths problem. They also use the linear time algorithm of Thorup [11] for the first part. For the second part, they use the transmuter [9,10] data structure described by Tarjan. Note that the technique described in this paper is simpler and easier to implement.

2   Replacement Paths in Case of Edge Failure

We assume that the graph $G(V,E)$ is undirected and edges have positive integer weights. For $s,t \in V$, we will denote the shortest path from s to t by $P = \{v_0, v_1, v_2...v_t\}$ with $s = v_0$ and $t = v_t$. So, for every edge $e = (v_k, v_{k+1})$, $k = 0...t-1$, we want to report the corresponding replacement path, that is the shortest $s-t$ path in the graph $G \setminus e$.

Like[6,3] we also find the shortest path trees X rooted at s and Y rooted at t respectively. Any edge $e = (x,y) \in P$ divides the tree X into two disjoint trees (X_s) and (X_t) such that $s \in X_s$ and $t \in X_t$.

Let the "reduced cost" of an edge (u,v) be

$$\varphi(u,v) = d(s,u) + c(u,v) + d(v,t)$$

Here $d(s,u)$ is the length of the shortest path from s to u, $c(u,v)$ is the cost of edge (u,v) and $d(v,t)$ is the length of the shortest path from v to t. Since G is an undirected graph, $d(v,t) = d(t,v)$, the length of the shortest path of a node v from t in the shortest path tree Y. Hershberger and Suri [3], use the property that the shortest path in the graph $G \setminus e$ is the path through a non-tree edge (u,v) such that $u \in X_s$, $v \in X_t$ and $\varphi(u,v)$ is minimum.

We say that such edge (u,v) is the *replacement edge* of e and $\varphi(u,v)$ is the *replacement cost* of e.

Note that the edge (u,v) belongs to the fundamental cut set [2] of edge e.

We preprocess non-tree edges in $O(m + d^2)$ time, so that for any given tree edge (v_k, v_{k+1}) on P, the corresponding replacement edge (u,v) can be found in $O(d)$ time. Since there can be $O(d)$ edges in P, we can compute the replacement cost for each edge in $O(d^2)$ time.

Let us carry out preorder traversal on the shortest path tree X. Let $pre(v)$ denote the preorder number of node v and $desc(v)$ denote the number of descendants of v, including v. Let

$$\alpha(v) = pre(v) + desc(v)$$

Let T_i be the set of nodes that are descendents of node v_i (including v_i) but not of v_{i+1}. Thus, the nodes of the tree are partitioned into sets $T_0, T_1...T_{t-1}$.

Let N_i be the set of non-tree edges, between T_i and T_{i+b} (for $b \geq 1$). Less informally:

$$N_i = \{(x,y) \mid (x \in T_i \text{ and } y \in T_j) \; \forall \; j \geq i+1\}$$

Thus, the non-tree edges are partitioned into sets $N_0, N_1, ..., N_{t-1}$.

2.1   Preprocessing

For graphs with integer weights, the shortest path trees can be obtained in $O(n+m)$ time using algorithm described by Thorup[11]. We can easily determine the preorder number and number of descendents of nodes in linear time. Hence the values $pre(v)$, $desc(v)$, $\alpha(v)$ and $\varphi(u,v)$ are determined in $O(n)$ time.

We use bucket sort to sort these non-tree edges (say (u,v)) on the $pre(v)$, the preorder number of the second entry v, in $O(n+m)$ time.

By post order traversal, we can easily determine the sets of tree edges $T_0, T_1, \ldots, T_{t-1}$ in linear time. Then by looking at each non-tree edge one by one, we can put it the sets N_j, in constant time. Thus, the sets $N_1, \ldots, N_{t-1}$ of non-trees edges can also be constructed in $O(m+n)$ time.

Berkman et al. [1] has shown that an array $A[1:n]$ can be preprocessed in linear time to answer range minima queries of the following form in constant time:

For $1 \le i, j \le n$ find the smallest item in $A[i], A[i+1]..., A[j-1], A[j]$.

We preprocess each set N_i independently to answer range minima query with reduced cost φ as the key. Since the total number of elements in all the sets together is $O(m)$ we need $O(m)$ time to perform this step.

For each set N_i and for every vertex v_j, $j \ge i+1$ on P, we maintain the rank of $pre(v_j)$ in N_i. In other words we find the "pointer" $Pre[i, j]$ which will points to the smallest element in N_i which is greater than or equal to $pre(v_j)$.

These pointers can be obtained by doing a binary search for each $pre(v_j)$ in N_i, but that will take $(t \log |N_i|)$ time.

Alternatively, we can also get the ranks by merging sorted array $pre(v_i), pre(v_{i+1}), \ldots, pre(v_t)$ with N_i. This will take $O(t + |N_i|)$ time. Or total time for all i's will be $\sum O(t + |N_i|) = O(\sum t + \sum |N_i|) = O(t^2 + m)$.

We similarly find pointer $A[i, j]$ which points to largest element in N_i which is smaller than or equal to $\alpha(v_j)$.

Thus the total time taken by algorithm preprocess is $O(m + d^2)$

2.2   Reporting Replacement Paths

To report the replacement path for edge (v_k, v_{k+1}) we are interested in all (non-tree) edges (u,v) which connect a non-descendant of v_{k+1} to a descendant of v_{k+1}. Because of preprocessing, we can assume that for any non tree edge (u,v), $pre(u) < pre(v)$

Case 1: (v is not a descendant of v_{k+1}) Let us first consider the case when u is a descendant of v_{k+1} and v is not.

As u is a descendant:

$$pre(v_{k+1}) \le pre(u) < \alpha(v_{k+1})$$

and as v is a non-descendant:

$$\alpha(v_{k+1}) \leq pre(v) \leq n$$

As u is a descendant of v_{k+1}, u will be in $T_{k+1} \bigcup \ldots \bigcup T_t$ and as v is a non descendant, v will be in $T_0 \bigcup T_1 \bigcup \ldots \bigcup T_k$. As the edges in N_i have one point in T_i, all these edges will be present in $N_0 \bigcup N_1 \bigcup \ldots \bigcup N_k$. Hence, we only need to look at edges in $N_0 \bigcup N_1 \bigcup \ldots \bigcup N_k$.

Case 2: (u is a non-descendant) In the other case, u is not a descendant of v_{k+1} but v is a descendant of v_{k+1}.
As u is not a descendant:

$$0 < pre(u) < pre(v_{k+1})$$

and as v is a descendant of v_{k+1},

$$pre(v_{k+1}) \leq pre(v) < \alpha(v_{k+1})$$

Further, u will be in $T_0 \bigcup T_1 \bigcup \ldots \bigcup T_k$ and v will be in $T_{k+1} \bigcup \ldots \bigcup T_t$. As edges in N_i have one point in T_i, all these edges will be present in $N_0 \bigcup N_1 \bigcup \ldots \bigcup N_k$. Hence, we only need look at edges in $N_0 \bigcup N_1 \bigcup \ldots \bigcup N_k$.

As the process of finding the edge in the two cases is similar, we will only discuss implementation of the first case.

Let $C_k = N_0 \cup N_1 \cup \ldots \cup N_k$ be the set of candidate edges which satisfy the conditions of Case 1. From this set we want to report the edge with the least reduced cost φ.

Because the way we constructed N_i during preprocessing, we can assume that edges in each N_i are sorted according according to $pre(v)$.

For each set T_i for $i \leq k$, we find the edge with minimum φ value between $pre(v_{k+1})$ and $\alpha(v_{k+1})$ by performing the following range minima query:

$$(N_i, \, Pre[i, k+1], \, A[i, k+1]).$$

These are edges in N_i with $pre(v)$ between $pre(v_{k+1})$ and $\alpha(v_{k+1})$; in other words we are only looking at those edges of N_i for which v is a descendant of v_{k+1}.

Each of these queries takes $O(1)$ time. As there are k queries, total time will be $O(k)$. But, since $k = O(d)$ time is $O(d)$.

The replacement edge in Case 1, for (v_k, v_{k+1}) is the edge corresponding to the minimum of these range minima queries.

By a similar procedure, we can find a replacement edge in the other case for (v_k, v_{k+1}). The required replacement edge will be the one with smaller φ-value.

We repeat the procedure, for each tree edge $(v_k, v_{k+1}) \in P$. The total time to find all these replacement edges is $O(dk) = O(d^2)$.

If (u, v) is the replacement edge with minimum reduced cost $\varphi(u, v)$ among all these replacement edges, then the replacement path R for P as the path $(P(s, v) \in X) + (u, v) + (P(v, t) \in Y)$.

Thus the total time required to find the replacement path for P including the preprocessing step is $O(m + d^2)$.

3   Conclusion

We have proposed an $O(m+d^2)$ time algorithm for the replacement paths problem in case of edge failures for undirected graphs with positive integer weights. If diameter d of the graph is $O(\sqrt{m})$ then, our algorithm runs in linear time. For undirected, integer weighted dense graphs or sparse graphs with small diameter, our algorithm performs better than the existing algorithms. For planar graphs, our algorithm takes $O(n+d^2)$ time. Thus we conclude that the replacement path problem can be solved as efficiently as the shortest path problem, if diameter of graph is $O(\sqrt{m})$

References

1. Berkman, O., Schieber, B., Vishkin, U.: Optimal doubly logarithmic parallel algorithms based on finding all nearest smaller values. J. Algorithms 14, 344–370 (1993)
2. Deo, N.: Graph Theory with Applications to Engineering and Computer Science. Prentice-Hall (1974)
3. Hershberger, J., Suri, S.: Vickrey prices and shortest paths: what is an edge worth? In: Proc. FOCS, pp. 252–259 (2001)
4. Hershberger, J., Suri, S.: Erratum to "Vickrey Pricing and Shortest Paths: What is an Edge Worth? In: FOCS, p. 809 (2002)
5. Henzinger, M.R., Klein, P., Rao, D., Subramanian, S.: Faster shortest-path algorithms for planar graphs. J. Comput. Syst. Sci. 55, 3–33 (1997)
6. Malik, K., Mittal, A.K., Gupta, S.K.: The most vital arcs in the shortest path problem. Oper. Res. Letters 8, 223–227 (1989)
7. Nardelli, E., Proietti, G., Widmayer, P.: Finding the most vital node of a shortest path. Theoretical Computer Science 296, 167–177 (2003)
8. Nardelli, E., Proietti, G., Widmayer, P.: A faster computation of the most vital edge of a shortest path. Inf. Process. Lett. 79(2), 81–85 (2001)
9. Tarjan, R.: Sensitivity Analysis of Minimum Spanning Trees and Shortest Path Trees. Information Processing Letters 14(1), 30–33 (1982)
10. Tarjan, R.: Applications of path compression on balanced trees. J. ACM 26, 690–715 (1979)
11. Thorup, M.: Floats, Integers, and Single Source Shortest Paths. In: Meinel, C., Morvan, M. (eds.) STACS 1998. LNCS, vol. 1373, pp. 14–24. Springer, Heidelberg (1998)

Acyclic Coloring with Few Division Vertices

Debajyoti Mondal[1,*], Rahnuma Islam Nishat[2],
Md. Saidur Rahman[3], and Sue Whitesides[2,**]

[1] Department of Computer Science, University of Manitoba, Winnipeg, MB, Canada
[2] Department of Computer Science, University of Victoria, Victoria, BC, Canada
[3] Department of Computer Science and Engineering,
Bangladesh University of Engineering and Technology, Dhaka-1000, Bangladesh
jyoti@cs.umanitoba.ca, rnishat@uvic.ca,
saidurrahman@cse.buet.ac.bd, sue@uvic.ca

Abstract. An acyclic k-coloring of a graph G is a mapping ϕ from the set of vertices of G to a set of k distinct colors such that no two adjacent vertices receive the same color and ϕ does not contain any bichromatic cycle. In this paper we prove that every triangulated plane graph with n vertices has a 1-subdivision that is acyclically 3-colorable (respectively, 4-colorable), where the number of division vertices is at most $2n - 5$ (respectively, $1.5n - 3.5$). On the other hand, we prove an $1.28n$ (respectively, $0.3n$) lower bound on the number of division vertices for acyclic 3-colorings (respectively, 4-colorings) of triangulated planar graphs. Furthermore, we establish the NP-completeness of deciding acyclic 4-colorability for graphs with the maximum degree 5 and for planar graphs with the maximum degree 7.

1   Introduction

A k-coloring of a graph G is a mapping ϕ from the set of vertices of G to a set of k distinct colors such that no two adjacent vertices receive the same color. We call ϕ an *acyclic k-coloring* if it does not contain any bichromatic cycle. The *acyclic chromatic number* of a graph G is the minimum number of colors required in any acyclic coloring of G.

Grünbaum [13] first introduced the concept of acyclic coloring in 1973 and proved that every planar graph admits an acyclic 9-coloring. Then Mitchem [17], Albertson and Berman [1], Kostochka [15] and finally Borodin [6] improved this upper bound to $8, 7, 6$ and 5, respectively. Since there exist planar graphs requiring 5 colors in any acyclic coloring [13], much research effort has been devoted to characterize planar graphs that are acyclically 3 or 4-colorable [5]. Both the problems of deciding acyclic 3 and 4-colorability are NP-hard for planar graphs with maximum degree 4 and 8, respectively [3,22].

---

* Work of the author is supported in part by the University of Manitoba Graduate Fellowship.
** Work of the author is supported by the Natural Sciences and Engineering Research Council of Canada (NSERC) and the University of Victoria.

S. Arumugam and B. Smyth (Eds.): IWOCA 2012, LNCS 7643, pp. 86–99, 2012.
© Springer-Verlag Berlin Heidelberg 2012

Grünbaum also considered acyclic colorings of non-planar graphs. He proved that any graph of maximum degree 3 admits an acyclic 4-coloring. Alon et al. [2] gave an $O(\Delta^{4/3})$ upper bound and an $\Omega(\Delta^{4/3}/(\log \Delta)^{1/3})$ lower bound on acyclic chromatic number for the graphs with maximum degree Δ. The currently best known upper bounds on acyclic chromatic numbers for the graphs with maximum degree $3, 4, 5$ and 6 are 4, 5, 7 and 11, respectively [23,7,16,14]. Little is known about the time complexity of deciding acyclic colorability for bounded degree graphs. Acyclic 3-colorability (respectively, 4-colorability) is NP-complete for bipartite planar graphs with maximum degree 4 (respectively, maximum degree 8) [22]. Recently, Mondal et al. [18] proved that acyclic 4-colorability is NP-complete for the graphs with maximum degree 7.

A *k-subdivision* of a graph G is a graph G' obtained by replacing every edge of G with a path that has at most k internal vertices. We call these internal vertices the *division vertices* of G. Wood [24] observed that every graph has a 2-subdivision that is acyclically 3-colorable. Angelini and Frati [3] proved that every triangulated planar graph with n vertices has a 1-subdivision with $3n - 6$ division vertices that is acyclically 3-colorable. This upper bound on the number of division vertices reduces to $2n - 6$ in the case of acyclic 4-coloring [18].

Acyclic colorings of graphs and their subdivisions find applications in diverse areas [10,11,12]. For example, an acyclic coloring of a planar graph has been used to obtain upper bounds on the volume of a 3-dimensional straight-line grid drawing of a planar graph [10]. Consequently, an acyclic coloring of a planar graph subdivision can give upper bounds on the volume of a 3-dimensional polyline grid drawing, where the number of division vertices gives an upper bound on the number of bends sufficient to achieve that volume. The acyclic chromatic number of a graph helps to obtain an upper bound on the size of "feedback vertex set" of a graph, which has wide applications in operating system, database system, genome assembly, and VLSI chip design [11]. Acyclic colorings are also used in efficient computation of Hessian matrix [12].

In this paper we examine acyclic colorings of 1-subdivisions of planar graphs. We also show some improvement over the previous NP-completeness results in terms of maximum degree. Our results are as follows.

- In Section 3 we prove that every triangulated plane graph with n vertices has a 1-subdivision with at most $2n - 5$ (respectively, $1.5n - 3.5$) division vertices that is acyclically 3-colorable (respectively, 4-colorable), which significantly improves the previously best known upper bounds $3n - 6$ and $2n - 6$ on the number of division vertices presented in [3,18].
- In Section 4 we establish a $1.28n$ (respectively, $0.3n$) lower bound on the number of division vertices for acyclic 3 and 4-colorings of triangulated planar graphs, respectively.
- In Section 5 we show that deciding acyclic 4-colorability is NP-complete for graphs with maximum degree 5 and for planar graphs with maximum degree 7. Our results improve the previously known NP-completeness results on acyclic 4-colorability for graphs with maximum degree 7 [18] and for planar graphs with maximum degree 8 [22].

2   Preliminaries

In this section we present some definitions and preliminary results that are used throughout the paper.

Let $G = (V, E)$ be a connected graph with vertex set V and edge set E. By $\deg(v)$ we denote the degree of the vertex v in G. The *maximum degree* Δ of G is the maximum of all $\deg(v), v \in V$. Let $P = u_0, u_1, u_2, \ldots, u_{l+1}$, $l \geq 1$, be a path of G such that $\deg(u_0) \geq 3$, $\deg(u_1) = \deg(u_2) = \ldots = d(u_l) = 2$ and $d(u_{l+1}) \geq 3$. Then we call the subpath $u_1, u_2, \ldots, u_l$ of P a *chain* of G. A *spanning tree* of G is a subgraph of G that is a tree and contains all the vertices of G. G is *k-connected* if the minimum number of vertices required to remove from G to obtain a disconnected graph or a single-vertex graph is k. The following remark is easy to verify.

Remark 1. *Let G be a graph and let G' be a graph obtained from G by adding a chain $w_1, \ldots, w_j$ between two distinct vertices u and v of G. Assume that G admits an acyclic 3-coloring, which can be extended to a 3-coloring ϕ of G' such that the vertices on the path $u, w_1, \ldots, w_j, v$ receive three different colors. Then ϕ is an acyclic 3-coloring of G'.*

A *plane graph* G is a planar graph with a fixed planar embedding on the plane. G delimits the plane into connected regions called faces. The unbounded face is the outer face of G and all the other faces are the inner faces of G. G is called *triangulated* (respectively, internally triangulated) if every face (respectively, every inner face) of G contains exactly three vertices on its boundary. The vertices on the outer face of G are called the *outer vertices* and all the remaining vertices are called the *inner vertices*. The edges on the outer face are called the *outer edges* of G.

Let $G = (V, E)$ be a triangulated plane graph with the outer vertices x, y and z in anticlockwise order on the outer face. Let $\pi = (v_1(= x), v_2(= y), \ldots, v_n(= z))$ be an ordering of all vertices of G. By G_k, $3 \leq k \leq n$, we denote the subgraph of G induced by $v_1 \cup v_2 \cup \ldots \cup v_k$ and by C_k the outer cycle (i.e., the boundary of the outer face) of G_k. We call π a *canonical ordering* of G with respect to the outer edge (x, y) if for each index k, $3 \leq k \leq n$, the following conditions are satisfied [21].

(a) G_k is 2-connected and internally triangulated.
(b) If $k + 1 \leq n$, then v_{k+1} is an outer vertex of G_{k+1} and the neighbors of v_{k+1} in G_k appears consecutively on C_k.

Assume that for some $k \geq 3$, the outer cycle C_k is $w_1(= x), \ldots, w_p, w_q(= v_k), w_r \ldots, w_t(= y)$, where the vertices appear in clockwise order on C_k. Then we call the edges (w_p, v_k) and (v_k, w_r) the *left-edge* and the *right-edge* of v_k, respectively. By $L(v_k)$ and $R(v_k)$ we denote the vertices w_p and w_r, respectively. Let E^* be the set of edges that does not belong to any C_k, $3 \leq k \leq n$. Assume that $V^* = V - \{v_1, v_2\}$. Then the graph $T_\pi = (V^*, E^*)$ is a tree. The graph induced by the right-edges (respectively, left-edges) of the vertices $v_k, 3 \leq k \leq n - 1$, is also a tree, which we denote by T_π^r (respectively, T_π^l). In fact, T_π, T_π^l and T_π^r form a "Schnyder's realizer" of G [25]. By G_π we denote the graph obtained from G by removing all the edges of T_π. Figure 1(a) illustrates π, T_π and G_π.

The existence of Schnyder's realizer implies that there exists another canonical ordering $\pi' = (u_1(= z), u_2(= x), ..., u_n(= y))$ of G with respect to the outer edge (z, x) such that the following properties hold [25].

(i) For each index $k, 3 \le k \le n-1$, the right-edge e of vertex u_k in π' coincides with the left-edge of that vertex in π, and hence both $G_{\pi'}$ and G_π contains e. On the other hand, the left-edge of vertex u_k belongs to $G_{\pi'}$, but does not belong to G_π.

(ii) $G_{\pi'}$ contains all the edges of G except the edges of T_π^r.

Figure 1(b) illustrates π' and T_π^r. The vertex v_6 in Figure 1(a) and the vertex u_3 in Figure 1(b) are the same, where the left-edge (v_6, v_1) of v_6 coincides with the right-edge (u_3, u_2) of u_3.

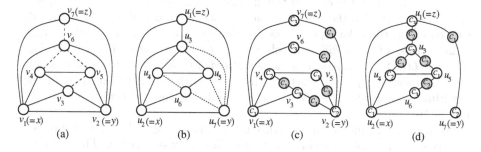

Fig. 1. (a) A graph G and its canonical ordering π. The edges of T_π are shown in dashed lines. All the remaining edges belong to G_π. (b) The canonical decomposition π' of G. The edges of T_π^r are shown in dotted lines. All the remaining edges belong to G'_π. (c) An acyclic 3-coloring of S. (d) An acyclic 3-coloring of S'. The division vertices are shown in gray color.

3   Acyclic Colorings of Planar Graph Subdivisions

In this section we prove that every triangulated plane graph with n vertices has a 1-subdivision with at most $2n - 5$ (respectively, $1.5n - 3.5$) division vertices that is acyclically 3-colorable (respectively, 4-colorable). To achieve our results we exploit the properties of canonical orderings of triangulated plane graphs.

Theorem 1. *Every triangulated plane graph G with $n \ge 3$ vertices has a 1-subdivision G' with at most $2n-5$ division vertices that is acyclically 3-colorable.*

Proof. Let x, y, z be the outer vertices of G in anticlockwise order on the outer face. Let $\pi = (v_1(= x), v_2(= y), \ldots, v_n(= z))$ and $\pi' = (u_1(= z), u_2(= x), \ldots, u_n(= y))$ be the canonical orderings of G as defined in Section 2. We use G_π and $G_{\pi'}$ to construct the required 1-subdivision G' and an acyclic 3-coloring of G'.

We first construct a 1-subdivision S of G_π and compute an acyclic 3-coloring ϕ of S with the colors c_1, c_2, c_3. We assign colors c_1 and c_2 to the vertices v_1 and v_2, respectively. For $3 \leq k \leq n$, we assign v_k a color other than the colors of $L(v_k)$ and $R(v_k)$. We then subdivide the right-edge $(v_k, R(v_k))$ with one division vertex and assign the division vertex the color different from its neighbors. The resulting 1-subdivision of G_π is the required subdivision S. It is now straightforward to prove inductively using Remark 1 that the resulting coloring ϕ of S is an acyclic 3-coloring[1]. Observe that every edge e of T_π^r contains a division vertex z in S and z along with the two end vertices of e receive three different colors in ϕ. This property also holds for the edge (y, z). Figure 1(c) illustrates S.

We then construct a 1-subdivision of $G_{\pi'}$ and color that subdivision acyclically with colors c_1, c_2, c_3 without changing the colors assigned to the original vertices by ϕ. For $3 \leq k \leq n - 1$, we subdivide the left-edge of u_k with one division vertex z'. If $col(u_k) = col(L(u_k))$, then we color z' with the color other than the colors of u_k and $R(u_k)$. Otherwise, we assign z' the color different from its neighbors. Finally, we subdivide the left-edge of u_n with a division vertex z'' and color it with the color of the division vertex on (y, z) in S. By Property (i) of $G_{\pi'}$ along with the computation of ϕ observe that $col(u_k) \neq col(R(u_k))$. Consequently, the coloring of z' ensures that the path $L(u_k), z', u_k, R(u_k)$ contains vertices of three different colors. This property holds when $k = n$, i.e., the path $L(u_n), z'', u_n, R(u_n)$ contains vertices of three different colors. It is now straightforward to prove inductively using Remark 1 that the coloring of the resulting 1-subdivision S' of $G_{\pi'}$ is an acyclic 3-coloring, which we denote by ϕ'. Figure 1(d) illustrates S'.

We now use S and S' to construct G'. For each edge e in G, we subdivide e if the corresponding edge in S or S' contains a division vertex. The resulting 1-subdivision is the required G'. Since ϕ and ϕ' do not contain any conflict, we can color the vertices of G' unambiguously. Suppose for a contradiction that the coloring we compute for G' contains a bichromatic cycle C. Since S' is colored acyclically, C must contain an edge e of G' that does not correspond to any edge in S'. By Property (ii) of $G_{\pi'}$, e must be an edge of T_π^r. Recall that every edge e of T_π^r contains a division vertex z in S and z along with the two end vertices of e receive three different colors in ϕ. Therefore, C cannot be a bichromatic cycle, a contradiction.

Observe that each of the subdivisions S and S' contains exactly $n - 2$ division vertices, where only the division vertex on (y, z) is common to both subdivisions. Therefore, the number of division vertices in G' is $2n - 5$.   □

Theorem 2. *Every triangulated plane graph G with $n \geq 3$ vertices has a 1-subdivision G' with at most $\lceil 3(n - 3)/2 \rceil \approx 1.5n - 3.5$ division vertices that is acyclically 4-colorable.*

---

[1] Note that the graph induced by v_1, v_2 in G_π is colored acyclically. Assume inductively that this holds for the graph H, which is a subgraph of G_π induced by $v_1, v_2, \ldots, v_z$, where $2 < z < n$. The graph induced by $v_1, v_2, \ldots, v_z, v_{z+1}$ in G_π is obtained by adding a chain to H and the colors assigned to the vertices on the chain satisfy the condition of Remark 1.

Proof. Zhang and He [25] proved that G has a canonical ordering π such that T_π contains $\lceil (n-3)/2 \rceil$ leaves. (The corresponding Schnyder's realizer is known as minimum Schnyder's realizer.) We use π to compute G'.

We first construct a 1-subdivision H of G_π and compute an acyclic 3-coloring of H as follows. We assign colors c_1 and c_2 to the vertices v_1 and v_2, respectively. For $3 \le k \le n$, we assign v_k a color other than the colors of $L(v_k)$ and $R(v_k)$. If $col(L(v_k)) = col(R(v_k))$, then we subdivide the edge $(v_k, R(v_k))$ with one division vertex and assign the division vertex the color different from its neighbors. The resulting 1-subdivision is the required subdivision H. It is now straightforward to prove inductively using Remark 1 that the resulting coloring of H is an acyclic 3-coloring.

We now count the number of division vertices in H. Observe that for each $v_k, 3 \le k \le n$, if v_k is a leaf in T_π or $k = n$, then the edge $(L(v_k), R(v_k))$ exists. Consequently, $col(L(v_k)) \ne col(R(v_k))$ and we do not add any division vertex in this situation. Since there are $\lceil (n-3)/2 \rceil$ leaves in T_π, the number of division vertices in H is at most $n - 2 - \lceil (n-3)/2 \rceil - 1 = \lceil (n-3)/2 \rceil$.

To construct G', we add the edges of T_π to H by subdividing each edge of T_π with one division vertex. We color all the new division vertices with color c_4. The resulting subdivision is the required subdivision G' of G. Suppose for a contradiction that the coloring we compute for G' contains a bichromatic cycle C. Since H is colored acyclically and T_π is a tree, C must contain at least one edge e from T_π and at least one edge e' from G_π. Since the division vertex on e is colored with c_4, and the end vertices of e' along with the division vertex on e' (if any) contribute two different colors to C other than c_4, C cannot be a bichromatic cycle, a contradiction.

Finally, the number of division vertices in G' is at most $n - 3 + \lceil (n-3)/2 \rceil = \lceil 3(n-3)/2 \rceil \le 1.5n - 3.5$. □

Since the canonical orderings of plane graphs used in Theorems 1 and 2 can be computed in linear time [25], the proofs of these theorems lead us to linear-time acyclic coloring algorithms.

Recently, Di Battista, Frati and Pach [4] have proved an $O(n \log^{16} n)$ and $O(n \log \log n)$ upper bound on volume of 3D straight-line and polyline drawings of planar graphs, respectively. However, to achieve the $O(n \log \log n)$ upper bound they need to allow $O(\log \log n)$ bends per edge. If we restrict each edge to have at most one bend, then a similar technique yields an $O(n \log^8 n)$ upper bound on volume of 3D polyline drawings as follows.

Every graph G with acyclic chromatic number c and 'queue number' q has 'track number' $t \le c(2q)^{c-1}$ [10]. The upper bound on the queue number of planar graphs is $O(\log^4 n)$ [4]. Since every planar graph G has a subdivision G' with $O(n)$ division vertices that is acyclically 3-colorable, the track number of G' is $t = O(\log^8 n)$. Dujmović and Wood [9] proved that every c-colorable graph with n vertices and track-number t has a 3D straight-line drawing with $O(nc^7 t)$ volume. Therefore, Theorems 1 and 2 imply the following.

Remark 2. *Every planar graph G admits an $O(n \log^8 n)$ volume (respectively, an $O(n \log^{12} n)$ volume) 3D polyline drawing with at most one bend per edge and at most $2n - 5$ bends (respectively, $1.5n - 3.5$ bends) in total.*

4 Lower Bounds on the Number of Division Vertices

In this section we present a triangulated planar graph G with n vertices such that any of its 1-subdivisions that is acyclically 3-colorable (respectively, 4-colorable), contains at least $1.28n$ (respectively, $0.3n$) division vertices.

In Figures 2(a) and (b) we exhibit two planar graphs $\mathcal{M}$ and $\mathcal{N}$ such that the following lemma holds.

Lemma 1. *Any 1-subdivision of $\mathcal{M}$ (respectively, $\mathcal{N}$) that is acyclically acyclically 3-colorable (respectively, 4-colorable), contains at least 9 (respectively, 3) division vertices.*

Lemma 1 can be verified by case study or by computer programs.

We use $\mathcal{M}$ and $\mathcal{N}$ along with a recursive graph structure $G_k, k \in \mathbb{Z}^+$, to construct the triangulated planar graphs that give rise to the lower bound. G_1 is the graph shown in Figure 2(c). A set of four edge disjoint empty triangles of G_1 are shown in gray, which we call *cells*. $G_k, k > 1$, is constructed by inserting a copy of G_1 into each cell of G_{k-1} and then identifying the outer cycle of each copy of G_1 with the boundary of the corresponding cell. Figure 2(d) shows G_2. The number of cells and the number of vertices in G_k is 4^k and $4^k + 2$, respectively.

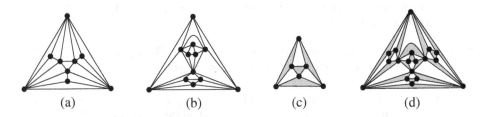

(a)   (b)   (c)   (d)

Fig. 2. Illustration for (a) $\mathcal{M}$, (b) $\mathcal{N}$, (c) G_1, and (d) G_2

Let $\mathcal{M}_k$ be the graph obtained by inserting a copy of $\mathcal{M}$ into each cell of G_k and then identifying the outer cycle of each copy of $\mathcal{M}$ with the boundary of the corresponding cell. Then the number of vertices in $\mathcal{M}_k$ is $4^k + 2 + 6 \cdot 4^k = 7 \cdot 4^k + 2$. The copies of $\mathcal{M}$ are edge disjoint in $\mathcal{M}_k$. Therefore by Lemma 1, any 1-subdivision of $\mathcal{M}_k$ that is acyclically acyclically 3-colorable contains at least $9 \cdot 4^k = (9t - 18)/7$ division vertices, where $t = 7 \cdot 4^k + 2$.

Similarly, for any $k \in \mathbb{Z}^+$, we use $\mathcal{N}$ to construct a triangulated planar graph with $t' = 4^k + 2 + 9 \cdot 4^k$ vertices such that any of its 1-subdivisions that is acyclically 4-colorable contains at least $3 \cdot 4^k = (3t' - 6)/10$ division vertices.

Theorem 3. *For every $k \in \mathbb{Z}^+$, there exists a triangulated planar graph with $t = 7 \cdot 4^k + 2$ vertices (respectively, $t' = 10 \cdot 4^k + 2$ vertices) such that any of its 1-subdivisions, which is acyclically 3-colorable (respectively, 4-colorable), contains at least $(9t - 18)/7$ (respectively, $(3t' - 6)/10$) division vertices.*

5   NP-Completeness

In this section we first prove that acyclic 4-colorability is NP-hard for the graphs with maximum degree 5. We then prove that the problem remains NP-hard for planar graphs with maximum degree 7.

5.1   Acyclic 4-Colorability of Graphs with $\Delta = 5$

To prove the NP-hardness of acyclic 4-colorability for maximum degree 5, we use the same technique as used in [18] to show the NP-hardness of 4-colorability for graphs with maximum degree 7. The crucial step is to construct a graph with low maximum degree such that in any acyclic 4-coloring of G a set of vertices of G receives the same color.

We use the graph shown in Figure 3(a) for this purpose. We call the graph of Figure 3(a) a *bead* and the vertices x, y the *poles*. A bead contains exactly one vertex s of degree 4, which we call the *center* of the bead.

Remark 3. *In any acyclic 4-coloring of a bead, the poles get different colors.*

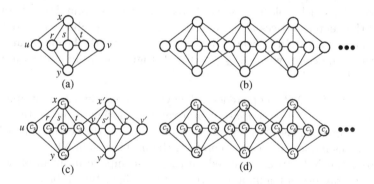

Fig. 3. Illustration for (a) a bead, (b) G_p, (c) a partial acyclic 4-coloring of G_2, and (d) an acyclic 4-coloring of G_p

For any $i \in \mathbb{Z}^+$, we now define a graph G_i with maximum degree 5 as follows.

(a) G_1 is a bead.
(b) $G_i, i > 1$, is constructed with an ordered sequence $B_1, B_2, \ldots, B_i$ of i beads by merging a vertex of degree two of B_j with a vertex of degree three of B_{j+1} and a vertex of degree three of B_j with a vertex of degree two of B_{j+1}, where $0 < j < i$. A construction for G_i is shown in Figure 3(b).

Observe that every bead in G_i contains exactly one vertex of degree 4. The following lemma gives some properties of acyclic 4-colorings of G_i.

Lemma 2. *For any $p \in \mathbb{Z}^+$, G_p contains an independent set $I(G_p)$ of size $\lfloor (p+1)/2 \rfloor$ such that every vertex of $I(G_p)$ is a vertex of degree 4 and in any acyclic 4-coloring of G_p, the vertices of $I(G_p)$ receive the same color.*

Proof. Let $B_1, B_2, \ldots, B_p$ be the ordered sequence of beads in G_p. It suffices to prove that in any acyclic 4-coloring of $G_p, p \geq 3$, the center of bead B_j and the center of bead B_{j+2} receive the same color, where $1 \leq j \leq p - 2$. To prove this claim we show that an acyclic 4-coloring of a single bead enforce the subsequent beads to follow some color pattern.

Figure 3(a) depicts a drawing of a single bead B. By Remark 3, in any acyclic 4-coloring ϕ of B the poles x, y receive different colors. Let the color of the poles be c_1 and c_2. Then all the vertices of B other than the poles are colored with c_3 or c_4. Without loss of generality assume that $col(x) = c_1, col(y) = c_2, col(u) = col(r) = col(t) = c_3$ and $col(s) = c_4$, as shown in Figure 3(c). Then the color of vertex v in ϕ can be c_3 or c_4.

Add another bead B' to B to form a G_2, as shown in Figure 3(c). Let the poles of B' be x' and y'. If $col(v) = c_3$, then both x' and y' must be colored with c_4 to avoid any bichromatic cycle. But by Remark 3, this partial coloring cannot be extended to an acyclic 4-coloring of G_2. Consequently, we have $col(v) = c_4$, which leaves us with the choice $\{col(x'), col(y')\} \subseteq \{c_1, c_2\}, col(t') = c_4$ and $col(v') = c_3$. It is now straightforward to verify that the resulting coloring is an acyclic 4-coloring of G_2.

Observe that each pole vertex in G_2 receives a color from $\{c_1, c_2\}$ and the colors of the center vertices alternate between c_3 and c_4. Since G_p is obtained from a repeated addition of beads, the center vertices of the beads B_j and B_{j+2} receive the same color, where $1 \leq j \leq p - 2$. Figure 3(d) illustrates an acyclic 4-coloring of G_p. $\qquad \square$

We now prove the NP-completeness of acyclic 4-colorability for graphs with maximum degree 5. Observe that given a valid 4-coloring of the vertices of the input graph, one can check in polynomial time whether the vertices of each pair of color classes induces a forest. Therefore, the problem is in NP.

To prove the NP-hardness we reduce the NP-complete problem of deciding acyclic 3-colorability of maximum degree 4 graphs [3] to the problem of deciding acyclic 4-colorability of maximum degree 5 graphs. Let G be an instance of acyclic 3-colorability problem, where G has n vertices and the maximum degree of G is 4. Take a copy of G_{2n-1} and connect each vertex of G with a distinct vertex of $I(G_{2n-1})$ by an edge. Let the resulting graph with maximum degree 5 be G', which is straightforward to construct in polynomial time. Using the proof technique of Theorem 3 of [18] we can show that G admits an acyclic 3-coloring if and only if G' admits an acyclic 4-coloring. However, we give a stand-alone proof in the following theorem.

Theorem 4. *It is NP-complete to decide whether a graph with maximum degree 5 admits an acyclic 4-coloring.*

Proof. Given a valid 4-coloring of the vertices of the input graph, we can check in polynomial time whether the vertices of each pair of color classes induces a forest. Therefore, the problem of deciding 4-colorability is in NP. To prove the NP-hardness we reduce the NP-complete problem of deciding acyclic 3-colorability of maximum degree 4 graphs [3] to the problem of deciding acyclic 4-colorability of maximum degree 5 graphs.

Let G be an instance of acyclic 3-colorability problem, where G has n vertices and the maximum degree of G is 4. Take a copy of G_{2n-1} and connect each vertex of G with a distinct vertex of $I(G_{2n-1})$ by an edge. Let the resulting graph with maximum degree 5 be G', which is straightforward to construct in polynomial time. By the *linkers* of G' we denote these edges that connect the vertices of $I(G_{2n-1})$ with the vertices in G. We now show that G admits an acyclic 3-coloring if and only if G' admits an acyclic 4-coloring.

We first assume that G admits an acyclic 3-coloring with the colors c_1, c_2, c_3 and then construct an acyclic 4-coloring of G'. For each vertex v in G, color the corresponding vertex in G' with $col(v)$. We color G_{2n-1} acyclically with the colors c_1, c_2, c_3 and c_4 such that the vertices of $I(G_{2n-1})$ receive color c_4. Which can be done in polynomial time by Lemma 2. If the resulting coloring of G' is not acyclic, then there is a bichromatic cycle C. Since G and G_{2n-1} are colored acyclically, C must contain a linker. Therefore, some vertex on C must be colored with color c_4. Since no two linkers have a common end vertex, C must contain an edge e of G. The end vertices of e must have two of the three colors c_1, c_2, c_3. Consequently, C cannot be a bichromatic cycle, a contradiction.

We now assume that G' admits an acyclic 4-coloring and then construct an acyclic 3-coloring of G. By Lemma 2, the vertices in $I(G_{2n-1})$ are colored with the same color. Since each vertex in G is adjacent to some vertex in $I(G_{2n-1})$, the vertices of G are colored with three colors. Since G' is colored acyclically, the coloring of G is acyclic. □

5.2   Acyclic 4-Colorability of Planar Graphs with $\Delta = 7$

In this section we prove the NP-completeness of acyclic 4-colorability of planar graphs with maximum degree 7. Ochem [22] proved the NP-hardness of acyclic 4-colorability of bipartite Planar Graphs with maximum degree 8. It does not seem straightforward to adapt his proof to show the NP-hardness of acyclic 4-colorability of planar graphs with maximum degree 7, even if we relax the 'bipartite' condition and try to replace his vertex gadget with another vertex gadget of low maximum degree that have the same functionality.

To define our vertex and edge gadgets, we first define some special graphs. A *jewel* is a graph obtained from a bead by connecting the vertices of degree 2 with distinct vertices of degree 3, as shown in Figure 4(a). By the *connectors of a jewel* J we denote the vertices of degree three in J. For any $i \in \mathbb{Z}^+$, a *necklace* N_i is a graph with maximum degree 6, which is constructed with an ordered sequence $J_1, J_2, \ldots, J_i$ of i jewels by merging a connector of J_q with a connector of J_{q+1}, where $0 < q < i$. We use the necklace N_{15} as the vertex gadget, as shown in Figure 4(c) inside the dashed rectangle.

We call the graph of Figure 4(b) a *link* L, where the vertex w is the *connector of* L and each of the vertices l, m, n is a *tail of* L. In a similar technique as we used in Lemma 2, we can prove the following lemma.

Lemma 3. *The following claims hold:*

(a) *Let ϕ be an acyclic 4-coloring of a necklace $N_i, i \in \mathbb{Z}^+$. Then all the connectors receive the same color c in ϕ. Let $c', c' \neq c$, be any color among the 4 colors used in ϕ. Then for any two connectors in N_i, there is a bichromatic path with colors c' and c in ϕ.*

(b) *In any acyclic 4-coloring of the link L, if $col(l) = col(m) = col(n) = c$ and there is no bichromatic path between any pair of the tails l, m, n, then $col(w) = c$. Furthermore, there exists an acyclic coloring ϕ of L such that there is no bichromatic path between any pair of the vertices w, l, m, n.*

We use two copies of N_{11} along with six copies of the link to construct the edge gadget. Figures 4(c)–(e) illustrate the edge gadget and its hypothetical representation. We call the vertices w_1, w_2, w_3 and w_4 the *free connectors* of the edge gadget. We now have the following theorem.

Theorem 5. *It is NP-complete to decide whether a planar graph with maximum degree 7 admits an acyclic 4-coloring.*

Proof. In a similar way as in Theorem 4, we can observe that the problem is in NP. To prove the NP-hardness we reduce the NP-complete problem of deciding 3-colorability of planar graphs with maximum degree 4 [8] to the problem of deciding acyclic 4-colorability of planar graphs with maximum degree 7.

Let G be an instance of 3-colorability problem, where G has n vertices and the maximum degree of G is 4. We now construct a graph G' by replacing the vertices and edges with appropriate gadgets, as illustrated in Figures 4(c)–(e). For every vertex gadget $\mathcal{X}$, we connect the edge gadgets incident to $\mathcal{X}$ by merging some of the free connectors such that the resulting graph remains planar and the maximum degree does not exceed 7. As a consequence, all the edge gadgets become connected, i.e., removal of all the vertex gadgets leaves a connected component. See Figures 4(f)-(g). Let the resulting planar graph be G', which is straightforward to construct in polynomial time. We now show that G admits a 3-coloring if and only if G' admits an acyclic 4-coloring.

We first assume that G admits a 3-coloring with the colors c_1, c_2, c_3 and then construct an acyclic 4-coloring of G'. For every vertex v in G, we color the connectors of the corresponding vertex gadget in G' with $col(v)$. We then color all the remaining connectors with color c_4. See Figures 4(h)-(i). Finally, we color the remaining vertices of G' according to the Figures 4(a)–(b). Suppose for a contradiction that the resulting coloring contains a bichromatic path C. It is straightforward to verify that every vertex gadget and edge gadget is colored acyclically. Moreover, we have colored every link L in such a way that there is no bichromatic path between the connector and any of the tails of L (See Lemma 3 and Figure 4(b)). Therefore, the cycle C must pass through at least one edge gadget $\mathcal{Y}$ and its two incident vertex gadgets. Since the connectors of $\mathcal{Y}$ are

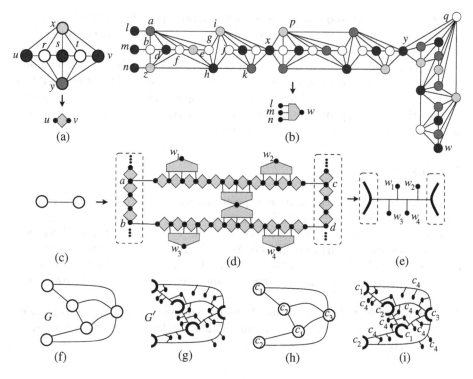

Fig. 4. (a) A jewel. (b) A link. (c) An edge e. (d) A vertex and edge gadgets replacing e. (e) A hypothetical representation of the gadgets. (f) A graph G. (g) G', which is obtained from G by first replacing the vertices and edges with appropriate gadgets and then merging the free connectors as necessary. (h) A 3-coloring ϕ of G. (i) An acyclic 4-coloring of G' that corresponds to ϕ, where a color associated with a vertex or edge gadget denotes the color of the connectors in that gadget.

colored with c_4 and the connectors of its two incident vertex gadgets are colored with two different colors other than c_4, the cycle C cannot be bichromatic, a contradiction. We now assume that G' admits an acyclic 4-coloring ϕ' and then construct a 3-coloring of G. By Lemma 3, all the connectors in each vertex gadget receive the same color in ϕ'. We assign the color associated to the connectors of a vertex gadget in G' to its corresponding vertex in G. Suppose for a contradiction that the resulting coloring ϕ of G is either a 4-coloring or contains two vertices with the same color that are adjacent in G.

By construction of G', all the edge gadgets are connected through the connectors. Therefore, the color of the connectors in all the edge gadgets must be the same. Without loss of generality let that color be c_4. Since every vertex gadget has a connector that is adjacent to some connector in some edge gadget in G', no connector of the vertex gadgets can receive color c_4. Therefore, ϕ contains only three different colors. We are now left with the case when ϕ contains two vertices with the same color z that are adjacent in G.

Let $\mathcal{Y}$ be the corresponding edge gadget and let $\mathcal{X}_1, \mathcal{X}_2$ be its incident vertex gadgets. Figure 4(d) illustrates an example, where $\mathcal{Y}$ meets $\mathcal{X}_1$ at the connectors a, b and $\mathcal{X}_2$ at the connectors c, d. If both the connectors of $\mathcal{X}_1$ and $\mathcal{X}_2$ are colored with color z, then by Lemma 3 we can construct a bichromatic cycle through a, b, c, d that is contained in the necklaces of $\mathcal{X}_1, \mathcal{X}_2$ and $\mathcal{Y}$. Therefore, any two connectors that lie in two different vertex gadgets must receive two different colors if those vertex gadgets are adjacent in G'. Hence, no two adjacent vertices of G can receive the same color in ϕ, a contradiction. $\square$

6   Conclusion

The question "What is the minimum positive constant c such that every triangulated planar graph with n vertices has an acyclic k-coloring, $k \in \{3, 4\}$, with at most cn division vertices?" was posed in the 22nd International Workshop on Combinatorial Algorithms (IWOCA 2011) [20]. Although we proved that $1.28 \le c \le 2$ and $0.3 \le c \le 1.5$ for $k = 3$ and $k = 4$, respectively, there is a gap between the upper bound and the lower bound leaving a scope for improvement.

Acknowledgment. We thank Bangladesh Academy of Sciences (BAS) for providing research travel grants to Md. Saidur Rahman for presenting the paper at IWOCA 2012.

References

1. Albertson, M.O., Berman, D.M.: Every planar graph has an acyclic 7-coloring. Israel Journal of Mathematics 28(1–2), 169–174 (1977)
2. Alon, N., McDiarmid, C., Reed, B.A.: Acyclic coloring of graphs. Random Struct. Algorithms 2(3), 277–288 (1991)
3. Angelini, P., Frati, F.: Acyclically 3-Colorable Planar Graphs. In: Rahman, M. S., Fujita, S. (eds.) WALCOM 2010. LNCS, vol. 5942, pp. 113–124. Springer, Heidelberg (2010)
4. Di Battista, G., Frati, F., Pach, J.: On the queue number of planar graphs. In: 51th Annual IEEE Symposium on Foundations of Computer Science (FOCS 2010), pp. 365–374 (2010)
5. Borodin, O.V., Ivanova, A.O.: Acyclic 5-choosability of planar graphs without adjacent short cycles. Journal of Graph Theory 68(2), 169–176 (2011)
6. Borodin, O.V.: On acyclic colorings of planar graphs. Discrete Mathematics 306(10–11), 953–972 (2006)
7. Burnstein, M.I.: Every 4-valent graph has an acyclic 5-coloring. Soobsc Akad. Nauk Grucin 93, 21–24 (1979)
8. Dailey, D.P.: Uniqueness of colorability and colorability of planar 4-regular graphs are NP-complete. Discrete Mathematics 30(3), 289 (1980)
9. Dujmović, V., Wood, D.R.: Three-dimensional grid drawings with sub-quadratic volume. In: Pach, J. (ed.) Towards a Theory of Geometric Graphs. Contemporary Mathematics. American Mathematical Society (2004)
10. Dujmović, V., Morin, P., Wood, D.R.: Layout of graphs with bounded tree-width. SIAM Journal of Computing 34, 553–579 (2005)

11. Fertin, G., Godard, E., Raspaud, A.: Minimum feedback vertex set and acyclic coloring. Inf. Process. Lett. 84(3), 131–139 (2002)
12. Gebremedhin, A.H., Tarafdar, A., Pothen, A., Walther, A.: Efficient computation of sparse hessians using coloring and automatic differentiation. INFORMS Journal on Computing 21, 209–223 (2009)
13. Grünbaum, B.: Acyclic colorings of planar graphs. Israel Journal of Mathematics 14(4), 390–408 (1973)
14. Hocquard, H.: Graphs with maximum degree 6 are acyclically 11-colorable. Inf. Process. Lett. 111(15), 748–753 (2011)
15. Kostochka, A.V.: Acyclic 6-coloring of planar graphs. Diskretn. Anal. 28, 40–56 (1976)
16. Kostochka, A.V., Stocker, C.: Graphs with maximum degree 5 are acyclically 7-colorable. Ars Mathematica Contemporanea 4(1), 153–164 (2011)
17. Mitchem, J.: Every planar graph has an acyclic 8-coloring. Duke Mathematical Journal 41(1), 177–181 (1974)
18. Mondal, D., Nishat, R.I., Whitesides, S., Rahman, M. S.: Acyclic Colorings of Graph Subdivisions. In: Iliopoulos, C.S., Smyth, W.F. (eds.) IWOCA 2011. LNCS, vol. 7056, pp. 247–260. Springer, Heidelberg (2011)
19. Mondal, D., Nishat, R.I., Whitesides, S., Rahman, M.S.: Acyclic colorings of graph subdivisions revisited. Journal of Discrete Algorithms (to appear, 2012)
20. Nishat, R.I.: Acyclic 3-colorings and 4-colorings of planar graph subdivisions, http://www.iwoca.org/main_iwocaproblems.php (accessed July 2012)
21. Nishizeki, T., Rahman, M.S.: Planar Graph Drawing. World Scientific (2004)
22. Ochem, P.: Negative results on acyclic improper colorings. In: European Conference on Combinatorics (EuroComb 2005), pp. 357–362 (2005)
23. Skulrattanakulchai, S.: Acyclic colorings of subcubic graphs. Information Processing Letters 92(4), 161–167 (2004)
24. Wood, D.R.: Acyclic, star and oriented colourings of graph subdivisions. Discrete Mathematics & Theoretical Computer Science 7(1), 37–50 (2005)
25. Zhang, H., He, X.: Canonical ordering trees and their applications in graph drawing. Discrete & Computational Geometry 33(2), 321–344 (2005)

Degree Associated Edge Reconstruction Number

S. Monikandan[1,*] and S. Sundar Raj[2]

[1] Department of Mathematics, Manonmaniam Sundaranar University,
Tirunelveli - 627 012, India
[2] Department of Mathematics, Vivekananda College,
Kanyakumari - 629 701, India
{monikandans,sundarrajvc}@gmail.com

Abstract. An edge-deleted subgraph of a graph G is called an *ecard* of G. An ecard of G with which the degree of the deleted edge is also given is called a *degree associated ecard* (or *da-ecard*) of G. The *edeck* (*da-edeck*) of a graph G is its collection of ecards (da-ecards). The *degree associated edge reconstruction number*, $dern(G)$, of a graph G is the size of the smallest collection of ecards of G uniquely determines G. The *adversary degree associated edge reconstruction number*, $adern(G)$, of a graph G is the minimum number k such that every collection of k da-ecards of G uniquely determines G. We prove that $dern(G) = adern(G) = 1$ for any regular graph G or any bidegreed graph G with exactly one vertex of different degree, which differs by at least three. We determine dern and adern for all complete bipartite graphs except $K_{1,3}$. We also prove that $dern(G) \leq 2$ and $adern(G) \leq 3$ for any complete 3-partite graph G with n vertices in which all partite sets are equal in size as possible and a few other results.

Keywords: reconstruction number, edge reconstruction number, card, dacard.

1 Introduction

All graphs considered are nonempty, simple, finite and undirected. We shall mostly follow the graph theoretic terminology of [1]. Graphs whose vertices all have one of two possible degrees are called *bidegreed graphs*. A *balanced complete m-partite graph* of order n, denoted by $T_{m,n}$, is one whose vertex set can be partitioned into m subsets $V_1, V_2, ..., V_m$ (called partite sets) such that each vertex in V_i is adjacent to every vertex in V_j if and only if $i \neq j$ and $||V_i| - |V_j|| \leq 1$. A tree T is a *bistar* if it contains exactly two vertices that are not endvertices. The bistar with central vertices of degrees $m + 1$ and $n + 1$ is denoted by $D_{m,n}$. A *vertex-deleted subgraph* or *card* $G - v$ of a graph G is the unlabeled graph obtained from G by deleting the vertex v and all edges incident to v. The ordered pair $(d(v), G - v)$ is called a *degree associated card* or *dacard* of the graph G, where $d(v)$ is the degree of v in G. The *deck* (*dadeck*) of a graph G is its

---

* Research is supported by the DST, Govt. of India, Grant No. SR/S4/MS:628/09.

S. Arumugam and B. Smyth (Eds.): IWOCA 2012, LNCS 7643, pp. 100–109, 2012.

collection of cards (dacards). Following the formulation in [2], a graph G is *reconstructible* if it can be uniquely determined from its deck.

For a reconstructible graph G, Harary and Plantholt [3] have defined the *reconstruction number* $rn(G)$ to be the minimum number of vertex-deleted subgraphs which can only belong to its deck and not to the deck of any other graph, thus uniquely identifying the graph G. Myrvold [4] has studied, for a reconstructible graph G, the *adversary reconstruction number*, which is the minimum number k such that every collection of k cards of G is not contained in the deck of any other graph H, $H \not\cong G$. For a reconstructible graph G from it's dadeck, Ramachandran [6] has defined the *degree associated reconstruction number* $drn(G)$ of a graph G to be the size of the smallest subcollection of the dadeck of G which is not contained in the dadeck of any other graph H, $H \not\cong G$. The *edge reconstruction number*, *degree associated edge reconstruction number* and *adversary degree associated edge reconstruction number* of a graph are defined similarly with edge deletions instead of vertex deletions.

The *degree* of an edge e, denoted by $d(e)$, is the number of edges adjacent to e. That is, if $e = uv$ is an edge, then $d(e) = d(u) + d(v) - 2$. An *edge-deleted subgraph* (or *ecard*) $G - e$ of a graph G is the unlabeled graph obtained from G by deleting the edge e. The ordered pair $(d(e), G - e)$ is called a *degree associated ecard* or *da-ecard* of the graph G. The *edeck* (*da-edeck*) of a graph G is its collection of ecards (da-ecards). For an edge reconstructible graph G, the *edge reconstruction number* $ern(G)$ is defined to be the size of the smallest subcollection of the edeck of G which is not contained in the edeck of any other graph H, $H \not\cong G$. For an edge reconstructible graph G from its da-edeck, the *degree associated edge reconstruction number* of a graph G, denoted by $dern(G)$, is the size of the smallest subcollection of the da-edeck of G which is not contained in the da-edeck of any other graph H, $H \not\cong G$. The *adversary degree associated edge reconstruction number* of a graph G, $adern(G)$, is the minimum number k such that every collection of k da-ecards of G is not contained in the da-edeck of any other graph H, $H \not\cong G$.

In this paper, we prove that $dern(G) = adern(G) = 1$ for any regular graph G or any bidegreed graph G with exactly one vertex of different degree, which differs by at least three. We also determine dern and adern for all complete bipartite graphs (except $K_{1,3}$), paths, wheels and bistars. Finally, we prove that $dern(G) \le 2$ and $adern(G) \le 3$ for any balanced complete 3-partite graph G.

2   dern and adern of Regular and Bidegreed Graphs

An *s-blocking* set of G is a family $\mathbb{F}$ of graphs such that $G \notin \mathbb{F}$ and each collection of s da-ecards of G will also appear in the da-edeck of some graph of $\mathbb{F}$. A graph non-isomorphic to G but having s da-ecards in common with G is called an *s-adversary-blocking graph* of G. The graphs $K_{1,3}$ and $K_3 \cup K_1$ are not edge reconstructible from their da-edeck. All other graphs G with $n \le 4$ vertices have $dern(G)=adern(G)=1$. The graphs $K_{1,3} \cup K_1$ and $K_3 \cup 2K_1$ are not edge reconstructible from their da-edeck. Most other graphs G with $n = 5$

Table 1. Graphs G on 5 vertices with $dern(G) = 3$ or $adern(G) = 4$

G	dern(G)	adern(G)	2-blocking set	3-adversary-blocking graph
$C_4 \cup K_1$	3			
$K_{2,3}$	3			
$K_{2,3} - e$		4		
		4		$K_{2,3} - e$
		4		
		4		

vertices have $dern(G)=adern(G)=1$. The exceptions are given in Table 1; the dashed edges of graphs given in the table denote the edges correspond to the common da-ecards.

A *generator* of a da-ecard $(d(e), G - e)$ of G is a graph obtained from the da-ecard by adding a new edge which joins two nonadjacent vertices whose degree sum is $d(e) - 2$ and it is denoted by $H(d(e), G - e)$.

For a graph G, to prove $dern(G) = k$ $(adern(G) = k)$, we proceed as follows.

(i) First find the da-edeck of G.
(ii) Determine next all possible generators of every da-ecard of G.
(iii) Finally, show that at least one generator other than G (every generator other than G) has at most $k - 1$ da-ecards in common with those of G, and that at least one generator has precisely $k - 1$ da-ecards in common with those of G.

Theorem 1. *If G is a bidegreed graph with exactly one vertex of different degree, which differs by at least three, then $dern(G) = adern(G) = 1$.*

Proof. Let G be a bidegreed graph of order n; let G have $n-1$ vertices of degree r and one vertex of degree s. Then G has $\frac{(n-1)r-s}{2}$ da-ecards with associated edge degree $2r-2$ and s da-ecards with associated edge degree $r+s-2$. If we join the vertices, which are the ends of the removed edge of the graph G, in the da-ecard $(2r-2, G-e)$, then $H(2r-2, G-e) \cong G$. To get a generator non-isomorphic to G, at least one of the two vertices to be joined must be different from these two ends. But then the degree sum of the two vertices to be joined is one of the four values namely $2r, 2r-1, r+s$ and $r+s-1$. Therefore, any graph non-isomorphic to G does not have the da-ecard taken as one of its da-ecards. Similarly, it can be easily proved that any da-ecard of G with associated edge degree $r+s-2$ uniquely determines G, which completes the proof.

Theorem 2. *If G is an r-regular non-empty graph, then $adern(G)$* $=$ *$dern(G) = 1$.*

Proof. Each da-ecard of G is of the form $(2r-2, G-e)$. If we join the vertices, which are the ends of the removed edge of the graph G, in the da-ecard $(2r-2, G-e)$, then $H(2r-2, G-e) \cong G$. To get a generator non-isomorphic to G, at least one of the two vertices to be joined must be different from these two ends. But then the degree sum of the two vertices to be joined is either $2r-1$ or $2r$. Therefore, any graph non-isomorphic to G does not have the da-ecard taken as one of its da-ecards. Thus, $dern(G) = 1$ and $adern(G) = 1$.

Corollary 1. *If $G \cong K_n$ or C_n, then $adern(G) = dern(G) = 1$ for $n > 1$.*

We now determine dern and adern for paths, wheels, bistars and complete bipartite graphs.

Theorem 3. *If P_n is the path with n vertices, then $dern(P_n) = 1$ and*
$$adern(P_n) = \begin{cases} 1 & \text{if } n \leq 4 \\ 3 & \text{if } n > 4 \end{cases}.$$

Proof. Since all graphs G (except $K_{1,3}$ and $K_3 \cup K_1$) of order at most four have $dern(G) = adern(G) = 1$ (Table 1), we assume that $n > 4$. Clearly, the generator $H(1, P_1 \cup P_{n-1})$ is isomorphic to P_n. The da-ecard of P_n with associated edge degree 2 has two components. If we join the vertices of different components, then the generator is isomorphic to P_n. If we join the vertices of same component (this is possible for $n > 4$), then the generator has two da-ecards in common with those of P_n with associated edge degree 2.

Theorem 4. *If W_n is the wheel with n (≥ 4) vertices, then $dern(W_n) = 1$ and*
$$adern(W_n) = \begin{cases} 3 & \text{if } n = 6 \\ 1 & \text{otherwise} \end{cases}.$$

Proof. For $n = 4$, the wheel W_n is isomorphic to K_n and $adern(W_n) = dern(W_n) = 1$. So, let us take that $n \geq 5$. Then there are $n-1$ isomorphic da-ecards of the form $(4, W_n - e)$ and there are $n-1$ isomorphic da-ecards of

the form $(n, W_n - e)$. There is one vertex of degree $n - 2$ $(\geq 3), n - 2$ vertices
of degree 3, and one vertex of degree 2 in the da-ecard $(n, W_n - e)$. Since the
$(n - 2)$-vertex is adjacent to every other vertex except the 2-vertex, the two
vertices to be joined must be different from the $(n - 2)$-vertex. Therefore, the
sum of degrees of the two vertices to be joined is 5 or 6. Thus, only for $n = 5$
or $= 6$, a graph non-isomorphic to W_n may have the da-ecard taken as one of
its da-ecards. Therefore, it suffices to consider these two cases. When $n = 5$,
there is only one 3-vertex non-adjacent to the 2-vertex in the da-ecard taken.
If we join these two vertices, then $H(n, W_n - e)$ is isomorphic to W_n. When
$n = 6$, the only graphs that have exactly one and two da-ecards in common
with those of W_n are, respectively, H_1 and H_2 shown in Fig. 1. There are only
one $(n - 1)$-vertex, two 2-vertices and $n - 3$ (≥ 2) vertices of degree 3 in the
da-ecard $(4, W_n - e)$. Here the sum of degrees of the two 2-vertices is 4 and the
generator $H(4, W_n - e)$ is isomorphic to W_n. The degree sum of all other two
vertices is greater than 4. Hence, for $n \geq 5$, any graph non-isomorphic to W_n
does not have the da-ecard in common with that of W_n with associated edge
degree 4, which completes the proof.

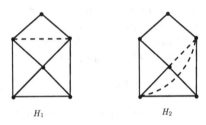

H_1 H_2

Fig. 1. The graphs H_1 and H_2

Theorem 5. *For a bistar* $D_{m,n}$ $(1 \leq m \leq n)$, $dern(D_{m,n}) = 1$ *and*

$$adern(D_{m,n}) = \begin{cases} 3 & if\ n = m + 2, m = 2\ or\ n = 2 \\ 1 & otherwise \end{cases}.$$

Proof. Denote the bistar $D_{m,n}$ simply by G. We consider two cases according to
whether $m = n$ or not.

Case 1. $m = n$
The graph G has $2n$ isomorphic da-ecards with associated edge degree n and
one da-ecard with associated edge degree $2n$. The da-ecard $(n, G - e)$ has only
one isolated vertex. To get a generator non-isomorphic to G, join two vertices
different from the isolated vertex; this is possible only for $n = 2$. When $n = 2$,
two endvertices are joined. If the two endvertices considered have a common
neighbor in the da-ecard $(n, G - e)$, then the generator $H(n, G - e)$ has only one
da-ecard in common with that of G with associated edge degree 2; Otherwise,
the generator has exactly two da-ecards in common with those of G with associated edge degree 2.

The da-ecard $(2n, G - e)$ has two isomorphic components, each of which is $K_{1,n}$. If we join the two vertices of degree n, then the generator $H(2n, G - e)$ is isomorphic to G. If at least one vertex considered in the da-ecard has degree different from n, then the degree sum of the two vertices is less than $2n$. Thus, no graph non-isomorphic to G has any da-ecard in common with that of G with associated edge degree $2n$. Thus,

$$dern(D_{n,n}) = 1 \text{ and } adern(D_{n,n}) = \begin{cases} 3 & \text{if } n = 2 \\ 1 & \text{otherwise} \end{cases}.$$

Case 2. $m \neq n$

The bistar G has m isomorphic da-ecards with associated edge degree m, n isomorphic da-ecards with associated edge degree n, one da-ecard with associated edge degree $m + n$. For $m = 2$ only, there exists a graph non-isomorphic to G having maximum of two da-ecards in common with those of G with associated edge degree 2 as similar to the case of $m = n$. The da-ecard $(n, G - e)$ has an isolated vertex. Suppose the da-ecard has a unique n-vertex. If the isolated vertex is joined with the n-vertex, then $H(n, G - e)$ is isomorphic to G. Suppose the da-ecard has two n-vertices (This is possible when $m + 1 = n$). If the isolated vertex is joined with any n-vertex, then the generator is isomorphic to G. To get a generator non-isomorphic to G, join two vertices different from the isolated vertex. This is possible only for two cases, namely $n = 2$ and $n = m + 2$. When $n = 2$ (here $m = 1$ as $m < n$), the two endvertices are joined. The generator has two components, one of which is a 4-cycle and the other one is an isolated vertex. The da-ecard corresponding to each edge of the generator is a da-ecard of G with associated edge degree 2. Thus, the generator has two da-ecards in common with those of G with associated edge degree 2. When $n = m + 2$, an endvertex is joined with the vertex of degree $m + 1$. The generator is disconnected with two components, one of which is an isolated vertex and the other has a triangle. Since no da-ecard of G has a triangle, the da-ecard of the generator corresponding to the edge of the triangle can only be in common with that of G. Here the da-ecards of the generator corresponding to the two edges other than the edge whose ends are of equal degree in the generator are in common with those of G with associated edge degree n. It is clear that no graph non-isomorphic to G has a da-ecard in common with that of G with associated edge degree $m + n$. Hence,

$$dern(D_{m,n}) = 1 \text{ and } adern(D_{m,n}) = \begin{cases} 3 & \text{if } n = m + 2, m = 2 \text{ or } n = 2 \\ 1 & \text{otherwise} \end{cases}.$$

Theorem 6. *If* $G = K_{m,n}, 1 \leq m \leq n$, *then*

$$adern(G) = dern(G) = \begin{cases} 3 & \text{if } m = 2, n = 3 \\ 2 & \text{if } m \geq 3, n = m + 1 \\ 2 & \text{if } m \geq 2, n = m + 2 \\ 1 & \text{otherwise (except when } m = 1 \text{ and } n = 3) \end{cases}.$$

Proof. Since all the da-ecards are isomorphic, $adern(K_{m,n}) = dern(K_{m,n})$. Let (A, B) be the bipartition of G, where $|A| = m$ and $|B| = n$. The graph G has mn da-ecards, all are isomorphic to $(m + n - 2, K_{m,n} - e)$. In any generator of the da-ecard $(m + n - 2, K_{m,n} - e)$, it holds that $m + n - 2 = 2n - 1, m + n - 2 = 2n, m + n - 2 = 2m - 1$ or $m + n - 2 = 2m$. Since $m \le n$, it reduces to the two cases namely, $n = m + 1$ or $n = m + 2$.

Case 1. $n = m + 1$

In this case, a vertex of degree m in B is joined with the vertex of degree $m - 1$ in B. Clearly, $m \ge 2$ (as otherwise the generator is isomorphic to G). Also, if $m = 2$, then G is isomorphic to $K_{2,3}$ and $dern(G) = 3$ (Table 1). Since all the da-ecards of $K_{2,3}$ are isomorphic, it follows that $\text{dern}(G) = \text{adern}(G) = 3$. Thus, we assume that $m \ge 3$. Now the generator has $m - 1$ (≥ 2) triangles with each triangle has the newly added edge as the base. Therefore, removal of the newly added edge can only give a da-ecard in common with that of G. Thus, the generator has only one da-ecard in common with that of G.

Case 2. $n = m + 2$

In this case, the two vertices of degree m in B are joined in the da-ecard taken. If $m = 1$, then $n = 3$, which is excluded in the hypothesis of the theorem. So, we assume that $m \ge 2$. Now, the generator has m (≥ 2) triangles such that each triangle has the newly added edge as the base. Therefore, removal of the newly added edge can only give a da-ecard in common with that of G. Thus, the generator has only one da-ecard in common with that of G, which completes the proof.

3   *dern* and *adern* of Balanced Complete Tripartite Graphs

Theorem 7. *(i) For $n = 3m$, $dern(T_{3,n}) = adern(T_{3,n}) = 1$.*

(ii) For $n = 3m + 1$, $dern(T_{3,n}) = \begin{cases} 1 & \text{if } m = 1 \\ 2 & \text{if } m \ge 2 \end{cases}$.

and $adern(T_{3,n}) = \begin{cases} 1 & \text{if } m = 1 \\ 3 & \text{if } m = 2 \\ 2 & \text{if } m \ge 3 \end{cases}$.

(iii) For $n = 3m + 2$, $dern(T_{3,n}) = 1$ and $adern(T_{3,n}) = \begin{cases} 1 & \text{if } m = 1 \\ 3 & \text{if } m = 2 \\ 2 & \text{if } m \ge 3 \end{cases}$.

Proof. We denote $T_{3,n}$ simply by G; we consider three cases depending on the fact that $n \equiv 0, 1, 2 \pmod 3$.

(i) $n \equiv 0 \pmod 3$

In this case, $n = 3m$ for some m and the graph is a $2m$-regular graph, and hence, by Theorem 2, $dern(T_{3,n}) = 1$ and $adern(T_{3,n}) = 1$.

(ii) $n \equiv 1 \pmod 3$

Now $n = 3m+1$ for some integer m. Let (A, B, C) be the tripartition of G, where $|A| = |C| = m$ and $|B| = m + 1$. The graph has $2m^2 + 2m$ isomorphic da-ecards with associated edge degree $4m - 1$ and m^2 isomorphic da-ecards with associated edge degree $4m$. When $m = 1$, the generator $H(4m, G - e)$ is isomorphic to G. So, we take that $m \geq 2$. If we join the two vertices of degree $2m$, each one is adjacent to none of the $m - 1$ vertices of degree $2m + 1$, then the generator is isomorphic to G. To get a graph non-isomorphic to G, join two vertices different from these vertices. We join two vertices of degree $2m$ from set B and let the newly added edge be x. Clearly the da-ecard of the generator $H(4m, G - e)$ corresponding to the edge x is a da-ecard of G with associated edge degree $4m$. Any other da-ecard of H corresponding to the edge of degree $4m$ is not a da-ecard of G, since in the da-ecard of G, there are two non-adjacent $2m$-vertices having no common $(2m + 1)$-neighbor, whereas the da-ecard of $H(4m, G - e)$ is not so. Also no da-ecard of H corresponding to the edge of degree $4m - 1$ is a da-ecard of G, since the $(2m - 1)$-vertex of the da-ecard of G is adjacent to each of the $(2m + 1)$-vertices of the da-ecard, whereas the da-ecard of $H(4m, G - e)$ is not so. Thus, for $m \geq 2$, $H(4m, G - e)$ has only one da-ecard in common with that of G with associated edge degree $4m$.

When $m = 1$, the generator $H(4m - 1, G - e)$ is isomorphic to G. When $m \geq 2$, the da-ecard $(4m - 1, G - e)$ has only one $(2m - 1)$-vertex and it is adjacent to every $(2m + 1)$-vertex in the da-ecard and all the $m + 1$ vertices of degree $2m$ induce a $K_{1,m}$. If we join the unique $(2m - 1)$-vertex with the $2m$-vertex, which is non-adjacent to exactly $m - 1$ vertices of degree $2m + 1$, then the generator is isomorphic to G. To get a generator non-isomorphic to G, join the $(2m - 1)$-vertex with a $2m$-vertex different from the vertex selected above. When $m = 2$, the generator $H(4m - 1, G - e)$ (Fig. 2; the dashed edges correspond to the common da-ecards) has exactly two da-ecards, corresponding to the newly added edge and an edge adjacent with this edge, in common with those of G with associated edge degree $4m - 1$. Any other da-ecard corresponding to the edge of degree $4m - 1$ contains a $(2m - 1)$-vertex adjacent with a $2m$-vertex or all the $2m$-vertices of the da-ecard are mutually adjacent, but which does not hold in the da-ecard of G.

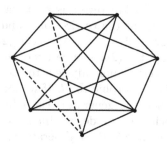

Fig. 2. The generator $H(4m - 1, G - e)$

When $m > 2$, the generator has only one da-ecard in common with that of G with associated edge degree $4m - 1$. Since any graph non-isomorphic to G having a da-ecard in common with that of G with associated edge degree $4m$ does not have any da-ecard in common with that of G with associated edge degree $4m - 1$, it follows that any graph non-isomorphic to G having a da-ecard in common with that of G with associated edge degree $4m - 1$ cannot have any da-ecard in common with that of G with associated edge degree $4m$. Hence

$$dern(T_{3,n}) = \begin{cases} 1 & \text{if } m = 1 \\ 2 & \text{if } m \geq 2 \end{cases} \text{ and } adern(T_{3,n}) = \begin{cases} 1 & \text{if } m = 1 \\ 3 & \text{if } m = 2 \\ 2 & \text{if } m \geq 3 \end{cases}.$$

(iii) $n \equiv 2 \pmod{3}$

Here $n = 3m + 2$ for some integer m. Let (A, B, C) be the tripartition of the graph, where $|A| = |C| = m + 1$ and $|B| = m$. The graph has $m^2 + 2m + 1$ isomorphic da-ecards with associated edge degree $4m$ and $2m^2 + 2m$ isomorphic da-ecards with associated edge degree $4m+1$. Clearly, the generator $H(4m, G-e)$ is isomorphic to G, since in the da-ecard exactly two non-adjacent vertices with each one is of degree $2m$ and all other vertices are of degree greater than $2m$. Thus, it follows that no graph non-isomorphic to G has a da-ecard in common with that of G with associated edge degree $4m$.   ...(E1)

If $m = 1$, then the generator $H(4m + 1, G - e)$ is isomorphic to G.

When $m = 2$, there is only one $2m$-vertex in the da-ecard $(4m+1, G-e)$ and all the $(2m + 1)$-vertices induce a tripartite subgraph $K_{m,m-1,m+1}$. Let D, E and F denote the set of $m, m - 1$ and $m + 1$ vertices in the tripartition of the induced subgraph, respectively. Then the $2m$-vertex is adjacent to all the vertices of the da-ecard $(4m + 1, G - e)$ except the vertices of E and F. The unique $(2m + 2)$-vertex in the da-ecard taken is adjacent to all the vertices of the da-ecard except the vertices of E. If we join the $2m$-vertex with the $(2m + 1)$-vertex, which is non-adjacent to that $(2m + 2)$-vertex, then the generator $H(4m + 1, G - e)$ is isomorphic to G. If we join the $2m$-vertex with any other $(2m + 1)$-vertex, then the generator has exactly two da-ecards in common with those of G with associated edge degree $4m + 1$.

If $m \geq 3$, then there is only one $2m$-vertex in the da-ecard $(4m + 1, G - e)$ and all the $(2m + 1)$-vertices induce a tripartite subgraph $K_{m,1,m+1}$. Let D, E and F denote, respectively, the set of $m, 1$ and $m + 1$ vertices in the tripartition of the induced subgraph. The $2m$-vertex is adjacent to all the vertices of the da-ecard $(4m + 1, G - e)$ except the vertices of E and F. Each $(2m + 2)$-vertex is adjacent to no other $(2m + 2)$-vertex. Also each $(2m + 2)$-vertex is adjacent to every other vertex except the $(2m + 1)$-vertex of the set E. If we join the $2m$-vertex with the $(2m + 1)$-vertex, non-adjacent to each of the $(2m + 2)$-vertices, then $H(4m + 1, G - e)$ is isomorphic to G. If we join the $2m$-vertex with any other $(2m + 1)$-vertex, then the generator has only one da-ecard in common with that of G with associated edge degree $4m + 1$ corresponding to the newly added edge. By (E1), when $m \geq 2$, these generators (non-isomorphic to G) have no da-ecard in common with that of G with associated edge degree $4m$. Hence,

$$\text{dern}(T_{3,n})= 1 \text{ and adern}(T_{3,n})= \begin{cases} 1 & \text{if } m = 1 \\ 3 & \text{if } m = 2 \\ 2 & \text{if } m \geq 3 \end{cases}.$$

4 Conclusion

It follows, from their definitions, that $\text{dern}(G) \leq \min \{\text{ern}(G), \text{adern}(G)\}$. However, $\text{ern}(G)$ and $\text{adern}(G)$ are not comparable in general. For instance, $\text{adern}(C_4 \cup 2K_1) = 3 = \text{ern}(C_4 \cup 2K_1)$, $\text{adern}(K_{1,4} \cup K_1) = 1 < 2 = \text{ern}(K_{1,4} \cup K_1)$ and $\text{adern}(K_3 \cup K_2 \cup K_1) = 4 > 2 = \text{ern}(K_3 \cup K_2 \cup K_1)$. Moreover, if all the da-ecards of a graph G are isomorphic, then it is clear that $dern(G) = adern(G)$. But the condition is not necessary. For instance, the graph $G = P_4 \cup 2K_1$ has non-isomorphic da-ecards and $\text{dern}(G) = \text{adern}(G) = 1$.

Acknowledgment. We are thankful to anonymous referees for their many valuable comments which largely improved the style of the paper and the proof of Theorem 7. The work reported here is supported by the Project SR/S4/MS:628/09 awarded to the first author by the Department of Science and Technology, Government of India, New Delhi.

References

1. Harary, F.: Graph Theory. Addison Wesley, Mass. (1969)
2. Harary, F.: On the reconstruction of a graph from a collection of subgraphs. In: Fieldler, M. (ed.) Theory of Graphs and its Applications, pp. 47–52. Academic Press, New York (1964)
3. Harary, F., Plantholt, M.: The graph reconstruction number. J. Graph Theory 9, 451–454 (1985)
4. Myrvold, W.J.: The ally and adversary reconstruction problems. Ph.D. Thesis, University of Waterloo (1988)
5. Myrvold, W.J.: The ally-reconstruction number of a disconnected graph. Ars Combin. 28, 123–127 (1989)
6. Ramachandran, S.: Degree associated reconstruction number of graphs and digraphs. Mano. Int. J. Mathematical Sciences 1, 41–53 (2000)
7. Ramachandran, S.: The Reconstruction number for Ulam's Conjecture. Ars Combin. 78, 289–296 (2006)
8. Barrus, M.D., West, D.B.: Degree-associated reconstruction number of graphs. Discrete Math. 310, 2600–2612 (2010)
9. Molina, R.: The Edge Reconstruction Number of a Disconnected Graph. J. Graph Theory 19(3), 375–384 (1995)

Touring Polygons: An Approximation Algorithm

Amirhossein Mozafari and Alireza Zarei

Department of Mathematical Sciences
Sharif University of Technology

Abstract. In this paper, we introduce a new version of the shortest path problem appeared in many applications. In this problem, there is a start point s, an end point t, an ordered sequence $\mathcal{S}=(S_0 = s, S_1, ..., S_k, S_{k+1} = t)$ of sets of polygons, and an ordered sequence $\mathcal{F}=(F_0, ..., F_k)$ of simple polygons named fences in $\Re^2$ such that each fence F_i contains polygons of S_i and S_{i+1}. The goal is to find a path of minimum possible length from s to t which orderly touches the sets of polygons of $\mathcal{S}$ in at least one point supporting the fences constraints. This is the general version of the previously answered Touring Polygons Problem (TPP). We prove that this problem is NP-Hard and propose a precision sensitive FPTAS algorithm of $O(k^2 n^2/\epsilon^2)$ time complexity where n is the total complexity of polygons and fences.

Keywords: Computational geometry, approximation algorithm, touring polygons, minimum link path.

1   Introduction

Finding a shortest path is a basic subroutine in computational geometry and appears in many applications in mathematics and engineering. There are several types of shortest path problems. In its most conventional form, we have a weighted graph and the problem is to obtain the shortest path (path of minimum weight) from a source node to a destination[5]. In this paper, we introduce a special version of the shortest path problem in which we have an ordered set $\mathcal{S}=(S_0, ..., S_{k+1})$ of sets of polygons, a start point s, an end point t, and an ordered set of polygonal fences $\mathcal{F}=(F_0, ..., F_k)$ in $\Re^2$, such that F_i contains S_i and S_{i+1}. In this notation we assume that S_0 is the start point s and S_{k+1} is the end point t. The goal is to find the shortest path P (path of minimum length) from s to t such that P intersects at least one point of a polygon from each set S_i according to their order in such a way that the portion of this path from S_i to S_{i+1} lies inside F_i. This problem can be considered as a general version of the known Touring Polygons Problem (TPP)[6]. In TPP we have only one polygon in each set S_i. We denote the general version in which each S_i may contain more than one polygon by TMP (Touring Multiple-polygons Problem). Fig. 1 shows an example of TPP and Fig. 2 shows an example of TMP. In the unconstrained version of TMP and TPP, denoted by UTMP and UTPP respectively, all fences F_i are assumed to be the whole plane which means that there

S. Arumugam and B. Smyth (Eds.): IWOCA 2012, LNCS 7643, pp. 110–121, 2012.

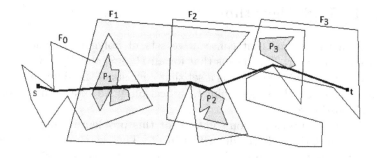

Fig. 1. An example of the TPP with polygons (P_1, P_2, P_3) and fences (F_0, F_1, F_2, F_3)

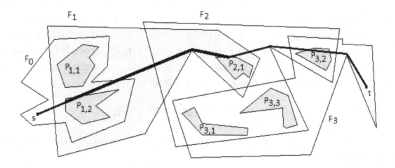

Fig. 2. The TMP with three sets of polygons $(\{P_{1,1}, P_{1,2}\}, \{P_{2,1}\}, \{P_{3,1}, P_{3,2}, P_{3,3}\})$ and fences (F_0, F_1, F_2, F_3)

is no constraint for the path from S_i to S_{i+1}. This problem has applications in several well-known problems in computational geometry including Watchman Route[4,11], Zookeeper[9], Safari[12], and Part Cutting[7] problems. Dror *et. al* [6] proved that TPP is NP-Hard when the polygons are non-convex and allowed to intersect each other. This implies that our problem is also NP-Hard when the polygons in $\mathcal{S}$ can intersect each other. In this paper, we prove that TMP is NP-Hard even if the polygons are disjoint and convex.

In this paper, we propose a precision sensitive ϵ-approximation algorithm for the TMP which is based on solving the shortest path problem on a graph built on the vertices of the polygons and some extra vertices put on the boundaries of these polygons. In the rest of this paper, in Section 2 we propose a precision sensitive FPTAS algorithm for TMP when the polygons are disjoint. In Section 3, we analyse the efficiency of this algorithm and show that the running time of this algorithm is the same as the running time of the best known approximation algorithm for TPP. In Section 4 we extend this algorithm to the overlapped situations. In Section 5, we prove the NP-Hardness of the TMP.

2   The FPTAS Algorithm

In this section we assume that consecutive sets of polygons S_i and S_{i+1} are disjoint from each other. This means that for all $P_r \in S_i$ and $P_q \in S_{i+1}$ we have $P_r \cap P_q = \emptyset$. Recall that we set S_0 as a set that consists of the single point s and S_{k+1} as a set that has the single point t. In Section 4, we extend this algorithm to the overlapped case.

To obtain an ϵ−approximation solution for this problem we use the *pseudo approximation technique* (PAT) described in [2]. The sketch of this method is as follows. If X is the space of all solutions of a problem and $x^* \in X$ is an optimal solution, X is classified into subsets $X_R \subseteq X$ for different values of a real parameter $R \geq 0$ which is called the search radius. This classification must satisfy three properties: (1) if $R_1 \leq R_2$, then $X_{R_1} \subseteq X_{R_2}$, (2) there exists $R^* \geq 0$ such that $X_{R^*} = X$. (3) if $length(x^*) \leq R$ then $length(x_R^*) = length(x^*)$ where x_R^* is an optimal solution in X_R. Having these properties, by constructing a pseudo approximation algorithm for the search radius R and iteratively running it for different values of R, an accurate ϵ−approximation algorithm is obtained. To be able to use this method the pseudo approximation algorithm must have this property that for each $R \geq 0$ and fixed $\epsilon \geq 0$

$$length(x_R^{apr}) \leq length(x_R^*) + \epsilon R,$$

where x_R^{apr} is the solution obtained by the pseudo approximation algorithm for the search radius R.

Now, we describe how to use PAT for solving the TMP. Denote by X the set of all solutions (acceptable paths) in our problem and use a search radius $R \geq 0$ to classify all solutions in X. We define X_R as the set of all solutions which are completely inside the disk of radius R with center s. Also, we denote x^* as an optimal solution (solution with minimum length) in X and x_R^* an optimal solution in X_R (note that if $X_R = \emptyset$ we set $length(x_R^*) = \infty$).

In order to use PAT, we need to check the three properties that the definition of the search radius must satisfy. It is simple to check that these properties are satisfied by the definition of our search radius and the classification method. Therefore, if we can obtain a pseudo approximation algorithm, we can use the PAT method to obtain an ϵ−approximation algorithm.

For fixed $R > 0$, we restrict our problem to this disk, i.e., we remove all parts of polygons and fences which are outside this disk. In this restriction, an edge e of a polygon is replaced by $e \cap D_R(s)$ where $D_R(s)$ is the disk of radius R and center s. We put $\lceil 4k/\epsilon \rceil$ points on each edge of polygons and divide each edge into $\lceil 4k/\epsilon \rceil + 1$ fragments of equal length. The length of each fragment is at most $2R\epsilon/4k$ (we call these points as extra points). We build a directed weighted visibility graph, DVG, which its vertex set is the set of vertices of polygons and fences and the extra points. An edge $\overrightarrow{uv}$ exists in DVG if and only if these conditions holds:

1. u corresponds to a vertex or extra point of S_i or a vertex of fence F_i.
2. v corresponds to a vertex or extra point of S_{i+1} or a vertex of fence F_i.
3. The corresponding points of u and v are visible from each other with respect to fence F_i, i.e.,their connecting segment lies completely inside fence F_i.

The weight of an edge $\vec{uv}$ is set to be the distance between the corresponding points of its vertices. We run a shortest path algorithm like Dijkstra[5] from s to t in this directed graph to obtain the path x_R^{apr}.

Lemma 1. If x_R^{apr} exists, it belongs to X_R.

Proof. According to its construction, all vertices of DVG lie inside $D_R(s)$. Therefore, the path x_R^{apr} entirely lies inside this disk. To complete the proof, We must show that x_R^{apr} is an acceptable path, i.e., it starts from s, ends at t, and intersects at least one polygon in each S_i in their correct order and satisfies the fences constraints. Trivially, x_R^{apr} starts from s and ends at t. Assume that $< p_r, p_{r+1}, ..., p_q >$ is a sub-path in x_R^{apr} where p_r and p_q are respectively the first vertices of x_R^{apr} belonging to the polygon sets S_i and S_{i+1} for $0 \le i \le k$. According to the direction of the edges in DVG, the outward edges from p_r is only to F_i and S_{i+1} vertices, and the inward edges to p_q is restricted to the vertices of F_i and S_i. Therefore, the vertices between p_{r+1} and p_{q-1} in path $< p_r, p_{r+1}, ..., p_q >$ only belong to the vertices of F_i (note that it is possible that p_r and p_q are directly connected by an edge which means that the path $< p_r, ..., p_q >$ is a single edge). Moreover, each edge $\vec{uv}$ which its start vertex belongs to S_i or F_i and its end vertex belongs to F_i or $Si + 1$ follows the visibility constraints of fence F_i. This implies that the sub-path $< p_r, ..., p_q >$ lies inside the fence F_i. Finally, we prove by contradiction that path X_R^{apr} intersects the polygon sets S_i according to their order. Assume that i is the smallest value which S_i polygons are not intersected by x_R^{apr} just after entering a polygon in S_{i-1}. According to the direction of the edges in DVG, after entering a polygon in S_{i-1} the path x_R^{apr} can only enter a vertex of F_{i-1} or S_i and to leave the vertices of F_{i-1} it must enter a vertex of S_i. Therefore, if x_R^{apr} ends at t, it must pass through a vertex of S_i after leaving S_{i-1}.   □

Now, we analyse the relation between x_R^{apr} and x_R^*. We can locate a sequence of p_i points on x_R^* where p_i is the first intersection point of x_R^* and the set of polygons S_i after visiting the polygon sets $S_1, ..., S_{i-1}$. Here, p_0 and p_{k+1} are respectively the start point s and the end point t. Therefore, the optimal path x_R^* can be divided into k sub-paths which the i'th sub-path ($0 \le i \le k$) starts from a point $p_i \in S_i$ and ends at a point $p_{i+1} \in S_{i+1}$. These sub-paths are denoted by $x_R^*(i)$ (Fig. 3).

Each point p_i lies on the boundary of a polygon in S_i. This boundary point may be a vertex of a polygon in S_i, an extra point on an edge or a point on a fragment of length at most $R\epsilon/2k$. According to the definition of p_i points, each sub-path $x_R^*(i)$ lies inside fence F_i. Moreover, each sub-path $x_R^*(i)$ lies inside a geometric structure called *hourglass* defined as bellow.

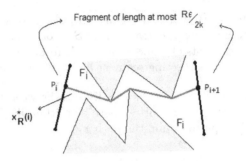

Fig. 3. Sub-path $x_R^*(i)$

Assume that b_s and b_e are respectively the fragments of length at most $R\epsilon/2k$ containing the points p_i and p_{i+1} of the sub-path $x_R^*(i)$. Note that in some cases b_s or b_e may be a single point. The corresponding hourglass of $x_R^*(i)$ is the region defined by these segments and the two shortest paths connecting the endpoints of b_s and b_e that $x_R^*(i)$ lies between them. Fig. 4 shows some configurations for hourglass shapes.

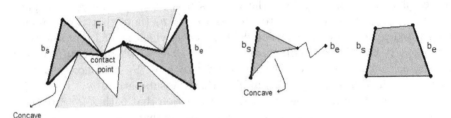

Fig. 4. Some configurations of hourglass shapes

We denote the sequence of hourglasses by $H_0, ..., H_k$ where each H_i contains $x_R^*(i)$. The end points of $x_R^*(i)$ lie on two edges of H_i that their lengths are at most $R\epsilon/2k$. These edges are shown as thick segments in Fig. 4 and we call them as the *base* edges of hourglass H_i. Let s_i and l_i be respectively the minimum and maximum length shortest paths that connect a point from one base edge of H_i to a point on the other base edge which completely lie inside H_i (or on its boundary). Fig. 5 shows some configurations for s_i and l_i paths. We define L_{min} and L_{max} as follows :

$$L_{min} = \sum_{i=0}^{k} |s_i|$$

$$L_{max} = \sum_{i=0}^{k} |l_i|$$

Lemma 2. $L_{max} \leq L_{min} + \epsilon R.$

Fig. 5. Some configurations for s_i and l_i

Proof. To prove the relation between L_{min} and L_{max} we first obtain the relation between $|s_i|$ and $|l_i|$ and extend it to L_{min} and L_{max}. Assume that for a hourglass H_i, the paths s_i and l_i are respectively composed of the sequence of points $< p_{l_1}, ..., p_{l_r} >$ and $< p_{s_1}, .., p_{s_q} >$. The points p_{l_1} and p_{s_1} lie on one base edge of H_i and p_{l_r} and p_{s_q} lie on the other base edge. Consider a new path $m_i = < p_{l_1}, p_{s_1}, p_{s_2}, ..., p_{s_q}, p_{l_r} >$. The path m_i connects the end points of l_i and completely lies inside H_i. While l_i is the shortest path between these points, $|l_i| \leq |m_i|$. On the other hand, the length of the base edges on which the segment $p_{l_1}p_{s_1}$ and $p_{s_q}p_{l_r}$ lie are at most $R\epsilon/2k$. Therefore,

$$|l_i| \leq |m_i| \leq |s_i| + 2(R\epsilon/2k) = |s_i| + R\epsilon/k.$$

From the above relation we prove the relation between L_{min} and L_{max} :

$$L_{max} = \sum_{i=0}^{k} |l_i| \leq \sum_{i=0}^{k} |s_i| + k(R\epsilon/k) = L_{min} + \epsilon R. \qquad \square$$

Now, we can prove the relation between x_R^{apr} and x_R^* which is required in the pseudo approximation algorithm of the PAT method.

Lemma 3. $length(x_R^{apr}) \leq length(x_R^*) + \epsilon R.$

Proof. It is trivial that $|s_i| \leq |x_R^*(i)| \leq |l_i|$. Therefore, we have :

$$L_{min} \leq length(x_R^*) \leq L_{max}.$$

Moreover, $length(x_R^*) \leq length(x_R^{apr})$ and $length(x_R^{apr}) \leq L_{max}$. The reason of the latter inequality is that if we use a path from s to t which lies only on the boundary of the sequence of H_i regions, its length is at most L_{max} and it is a valid path in DVG. Therefore, the length of x_R^{apr} cannot be greater than the length of this path. Hence, we have :

$$L_{min} \leq length(x_R^*) \leq length(x_R^{apr}) \leq L_{max}.$$

Combining this relation with the result of Lemma 2 we obtain the final result:

$$L_{min} \leq length(x_R^*) \leq length(x_R^{apr}) \leq L_{max} \leq L_{min} + \epsilon R \leq length(x_R^*) + \epsilon R.$$

$$\implies length(x_R^*) \leq length(x_R^{apr}) \leq length(x_R^*) + \epsilon R. \qquad \square$$

Now, we have all of the requirements of PAT and we can use this method to have the correct FPTAS algorithm. We assume that all inputs are rational numbers. If we set R^* as 2^L where L is the maximum bit length of the input integers, we can use the conversion procedure of PAT to obtain a *precision sensitive* ϵ−approximation algorithm. In the next section we analyse the efficiency of this algorithm.

3   Efficiency of the Algorithm

The running time of this algorithm depends on the size of the built graph and running time of finding the shortest path from s to t in this graph. We first obtain the complexity of computing a pseudo approximation path for fixed ϵ and R. We have $O(k/\epsilon)$ points on each edge and if n is the complexity of our problem (number of vertices of all polygons and fences) we have $O(nk/\epsilon)$ vertices in the graph. Let f_i $(0 \leq i \leq k)$ be the number of these vertices inside F_i (vertices of F_i and vertices and extra points of S_i and S_{i+1}). We can construct visibility graph for each F_i in f_i^2 time[1]. The sum of f_is is $O(nk/\epsilon)$ so we can construct entire visibility graph in $O(n^2k^2/\epsilon^2)$ time. The number of vertices of this graph is $O(nk/\epsilon)$. Therefore, running Dijkstra algorithm on this graph takes $O(n^2k^2/\epsilon^2)$ time. Hence, we can obtain a pseudo approximation path in $O(n^2k^2/\epsilon^2)$ time. To obtain the ϵ-approximation path with the geometric search of the PAT method[2], we use the pseudo approximation algorithm $O(\log\log(R^*/length(x^*)))$ times. We assume that all inputs are rational numbers each of which consists of integer numerator and denominator. If we set R^* as 2^L where L is the maximum bit length of the input integer in our system, the maximum value of R^* is 2^L and minimum value of $length(x^*)$ is 2^{-L}. Then we need to run the pseudo approximation algorithm for $O(\log\log(2^{2L}))$ times. While on a typical machine L is constant, we must run the pseudo approximation algorithm a constant number of times. This means that the total running time of this algorithm is $O(k^2n^2/\epsilon^2)$.

4   Extending to the Overlapped Cases

In Section 2, we proposed an ϵ-approximation algorithm for solving the TMP when polygons in S_i are disjoint from polygons in S_{i+1}. In this section, we extend this algorithm to the cases where polygons in S_i are allowed to intersect polygons in S_{i+1}. Fig. 6 shows an example where the approximation factor of our algorithm is not depend on the value of ϵ and for arbitrarily small value of ϵ it remains large. In this example, we have three sets of polygons ($S_1 = \{P_{1,1}\}, S_2 = \{P_{2,1}\}, S_3 = \{P_{3,1}\}$) each of which has one polygon. For this configuration, the approximation factor of the algorithm is approximately 2 even for infinitely small value of ϵ.

This problem happens because in our algorithm the touring path is forced to touch the polygons in their boundaries. But, as seen in this example, we can obtain better approximation by touching some polygons ($P_{2,1}$ in this example)

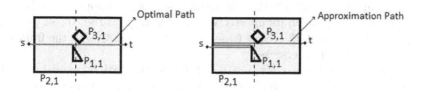

Fig. 6. A negative example for the first algorithm on intersecting polygons

in its interior. In order to solve problem we need to built DVG in such a way that handles such situations. We built DVG as follows. The vertex set of DVG is the same as before, but, there is a directed edge $\vec{uv}$ in this graph if and only if any one of the following conditions holds:

1. u and v are visible vertices of a fence with respect to that fence.
2. u is a vertex or an extra point on the boundary of a polygon of S_i which lies inside at least one polygon from each set $S_{i+1}, ..., S_{i+j}$ and does not lie inside a polygon of S_{i+j+1}, and v is a point of some polygon in S_{i+r} for $1 \leq r \leq j + 1$ which u and v are visible from each other with respect to F_{i+r-1}, or,
   v is a vertex of F_{i+r} for $0 \leq r \leq j$ which is visible from u with respect to this fence.
3. u is a vertex of fence F_i and v is a vertex or extra point of a polygon of S_{i+1} and u and v are visible from each other with respect to F_i.

Lemma 4. Running Dijkstra algorithm from s to t on this graph returns a valid pseudo approximation path.

Proof. (Sketch) Let x_R^{apr} be this path. By the same argument as Lemma 1 and according to the construction of the graph, it is simple to show that each path from s to t in this graph touches all polygon sets in correct order supporting the fences constraints. This means that x_R^{apr} is a valid touring path.
To satisfy the pseudo approximation path requirement, we must show that $length(x_R^{apr}) \leq length(x_R^*) + \epsilon R$. Consider x_R^* as a sequence of $k + 1$ sub-paths $(0 \leq i \leq k)$ such that $x_R^*(i)$ starts from the first point of x_R^* that lies on S_i after touring S_{i-1} and ends at the first point of x_R^* that lies on S_{i+1}. While S_i polygons may have intersections, a sub-path $x_R^*(i)$ may have zero length. It is simple to prove that the start point (end point) of $x_R^*(i)$ lies on the boundary of a polygon of a set S_j (S_l) $0 \leq j \leq i$ ($j \leq l \leq k + 1$). We denote the fragments that contain the start and end points of $x_R^*(i)$ by f_i^s and f_i^e, respectively. Trivially, $f_0^s = s$ and $f_{k+1}^e = t$. Moreover, $f_{i+1}^s = f_i^e$. We build a path P from s to t in DVG where $length(P) \leq length(x_R^*) + \epsilon R$. This path follows x_R^* in such a way that for each sub-path $x_R^*(i)$ which ends at a point on f_i^e, its corresponding sub-path in P, denoted by P_i, either (Case 1) ends at an endpoint of f_i^e or (Case 2) another fragment $g_i^e \in S_{i+1}$ which intersects f_i^e. Case 1 happens when the segment $x_i a$ lies inside F_{i+1} where x_i is the endpoint of $x_R^*(i)$ on f_i^e and a

is an endpoint of f_i^e at which P_i ends. Otherwise, Case 2 happens in where g_i^e is a fragment of S_{i+1} which its intersection point with f_i^e is the closest to x_i. It is simple to show that if $f_i^e \in S_{i+1}$ Case 1 happens and in Case 2 the fragment g_i^e always exists which intersects f_i^e.

Now, we inductively on i for $0 \leq i \leq k$ follow the path x_R^* and build the path P and show that in each step $0 \leq i \leq k$, $length(P)$ increases by at most $length(x_R^*(i)) + R\epsilon/k$. For $i = 0$, we start from s and if $s \in S_1$ then $x_R^*(0)$ and hence P_0 have zero length. Otherwise, $x_R^*(0)$ lies inside a hourglass with s and $f_0^e \in S_1$ as its bases. By the same argument as for the non-overlapping polygons, we can build P_0 from s to an endpoint of f_0^e with length at most $length(x_R^*(0)) + R\epsilon/k$. For $i > 0$, we consider two cases, $length(x_R^*(i)) = 0$ and $length(x_R^*(i)) \neq 0$ separately.

Assume that $length(x_R^*(i)) = 0$. Based on the cases applied on P_{i-1} and P_i, four options may happens which are shown in Fig. 7.

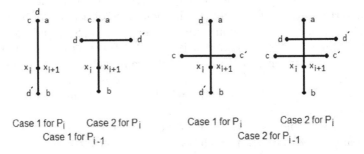

Fig. 7. Building P_i when $length(x_R^*(i)) = 0$

In this figure $f_i^s = f_i^e = f_{i-1}^e = ab$, x_i is the end point of $x_R^*(i)$, c is the endpoint of P_{i-1} and d is the endpoint of P_i. In all options we can find the union of paths P_{i-1} and P_i which starts from a and ends at an endpoint of dd' in such a way that their total length is at most $length(x_R^*(i-1)) + length(x_R^*(i)) + 2R\epsilon/k$. Now assume that $length(x_R^*(i)) \neq 0$. Hence, we also have four options based on the cases applied on P_{i-1} and P_i shown in Fig. 8 and we can build proper paths as well.   □

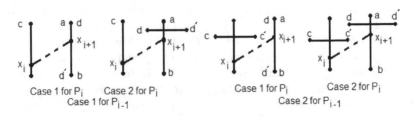

Fig. 8. Building P_i when $length(x_R^*(i)) \neq 0$

5   Complexity of the Problem

In this section, we show that UTMP (Unconstrained Touring Multiple-polygons Problem) is NP-Hard even for the L_1 norm and the case that all polygons are convex and disjoint from each other. Our proof is similar to the NP-Hardness proof of UTPP in Section 6 of [6] which itself is based on the NP-Hardness proof of three-dimensional shortest path problem[3]. This proof is a reduction from 3-SAT. Suppose that we have an instance of the 3-SAT problem with n variables $b_1, ..., b_n$ and m clauses $C_i = (l_{i1} \lor l_{i2} \lor l_{i3})$. For fixed points s and t we construct a sequence of sets of polygons with total complexity $O(n + m)$ for which solving the UTMP enables us to determine whether our 3-SAT problem has a satisfying truth assignment. We construct five kinds of gadgets : 2-way path splitter that doubles the number of shortest paths, 3-way path splitter that triples the number of shortest paths without changing their order (Fig. 9), path shuffler that performs a perfect shuffle of the input paths, literal filter that selects paths whose encodings have 0 or 1 in the ith bit (literal filters consisting of n shufflers and one horizontal segment to stretch all paths having special bit equal to 0 or 1) (Fig. 10), clause filter that determines whether a specific clause has satisfying truth assignment (Fig 11).

In these gadgets we have only line segment polygons which are convex and disjoint from each other and their angle with the x-axis is $0, \pm 45$ and 90. In this construction, we use n 2-way splitters to create 2^n distinct paths each of which encodes a truth assignment for n variables. Then, we use sequence of m clause filters each consisting of three literal filters contained between two 3-way splitters. This permits us to filter those paths fail to satisfy each clause. Fig. 12 shows how we can select polygonal sets to build three parallel literal filters inside a clause filter. We need to put a blocker set after each shuffler. Each blocker set has a segment that determines whether the output paths of the shuffler need to be stretched. As shown in Fig. 10, if this segment is full no output path is forced to be stretched.

Note that we can always modify size and position of these gadgets to enforce that each input path directly goes to one of the segments without bending or intersecting endpoint of segments. Finally, we use 2-way splitters to collect all paths back to a single path that terminates at t. In this construction, all segments are in the plane and disjoint from each other. So, the initial 3-SAT problem has

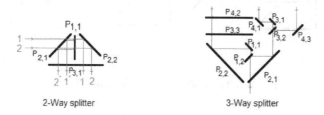

Fig. 9. Two and Three way path splitter

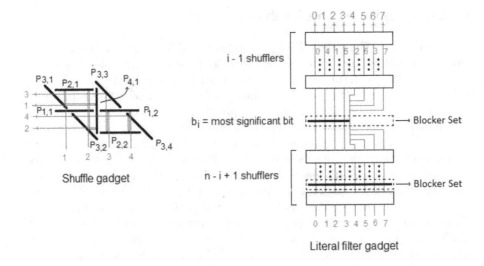

Fig. 10. Shuffle and Literal filter

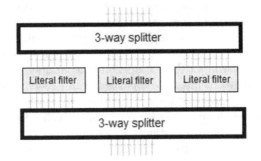

Fig. 11. Clause filter gadget

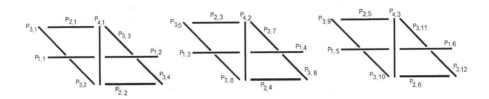

Fig. 12. Three shuffle gadgets in parallel

a satisfying truth assignment if and only if the solution of this UTMP problem is equal to the distance from s to t.

References

1. Asano, T., Asano, T., Guibas, L., Hershberger, J., Imai, H.: Visibility-polygon search and Euclidean shortest paths. In: Proc. 26th IEEE Symposium on Foundations of Computer Science, pp. 155–164 (1985)
2. Asano, T., Kirkpatrick, D., Yap, C.: Pseudo approximation algorithms, with applications to optimal motion planning. In: Proc. 18th Annu. ACM Sympos. Comput. Geom., pp. 170–178 (2002)
3. Canny, J., Reif, J.H.: New lower bound techniques for robot motion planning problems. In: Proc. 28th Annu. IEEE Sympos. Found. Comput. Sci, pp. 49–60 (1987)
4. Chin, W., Ntafos, S.: Shortest Watchman Routes in Simple Polygons. Discrete and Computational Geometry 6(1), 9–31 (1991)
5. Cormen, T.H., Leiserson, C.E., Rivest, R.L., Stein, C.: Introduction to Algorithms. MIT press (2009) ISBN 978-0-262-03384-8
6. Dror, M., Efrat, A., Lubiw, A., Mitchell, J.: Touring a sequence of polygons. In: Proc. STOC 2003, pp. 473–482 (2003)
7. Dror, M.: Polygon plate-cutting with a given order. IIE Transactions 31, 271–274 (1999)
8. Guibas, L.J., Hershberger, J.: optimal shortest path queries in simple polygon. J. Comput. Syst. Sci. 39, 126–152 (1989)
9. Hershberger, J., Snoeyink, J.: An efficient solution to the zookeeper's problem. In: Proc. 6th Canadian Conf. on Comp. Geometry, pp. 104–109 (1994)
10. Li, F., Klette, R.: Rubberband algorithms for solving various 2D or 3D shortest path problems. In: Proc. Computing: Theory and Applications, The Indian Statistical Institute, Kolkata, pp. 9–18. IEEE (2007)
11. Tan, X., Hirata, T.: Constructing Shortest Watchman Routes by Divide and Conquer. In: Ng, K.W., Balasubramanian, N.V., Raghavan, P., Chin, F.Y.L. (eds.) ISAAC 1993. LNCS, vol. 762, pp. 68–77. Springer, Heidelberg (1993)
12. Tan, X., Hirata, T.: Shortest Safari Routes in Simple Polygons. In: Du, D.-Z., Zhang, X.-S. (eds.) ISAAC 1994. LNCS, vol. 834, pp. 523–531. Springer, Heidelberg (1994)

Super Connectivity of the Generalized Mycielskian of Graphs

S. Francis Raj

Department of Mathematics, Pondicherry University, Puducherry-605014, India
francisraj_s@yahoo.com

Keywords: Mycielskian, Generalized Mycielskian, Vertex-connectivity, Edge- connectivity, Super connectivity, Super edge connectivity.

1 Introduction

All graphs considered in this paper are simple, finite, nontrivial and undirected.

Let G be a graph with vertex set $V^0 = \{v_0^0, v_1^0, \ldots, v_{n-1}^0\}$ and edge set E^0. Given an integer $m \geq 1$, the m-Mycielskian (also known as the generalized Mycielskian) of G, denoted by $\mu_m(G)$, is the graph whose vertex set is the disjoint union

$$V^0 \cup V^1 \cup \ldots \cup V^m \cup \{u\},$$

where $V^i = \{v_j^i; v_j^0 \in V^0\}$ is the i-th copy of V^0, $i = 1, 2, \ldots, m$, and edge set

$$E^0 \cup \left(\overset{m-1}{\underset{i=0}{\cup}} \{v_j^i v_{j'}^{i+1} : v_j^0 v_{j'}^0 \in E^0\} \right) \cup \{v_j^m u : v_j^m \in V^m\}.$$

For every pair $i, j \in \{0, 1, \ldots, m\}$, $i \neq j$, and $s \in \{0, 1, \ldots, n-1\}$, the vertices $v_s^i \in V^i$ and $v_s^j \in V^j$ are considered as twins of each other. Also if $S \subseteq V^0$, then $S^i \subseteq V^i$ denotes the twins of the vertices of S in V^i. The Mycielskian of G, denoted by $\mu(G)$, is simply $\mu_1(G)$.

The Mycielskian and generalized Mycielskians have fascinated graph theorists a great deal. This has resulted in studying several graph parameters of these graphs [see for instance [1], [3], [4], [7], [8]].

The connectivity $\kappa(G)$ of a connected graph G is the least positive integer k such that there exists $S \subseteq V(G), |S| = k$ and $G \backslash S$ is disconnected or becomes trivial. The edge connectivity $\kappa'(G)$ of a connected graph G is defined similarly. A graph G is super connected, or simply super-κ, if every minimum vertex cut is the set of neighbors of a vertex of G, that is, every minimum vertex cut isolates a vertex. Similarly, we can define super-κ' graphs (Refer [5], [6]).

The vertex-connectivity and edge-connectivity of the generalized Mycielskian of any digraph has already been determined in [4]. Also the super connectivity and super edge connectivity of the Mycielskian of graphs have been discussed in [5]. In this paper, we study the super connectivity and super edge connectivity of the generalized Mycielskian of graphs. This turns out to be a generalization of the result due to L. Guo et.al. [5].

S. Arumugam and B. Smyth (Eds.): IWOCA 2012, LNCS 7643, pp. 122–124, 2012.

2 Super Connectivity of the Generalized Mycielskian

For generalized Mycielskian, the following results have already been determined.

Theorem 21 ([4]). *If G is a connected graph, then $\kappa(\mu_m(G)) \geq \kappa(G) + 1$.*

Theorem 22 ([4]). *If G is a connected graph, then $\kappa(\mu_m(G)) = \kappa(G) + i + 1$ iff $\delta(G) = \kappa(G) + i$ for each i, $0 \leq i < m\kappa(G)$.*

Remark 23. *If S is a minimum vertex cut of G with $|S| = \kappa(G)$ and S^i is the corresponding set of twins in V^i, then $S \overset{m}{\underset{i=1}{\cup}} S^i \cup \{u\}$ is a vertex cut of $\mu_m(G)$. Therefore $\kappa(G) + 1 \leq \kappa(\mu_m(G)) \leq (m+1)\kappa(G) + 1$.*

Theorem 24 ([4]). *If G is a connected graph, then (i) $\kappa(\mu_m(G)) = (m+1)\kappa(G) + 1$ iff $\delta(G) \geq (m+1)\kappa(G)$ and (ii) $\kappa(\mu_m(G)) = min\{\delta(G) + 1, (m+1)\kappa(G) + 1\}$.*

With the help of these results, let us discuss about the super connectivity of the generalized Mycielskian. First, let us consider the following cases.

(a) Let $G = K_n$, $n \geq 2$. In this case, $\mu_m(K_n)$, $m \geq 2$ is not super connected, as one can see that $\kappa(G) = \delta(G) = n - 1$ and $\kappa(\mu_m(G)) = n$. Thus the set $S = V^1$ forms a vertex cut and this S does not isolate a vertex.

(b) Let $G = K_{a,b} = \{X, Y\}, |X| = a$, $|Y| = b$, $b \geq a$. In this case $\mu_m(K_{a,b})$, $m \geq 3$ is not super connected, as one can see that $\kappa(G) = \delta(G) = a$ and $\kappa(\mu_m(G)) = a + 1$. The set $S = X^1 \cup \{u\}$ is a minimum vertex which does not isolates a vertex.

Let us find a necessary and sufficient condition for graphs G other than K_n and $K_{a,b}$ for which $\mu_m(G)$ is super connected.

Theorem 25. *Let G be a connected graph where $G \notin \{K_n, K_{a,b}\}$ and $n \geq 2$. Then $\mu_m(G)$ is super-κ iff $\delta(G) < (m+1)\kappa(G)$.*

Proof. Let us first assume that for $G \notin \{K_n, K_{a,b}\}$ and $n \geq 2$, $\mu_m(G)$ is super-κ. Suppose $\delta(G) \geq (m+1)\kappa(G)$. Then by Theorem 24, $\kappa(\mu_m(G)) = (m+1)\kappa(G)+1$. Thus by Remark 23, if S is a minimum vertex cut of G then $S \overset{m}{\underset{i=1}{\cup}} S^i \cup \{u\}$ is a minimum vertex cut of $\mu_m(G)$ which does not isolate any vertex. Hence $\mu_m(G)$ is not super-κ, a contradiction.

Secondly assume that $\delta(G) < (m+1)\kappa(G)$. Then by Theorem 22, $\kappa(\mu_m(G)) = \delta(G)+1$. Suppose $\mu_m(G)$ is not super-κ, then there exist a minimum vertex cut S of $\mu_m(G)$ such that S does not isolate a vertex, that is, $S \neq N_{\mu_m(G)}(v)$, for any $v \in V(\mu_m(G))$. We then divide the proof into two cases: (i) $u \notin S$ (ii) $u \in S$. Proving Case (i) is not very tedious. But Case (ii) requires a little involvement to get through. In both the case we would show that $\mu_m(G) \backslash S$ is connected, a contradiction.   ∎

A consequence of the above result is the following result due to L.Guo et.al [5].

Corollary 26 *For a connected graph G with $|V(G)| \geq 2$, $\mu(G)$ is super-κ if and only if $\delta(G) < 2\kappa(G)$.*

Proof. For any graphs $G \notin \{K_n, K_{a,b}\}$, put $m = 1$ in Theorem 25. For $G = K_n$ or $K_{a,b}$, we can directly verify that $\mu(G)$ is super-κ.
∎

We next consider the super edge-connectivity of the generalised Mycielskian.
Let us recall the result proved in [4].

Theorem 27 ([4]). *If G is a connected graph, then $\kappa'(\mu_m(G)) = \delta(\mu_m(G)) = \delta(G) + 1$.*

Let us first consider the case when $G = K_2$. In this case, $\mu_m(K_2)$, $m \geq 1$ is not super connected, as one can see that $\kappa(G) = \delta(G) = 1$ and $\kappa(\mu_m(G)) = 2$. Thus if $V(K_2) = \{v, w\}$, then the set $F = \{vw^1, v^1w\}$ forms a edge cut and this F does not isolate any vertex.

We have shown that for any connected graph G with $|V(G)| \geq 2$, $\mu_m(G)$ is super-κ' if and only if $G \neq K_2$. This generalise the result proved by L. Guo et.al. [5] that for any connected graph G with $|V(G)| \geq 2$, $\mu(G)$ is super-κ' if and only if $G \neq K_2$.

References

1. Balakrishanan, R., Francis Raj, S.: Connectivity of the Mycielskian of a graph. Discrete Math. 308, 2607–2610 (2007)
2. Balakrishnan, R., Ranganathan, K.: A Textbook of Graph Theory. Springer, New York (2000)
3. Fisher, D.C., McKena, P.A., Boyer, E.D.: Hamiltonicity, diameter, domination, packing and biclique partitions of the Mycielski's graphs. Discrete Appl. Math. 84, 93–105 (1998)
4. Francis Raj, S.: Connectivity of the generalised Mycielskian of digraphs, Graphs and Combin., doi: 10.1007/s00373-012-1151-5
5. Guo, L., Liu, R., Guo, X.: Super Connectivity and Super Edge Connectivity of the Mycielskian of a Graph. Graphs and Combin. 28, 143–147 (2012)
6. Liu, J., Meng, J.: Super-connected and super-arc-connected Cartesian product of digraphs. Inform. Process. Lett. 108, 90–93 (2008)
7. Lam, P.C.B., Gu, G., Lin, W., Song, Z.: Circular Chromatic Number and a generalization of the construction of Mycielski. J. Combin. Theory, Ser. B 89, 195–205 (2003)
8. Lin, W., Wu, J., Lam, P.C.B., Gu, G.: Several parameters of generalised Mycielskians. Discrete Appl. Math. 154, 1173–1182 (2006)
9. Mycielski, J.: Sur le colouriage des graphes. Colloq. Math. 3, 161–162 (1955)

A Graph Radio k-Coloring Algorithm

Laxman Saha* and Pratima Panigrahi

Department of Mathematics, Indian Institute of Technology Kharagpur,
Kharagpur 721302, India
laxman.iitkgp@gmail.com,
pratima@maths.iitkgp.ernet.in

Abstract. For a positive integer k, a radio k-coloring of a simple connected graph $G = (V(G), E(G))$ is a mapping $f \colon V(G) \to \{0, 1, 2, \ldots\}$ such that $|f(u) - f(v)| \geqslant k + 1 - d(u, v)$ for each pair of distinct vertices u and v of G, where $d(u, v)$ is the distance between u and v in G. The *span* of a radio k-coloring f, $rc_k(f)$, is the maximum integer assigned by it to some vertex of G. The *radio k-chromatic number*, $rc_k(G)$ of G is $\min\{rc_k(f)\}$, where the minimum is taken over all radio k-colorings f of G. If k is the diameter of G, then $rc_k(G)$ is known as the radio number of G. In this paper, we give an algorithm to find an upper bound of $rc_k(G)$. We also give an algorithm that implement the result in [16,17] for lower bound of $rc_k(G)$. We check that for cycle C_n, upper and lower bound obtained from these algorithms coincide with the exact value of radio number, when n is an even integer with $4 \leqslant n \leqslant 400$. Also applying these algorithms we get the exact value of the radio number of several circulant graphs.

Keywords: Channel assignment, Radio k-coloring, Radio k-chromatic number, Span.

1 Introduction

A number of graph coloring problems have their roots in a communication problem known as the channel assignment problem. The channel assignment problem is the problem of assigning channels (non-negative integers) to the stations in an optimal way such as the interference is avoided, see Hale [3]. The interference is closely related to the location of the stations. Radio k-coloring of graphs is a variation of this channel assignment problem. For a positive integer k, a *radio k-coloring* f of a simple connected graph G is an assignment of non-negative integers to the vertices of G such that for every two distinct vertices u and v of G, $|f(u) - f(v)| \geqslant k + 1 - d(u, v)$. The *span* of a radio k-coloring f, $rc_k(f)$, is the maximum integer assigned by f to some vertex of G. The *radio k-chromatic number*, $rc_k(G)$ of G is $\min\{rc_k(f)\}$, where the minimum is taken over all radio k-colorings f of G. If k is the diameter of G, then $rc_k(G)$ is known as the radio number and is denoted by $rn(G)$.

---

* Corresponding author.

S. Arumugam and B. Smyth (Eds.): IWOCA 2012, LNCS 7643, pp. 125–129, 2012.
© Springer-Verlag Berlin Heidelberg 2012

The problem of finding radio k-chromatic number of a graph is of great interest for its widespread applications to channel assignment problem. So far, radio k-chromatic number is known for very limited families of graphs and specific values of k. Radio number of C_n and P_n [12], C_n^2[11], P_n^2[10], Q_n [4] and complete m-ary trees [13] are determined. Ortiz et al. [15] have studied the radio number of generalized prism graphs and have computed the exact value for some particular cases. A lower bound for radio number of any tree have been given in [9] by Liu. In [16,17], Saha et al. have given a lower bound for $rc_k(G)$ of an arbitrary graph G.

The objective of this paper is to provide an algorithm to find an upper bound of $rc_k(G)$ of an arbitrary graph G and implementation of the results in [16,17] for a lower bound of the same.

2   Algorithm to Find a Lower Bound of $rc_k(G)$

In this section, we give the Algorithm 1 to find a lower bound of $rc_k(G)$.

Algorithm 1. Finding a lower bound of $rc_k(G)$

Data: Positive integer k and G be an n-vertex graph.
Result: Lower bound of $rc_k(G)$.
Using Floyed-Warshall's algorithm compute the distance matrix $D[n][n]$
$=(d[i][j])_{n \times n}$, where $d[i][j]$ represents the distance between the vertex i and j of G.
for $l = 0$ *to* $n - 1$ **do**
    for $i = 0$ *to* $n - 1$ **do**
        for $j = 0$ *to* $n - 1$ **do**
            $s[l][i][j] = d[l][j] + a[i][j] + a[j][i]$
            if $b \leqslant s[l][i][j]$ *and* $i \notin \{l, j\}$ *and* $j \neq l$ **then**
                $b = s[l][i][j]$
            end
        end
    end
end
if $k = \text{diam}(G)$ *and n is an even integer* **then**
    Lower bound $= \left\lceil \frac{3(k+1)-b}{2} \right\rceil \left(\frac{n-2}{2} \right) + 1$
end
else
    Lower bound $= \left\lceil \frac{3(k+1)-b}{2} \right\rceil \left\lfloor \frac{n-2}{2} \right\rfloor$
end
Print Lower bound.

---

3   Radio k-Coloring Algorithm

In this section, we developed an algorithm that gives a radio coloring of an arbitrary graph G and hence finds an upper bound of $rc_k(G)$. The time complexity of this algorithm is $O(n^3)$.

Algorithm 2. Finding a radio k-coloring of a graph

---

input : G be an n-vertex simple connected graph and k be a positive integer.
output: A radio k-coloring of G.
begin

    Compute the adjacency matrix $a[n][n]$ of G.
    Using Floyed-Warshall's algorithm compute the distance matrix
    $D[n][n] = (d[i][j])_{n \times n}$, where $d[i][j]$ represents the distance between the
    vertex i and j of G.
    Initialization : Choose a vertex r, c be a two dimensional matrix
    for $i = 0$ *to* $n - 1$ **do**
        for $j = 0$ *to* $n - 1$ **do**
            if $k + 1 \geqslant d[i][j]$ **then**
                $c[i][j] = k + 1 - d[i][j]$
            else
                $c[i][j] = 0$
            $c[j][i] = c[i][j]$
        $c[i][i] = \infty$, a large number
    $l = r$ `/* Use r as the initialization vertex */`
    print : l and its color is zero   `/* Use zero is the color of r */`
    for $i = 1$ *to* $n - 1$ **do**
        $min = \infty$, a large number
        for $j = 0$ *to* $n - 1$ **do**
            if $min \geqslant c[l][j]$ **then**
                $min = c[l][j]$
                $p = j$
        for $j = 0$ *to* $n - 1$ **do**
            $c[p][j] = c[p][j] + min$
        for $j = 0$ *to* $n - 1$ **do**
            if $c[p][j] < c[l][j]$ **then**
                $c[p][j] = c[l][j]$
        print : p and min `/* Use min is the color of the vertex p */`
        $l = p$

Example 1. In this example, we explain the intermediate stages of Algorithm 2 considering G as the graph in Fig. 1.

Let $k = 4$. Here D is the distance matrix and C_0 is the k-labeling matrix. Let R_j^p be the jth-row of a matrix C_p. Here we consider $r = x_1$ is the initial vertex.

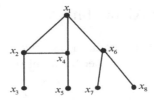

Fig. 1. The graph G

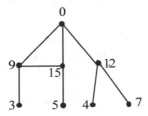

Fig. 2. Radio 4-coloring of G

Give color 0 to the vertex x_1. Minimum element of 1st- row of the matrix C_0 is 3 and a position of this minimum element is at 3rd-column of the matrix C_0. We assign color 3 to the vertex x_3. Replace the 3rd-row of the matrix C_0 by $\max\{R_3^0 + 3, R_1^0\}$, let this new matrix be C_1. The minimum element in 3rd-row of the matrix C_1 is 4 and a position of this minimum element is at 7th column. We give color 4 to the vertex x_7. Replace the 7th-row of the matrix C_1 by $\max\{R_7^1 + 4, R_3^1\}$ and we obtained a new matrix C_2. The minimum element in 7th-row of the matrix C_2 is 5 and a position of this minimum element is at 5th-column. We assign color 5 to the vertex x_5. By similar way we can give a coloring of the vertices of G as shown in Fig.2.

$$
D : \begin{pmatrix}
0 & 1 & 2 & 1 & 2 & 1 & 2 & 2 \\
1 & 0 & 1 & 1 & 2 & 2 & 3 & 3 \\
2 & 1 & 0 & 2 & 3 & 3 & 4 & 4 \\
1 & 1 & 2 & 0 & 1 & 2 & 3 & 3 \\
2 & 2 & 3 & 1 & 0 & 3 & 4 & 4 \\
1 & 2 & 3 & 2 & 3 & 0 & 1 & 1 \\
2 & 3 & 4 & 3 & 4 & 1 & 0 & 2 \\
2 & 3 & 4 & 3 & 4 & 1 & 2 & 0
\end{pmatrix}
\quad
C_0 : \begin{pmatrix}
\infty & 4 & \boxed{3} & 4 & 3 & 4 & 3 & 3 \\
4 & \infty & 4 & 4 & 3 & 3 & 2 & 2 \\
3 & 4 & \infty & 3 & 2 & 2 & 1 & 1 \\
4 & 4 & 3 & \infty & 4 & 3 & 2 & 2 \\
3 & 3 & 2 & 4 & \infty & 2 & 1 & 1 \\
4 & 3 & 2 & 3 & 2 & \infty & 4 & 4 \\
3 & 2 & 1 & 2 & 1 & 4 & \infty & 3 \\
3 & 2 & 1 & 2 & 1 & 4 & 3 & \infty
\end{pmatrix}.
$$

$$
C_1 : \begin{pmatrix}
\infty & 4 & 3 & 4 & 3 & 4 & 3 & 3 \\
4 & \infty & 4 & 4 & 3 & 3 & 2 & 2 \\
\infty & 7 & \infty & 6 & 5 & 5 & \boxed{4} & 4 \\
4 & 4 & 3 & \infty & 4 & 3 & 2 & 2 \\
3 & 3 & 2 & 4 & \infty & 2 & 1 & 1 \\
4 & 3 & 2 & 3 & 2 & \infty & 4 & 4 \\
3 & 2 & 1 & 2 & 1 & 4 & \infty & 3 \\
3 & 2 & 1 & 2 & 1 & 4 & 3 & \infty
\end{pmatrix}
\quad
C_2 : \begin{pmatrix}
\infty & 4 & 3 & 4 & 3 & 4 & 3 & 3 \\
4 & \infty & 4 & 4 & 3 & 3 & 2 & 2 \\
\infty & 7 & \infty & 6 & 6 & 5 & 4 & 4 \\
4 & 4 & 3 & \infty & 4 & 3 & 2 & 2 \\
3 & 3 & 2 & 4 & \infty & 2 & 1 & 1 \\
4 & 3 & 2 & 3 & 2 & \infty & 4 & 4 \\
\infty & 7 & \infty & 6 & \boxed{5} & 8 & \infty & 7 \\
3 & 2 & 1 & 2 & 1 & 4 & 3 & \infty
\end{pmatrix}
$$

References

1. Chartrand, G., Erwin, D., Harrary, F., Zhang, P.: Radio labeling of graphs. Bull. Inst. Combin. Appl. 33, 77–85 (2001)
2. Chartrand, G., Erwin, D., Zhang, P.: A graph labeling problem suggested by FM channel restrictions. Bull. Inst. Combin. Appl. 43, 43–57 (2005)
3. Hale, W.K.: Frequency assignment, Theory and application. Proc. IEEE 68, 1497–1514 (1980)
4. Khennoufa, R., Togni, O.: The radio antipodal and radio numbers of the hypercube. Ars Combin. 102, 447–461 (2011)
5. Khennoufa, R., Togni, O.: A note on radio antipodal colorigs of paths. Math. Bohem. 130(1), 277–282 (2005)
6. Kola, S.R., Panigrahi, P.: Nearly antipodal chromatic number $ac'(P_n)$ of a path P_n. Math. Bohem. 134(1), 77–86 (2009)
7. Kola, S.R., Panigrahi, P.: On Radio $(n-4)$-chromatic number the path P_n. AKCE Int. J. Graphs Combin. 6(1), 209–217 (2009)
8. Kola, S.R., Panigrahi, P.: An improved Lower bound for the radio k-chromatic number of the Hypercube Q_n. Comput. Math. Appl. 60(7), 2131–2140 (2010)
9. Liu, D.D.-F.: Radio number for trees. Discrete Math. 308, 1153–1164 (2008)
10. Liu, D.D.-F., Xie, M.: Radio Number for Square Paths. Ars Combin. 90, 307–319 (2009)
11. Liu, D.D.-F., Xie, M.: Radio number for square of cycles. Congr. Numer. 169, 105–125 (2004)
12. Liu, D., Zhu, X.: Multi-level distance labelings for paths and cycles. SIAM J. Discrete Math. 19(3), 610–621 (2005)
13. Li, X., Mak, V., Zhou, S.: Optimal radio labellings of complete m-ary trees. Discrete Appl. Math. 158, 507–515 (2010)
14. Morris-Rivera, M., Tomova, M., Wyels, C., Yeager, Y.: The radio number of $C_n DC_n$. Ars Combin. (to appear)
15. Ortiz, J.P., Martinez, P., Tomova, M., Wyels, C.: Radio numbers of some generalized prism graphs. Discuss. Math. Graph Theory 31(1), 45–62 (2011)
16. Saha, L., Panigrahi, P.: Antipodal number of some powers of cycles. Discrete Math. 312, 1550–1557 (2012)
17. Saha, L., Panigrahi, P.: On Radio number of power of cycles. Asian-European J. Math. 4, 523–544 (2011)

Maximum Order of a Planar Oclique Is 15

Sagnik Sen[1,2,*]

[1] Univ. Bordeaux, LaBRI, UMR5800, F-33400 Talence, France
[2] CNRS, LaBRI, UMR5800, F-33400 Talence, France
sen@labri.fr

Abstract. An oclique is an oriented graph where every pair of distinct non-adjacent vertices are connected by a directed path of length 2. Klostermeyer and MacGillivray conjectured that the maximum order of a planar oclique is 15. In this article we settle that conjecture.

1 Introduction

An *oriented graph* $\vec{G}$ is a directed graph obtained by replacing each edge uv of a simple graph G with an arc (ordered pair of vertices) $\vec{uv}$ or $\vec{vu}$. The graph $\vec{G}$ is an *orientation* of G and G is the *underlying graph* of $\vec{G}$, denoted by $und(\vec{G})$. We denote by $V(\vec{G})$ and $A(\vec{G})$ respectively the set of vertices and arcs of $\vec{G}$. Similarly, $V(G)$ and $E(G)$ denote respectively the set of vertices and edges of G. In this article, by graph we will mean either a simple undirected graph or an oriented graph.

The set of all vertices adjacent to a vertex v in a graph is the set of *neighbours* and is denoted by $N(v)$. For oriented graphs, if there is an arc $\vec{uv}$, then u is an *in-neighbour* of v and v is an *out-neighbour* of u. The sets of all in-neighbours and out-neighbours of v are denoted by $N^-(v)$ and $N^+(v)$ respectively. Two vertices u and v *agree* or *disagree* with each other on another vertex w if $w \in N^\alpha(u) \cap N^\alpha(v)$ or $N^\alpha(u) \cap N^{\overline{\alpha}}(v)$ respectively, where $\{\alpha, \overline{\alpha}\} = \{+, -\}$. A path obtained by two consecutive arcs $\vec{uv}$ and $\vec{vw}$ is called a *2-dipath*.

An *oriented k-colouring* [1] of an oriented graph $\vec{G}$ is a mapping f from the vertex set $V(\vec{G})$ to the set $\{1, 2, ..., k\}$ such that, (i) $f(u) \neq f(v)$ whenever u and v are adjacent and (ii) if $\vec{xy}$ and $\vec{uv}$ are two arcs in $\vec{G}$, then $f(x) = f(v)$ implies $f(y) \neq f(u)$. The *oriented chromatic number* $\chi_o(\vec{G})$ of an oriented graph $\vec{G}$ is the smallest integer k for which $\vec{G}$ has an oriented k-colouring. An *oriented clique* or simply *oclique* is an oriented graph $\vec{G}$ for which $\chi_o(\vec{G}) = |\vec{G}| = |V(\vec{G})|$. Note that, ocliques can hence be characterized as those oriented graphs whose any two distinct vertices are either adjacent or connected by a 2-dipath. Note that an oriented graph with an oclique of order n as a subgraph has oriented chromatic number at least n, where the *order* of a graph G is the number of its vertices, denoted by $|G|$.

For the family $\mathcal{P}$ of planar graphs we have $17 \leq \chi_o(\mathcal{P}) \leq 80$ where the lower bound is due to Marshall [2] and the upper bound to Raspaud and Sopena [3].

* This work is supported by ANR GRATEL project ANR-09-blan-0373-01.

S. Arumugam and B. Smyth (Eds.): IWOCA 2012, LNCS 7643, pp. 130–142, 2012.

Tightening these bounds are challenging problems in the domain of oriented colouring. A naturally related question to this problem is: What is the maximum order of a planar oclique? In order to find the answer to this question, Sopena [4] found a planar oclique of order 15 while Klostermeyer and MacGillivray [5] showed that there is no planar oclique of order more than 36 and conjectured that the maximum order of a planar oclique is 15. In this paper we settle this conjecture by proving the following result:

Theorem 1. *If $\vec{G}$ is a planar oclique, then $|\vec{G}| \leq 15$.*

The *distance* $d(u, v)$ between two vertices u, v of a graph G is the length (number of edges or arcs) of a shortest path joining u and v. Similarly, the *directed distance* $\vec{d}(u, v)$ between two vertices u, v of an oriented graph $\vec{G}$ is the length (number of arcs) of a shortest directed path joining u and v. The diameter of a graph G is the maximum of $d(u, v)$ taken over all $(u, v) \in V(G) \times V(G)$. Clearly any oclique has diameter at most 2.

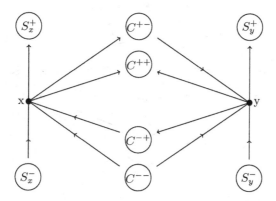

Fig. 1. Structure of $\vec{G}$ (not a planar embedding)

2   Proof of Theorem 1

For a graph G, $D \subseteq V(G)$ *dominates* G if any vertex of G is either in D or adjacent to a vertex in D. The *domination number* of a graph G is the minimum cardinality of a dominating set. Goddard and Henning [6] showed that every planar graph of diameter 2 has domination number at most 2 except for a particular graph on eleven vertices.

Let $\vec{B}$ be a planar oclique dominated by the vertex v. Sopena [1] showed that any oriented outerplanar graph has an oriented 7-colouring. Hence let c be an oriented 7-colouring of the oriented outerplanar graph obtained from $\vec{B}$ by deleting the vertex v. Now for $u \in N^\alpha(v)$ let us assign the colour $(c(u), \alpha)$ to u for $\alpha \in \{+, -\}$ and the colour 0 to v. It is easy to check that this is an oriented 15-colouring of $\vec{B}$. Hence any planar oclique dominated by one vertex has order at most 15.

Now let $\vec{G}$ be a planar oclique with $|\vec{G}| > 15$. Clearly $\vec{G}$ has diameter 2. Then by the above discussion, the domination number of $\vec{G}$ is 2. Without loss of generality, we may assume that $\vec{G}$ is triangulated.

We define the partial order $\preccurlyeq$ for the set of all dominating sets of order 2 of $\vec{G}$ as follows: for any two dominating sets $D = \{x, y\}$ and $D' = \{x', y'\}$ of order 2 of $\vec{G}$, $D' \preccurlyeq D$ if and only if $|N(x') \cap N(y')| \le |N(x) \cap N(y)|$.

Let $D = \{x, y\}$ be a maximal dominating set of order 2 of $\vec{G}$ with respect to $\preccurlyeq$. Also for the rest of this article, $t, t', \alpha, \overline{\alpha}, \beta, \overline{\beta}$ are variables satisfying $t = \{x, y\}$ and $\{\alpha, \overline{\alpha}\} = \{\beta, \overline{\beta}\} = \{+, -\}$.

Now, we fix the following notations (Fig: 1): $C = N(x) \cap N(y)$, $C^{\alpha\beta} = N^\alpha(x) \cap N^\beta(y)$, $C_t^\alpha = N^\alpha(t) \cap C$, $S_t = N(t) \setminus C$, $S_t^\alpha = S_t \cap N^\alpha(t)$, $S = S_x \cup S_y$.

Hence we have,

$$16 \le |\vec{G}| = |D| + |C| + |S|. \tag{1}$$

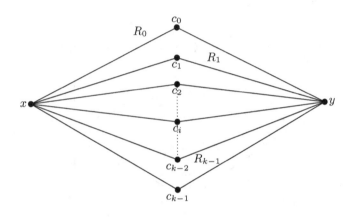

Fig. 2. A planar embedding of $und(\vec{H})$

Let $\vec{H}$ be the oriented graph obtained from the induced subgraph $\vec{G}[D \cup C]$ of $\vec{G}$ by deleting all the arcs between the vertices from D and between the vertices from C. Note that it is possible to extend the planar embedding of $und(\vec{H})$ given in Fig 2 to a planar embedding of $und(\vec{G})$ for some particular ordering of the elements of, say $C = \{c_0, c_1, ..., c_{k-1}\}$.

Notice that $und(\vec{H})$ has k faces, namely the unbounded face F_0 and the faces F_i bounded by edges $xc_{i-1}, c_{i-1}y, yc_i, c_ix$ for $i \in \{1, ..., k-1\}$. Geometrically, $und(\vec{H})$ divides the plane into k connected components. The *region* R_i of $\vec{G}$ is the ith connected component (corresponding to the face F_i) of the plane. *Boundary points* of a region R_i are c_{i-1} and c_i for $i \in \{1, ..., k-1\}$ and, c_0 and c_{k-1} for $i = 0$. Two regions are adjacent if and only if they have at least one common boundary point.

Now for the different possible values of $|C|$, we want to show that $und(\vec{H})$ cannot be extended to a planar oclique of order at least 16. Note that, for extending $und(\vec{H})$ to $\vec{G}$ we can add new vertices only from S. Any vertex $v \in S$ will be inside one of the regions R_i. If there is at least one vertex of S in a region R_i, then R_i is *non-empty* and *empty* otherwise. In fact when there is no chance of confusion, R_i might represent the set of vertices contained in the region R_i.

As any two distinct non-adjacent vertices of $\vec{G}$ must be connected by a 2-dipath, we have the following three lemmas:

Lemma 1. *(a) If $(u,v) \in S_x \times S_y$ or $(u,v) \in S_t^\alpha \times S_t^\alpha$, then u and v are in adjacent regions.*

(b) If $(u,c) \in S_t^\alpha \times C_t^\alpha$, then c is a boundary point of a region adjacent to the region containing u.

Lemma 2. *Let R, R^1, R^2 be three distinct regions such that R is adjacent to R^i with common boundary point c^i while the other boundary points of R^i is $\overline{c^i}$ for all $i \in \{1,2\}$. If $v \in S_t^\alpha \cap R$ and $u^i \in ((S_t^\alpha \cup S_{t'}) \cap R^i) \cup (\{\overline{c^i}\} \cap C_t^\alpha)$, then v disagrees with u^i on c^i, where $i \in \{1,2\}$. If both u^1 and u^2 exist, then $|S_t^\alpha \cap R| \le 1$.*

Lemma 3. *For any arc $\vec{uv}$ in an oclique, we have $|N^\alpha(u) \cap N^\beta(v)| \le 3$.*

Lemma 4. $|C| \ge 2$.

Proof. We know that x and y are either connected by a 2-dipath or by an arc. If x and y are adjacent, then as $\vec{G}$ is triangulated, we have $|C| \ge 2$. If x and y are non-adjacent, then $|C| \ge 1$. Hence it is enough to show that we cannot have $|C| = 1$ while x and y are non-adjacent.

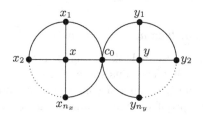

Fig. 3. For $|C| = 1$ while x and y are non-adjacent

If $|C| = 1$ and x and y are non-adjacent, then the triangulation will force the configuration depicted in Fig 3 as a subgraph of $und(\vec{G})$, where $C = \{c_0\}$, $S_x = \{x_1, ..., x_{n_x}\}$ and $S_y = \{y_1, ..., y_{n_y}\}$. Without loss of generality we may assume $|S_y| \ge |S_x|$. Then by (1) we have $n_y = |S_y| \ge \lceil (16 - 2 - 1)/2 \rceil = 7$.

Clearly $n_x \ge 3$ as otherwise $\{c_0, y\}$ is a dominating set with at least two common neighbours $\{y_1, y_{n_y}\}$ which contradicts the maximality of D.

For $n_x = 3$, we know that c_0 is not adjacent to x_2 as otherwise $\{c_0, y\}$ is a dominating set with at least two common neighbours $\{y_1, y_{n_y}\}$ contradicting the maximality of D. But then, x_2 should be adjacent to y_i for some $i \in \{1, ..., n_y\}$ as otherwise $d(x_2, y) > 2$. Now the triangulation will force x_2 and y_i to have at least two common neighbours. Also x_2 cannot be adjacent to y_j for any $j \neq i$, as it will create a dominating set $\{x_2, y\}$ with at least two common neighbours $\{y_i, y_j\}$ contradicting the maximality of D. Hence, x_2 and y_i are adjacent to both x_1 and x_3. Note that, t_{ℓ_t} and t_{ℓ_t+k} are adjacent if and only if $k = 1$, as otherwise $d(t_{\ell_t+1}, t') > 2$ for $1 \leq \ell_t < \ell_t + k \leq n_t$. In this case, by (1) we have $n_y = |S_y| \geq 16 - 2 - 1 - 3 = 10$. Assume $i \geq 6$. Hence, c_0 is adjacent to y_j for all $j = 1, 2, 3, 4$, as otherwise $d(y_j, x_3) > 2$. This implies $d(y_2, x_2) > 2$, a contradiction. Similarly $i \leq 5$ will force, (something), a contradiction. Hence $n_x \geq 4$.

For $n_x = 4$, c_0 cannot be adjacent to both x_3 and $x_{n_x-2} = x_2$ as it creates a dominating set $\{c_0, y\}$ with at least two common neighbours $\{y_1, y_{n_y}\}$ contradicting the maximality of D. For $n_x \geq 5$, c_0 is adjacent to x_3 implies, either for all $i \geq 3$ or for all $i \leq 3$, x_i is adjacent to c_0, as otherwise $d(x_i, y) > 2$. Either of these cases will force c_0 to become adjacent to y_j, as otherwise we will have either $d(x_1, y_j) > 2$ or $d(x_{n_x}, y_j) > 2$ for all $j \in \{1, 2, ..., n_y\}$. But then we will have a dominating set $\{c_0, x\}$ with at least two common vertices contradicting the maximality of D. Hence for $n_x \geq 5$, c_0 is not adjacent to x_3. Similarly we can show, for $n_x \geq 5$, that c_0 is not adjacent to x_{n_x-2}.

So for $n_x \geq 4$, without loss of generality we can assume that c_0 is not adjacent to x_3. We know that $d(y_1, x_3) \leq 2$. We have already noted that, t_{ℓ_t} and t_{ℓ_t+k} are adjacent if and only if $k = 1$ for any $0 \leq \ell_t < \ell_t + k \leq n_t$. Hence to have $d(y_1, x_3) \leq 2$, we must have one of the following edges: $y_1 x_2$, $y_1 x_3$, $y_1 x_4$ or $y_2 x_3$. The first edge will imply the edges $x_2 y_j$ as otherwise $d(x_1, y_j) > 2$ for all $j = 3, 4, 5$. These three edges will imply $d(x_4, y_3) > 2$. Hence we do not have $y_1 x_2$. We also cannot have the other three edges as they will imply $d(x_1, y_4) > 2$. Hence we are done. □

Lemma 5. *If $2 \leq |C| \leq 5$, then at most one region of $\vec{G}$ is non-empty.*

Proof. For pictorial help one can look at Fig 2. For $|C| = 2$, if x and y are adjacent, then the region that contains the edge xy is empty, as otherwise triangulation will force x and y to have a common neighbour other than c_0 and c_1. So for the rest of the proof we can assume x and y to be non-adjacent for $|C| = 2$.

First we shall show that it is not possible to have either $S_x = \emptyset$ or $S_y = \emptyset$ and have at least two non-empty regions. Without loss of generality assume that $S_x = \emptyset$. Then x and y are non-adjacent, as otherwise y will be a dominating vertex which is not possible. For $|C| = 2$, if both $S_y \cap R_0$ and $S_y \cap R_1$ are non-empty, then triangulation will force, either multiple edges $c_0 c_1$ (one in each region) or a common neighbour of x, y other than c_0, c_1, a contradiction. For $|C| = 3, 4$ and 5, triangulation implies the edges $c_0 c_1, ..., c_{k-2}c_{k-1}, c_{k-1}c_0$. Hence every $v \in S_y$ must be connected to x by a 2-dipath through c_i for some $i \in \{1, 2, ..., k-1\}$. Now assume $|S_y^\alpha| \geq |S_y^{\bar{\alpha}}|$ for some $\alpha \in \{+, -\}$. Then by (1), we have $|S_y^\alpha| \geq \lceil (16 - 2 - 5)/2 \rceil = 5$. Now by Lemma 1, we know that the vertices of S_y^α would be contained in two adjacent regions for $|C| = 4, 5$. For $|C| = 3$, $S_y^\alpha \cap R_i$ for all $i \in \{0, 1, 3\}$ implies $|S_y^\alpha| \leq 3$ by Lemma 2. Hence without loss of generality we may assume $S_y^\alpha \subseteq R_1 \cup R_2$. If both $S_y^\alpha \cap R_1$ and $S_y^\alpha \cap R_2$ are non-empty, then, by Lemma 2, each vertex of $S_y^\alpha \cap R_1$ disagrees with each vertex of $S_y^\alpha \cap R_2$ on c_1. Hence by Lemma 3 we have $|S_y^\alpha \cap R_1|, |S_y^\alpha \cap R_2| \le 3$. That implies we have at least two vertices in each of the sets. Without loss of generality assume that the vertices of $S_y^\alpha \cap R_1$ are connected to x by 2-dipaths through c_1. Hence vertices (at least two) of $S_y^\alpha \cap R_2$ must be connected to x by 2-dipaths through c_2. But it is not possible to have two vertices of S_y adjacent to both c_1 and c_2 keeping the graph planar. Hence both S_x and S_y are non-empty.

Now we prove that at most four sets out of the $2k$ sets $S_t \cap R_i$ can be non-empty, for all $t \in \{x, y\}$ and $i \in \{0, 1, ..., k-1\}$. It is trivial for $|C| = 2$. For $|C| = 4$ and 5, the statement follows from Lemma 1. For $|C| = 3$, we consider the following two cases:

(i) Assume $S_t \cap R_i \ne \emptyset$ for all $t \in \{x, y\}$ and for all $i \in \{0, 1, 2\}$. Then by Lemma 2 we have, $|S_t \cap R_i| \le 1$ for all $t \in \{x, y\}$ and for all $i \in \{0, 1, 2\}$. Then by (1) we have, $16 \le |\vec{G}| = 2 + 3 + 4 = 9$, a contradiction.

(ii) Assume that five out of the six sets $S_t \cap R_i$ are non-empty and the other one is empty, where $t \in \{x, y\}$ and $i \in \{0, 1, 2\}$. Without loss of generality we can assume $S_x \cap R_0 = \emptyset$. By Lemma 2 we have $|S_t \cap R_i| \le 1$ for all $(t, i) \in \{(x, 1), (x, 2), (y, 0)\}$. Now let $u \in S_y \cap R_0$. Assume $u \in N^\alpha(y)$, where $\{\alpha, \overline{\alpha}\} = \{+, -\}$. So $|S_y \cap R_i \cap N^\alpha(y)| \le 1$ for $i \in \{1, 2\}$, because any $w_i \in S_y \cap R_i \cap N^\alpha(y)$ will have an arc with c_{i-1} and c_i in order to satisfy $\vec{d}(w_i, u) \le 2$ and $d(w_i, v_i) \le 2$, where $v_i \in S_x \cap R_{3-i}$, for all $i \in \{1, 2\}$. Any $w_i \in S_y \cap R_i$ will disagree on c_1 with $v_i \in S_x \cap R_{3-i}$, for all $i \in \{1, 2\}$ to satisfy $\vec{d}(w_i, v_i) \le 2$. Then by Lemma 3 we have $|S_y \cap R_i \cap N^{\overline{\alpha}}(y)| \le 3$ for all $i \in \{1, 2\}$. Hence, $|S_y \cap N^{\overline{\alpha}}(y)| \le 6$. Also we have $|S_y \cap N^\alpha(y)| \le 3$ and $|S_x| \le 2$. Therefore by (1) we have, $16 \le |\vec{G}| = 2 + 2 + 11 = 15$, a contradiction.

Hence at most four sets out of the $2k$ sets $S_t \cap R_i$ can be non-empty, where $t \in \{x, y\}$ and $i \in \{0, 1, ..., k-1\}$.

Now assume that exactly four sets out of the sets $S_t \cap R_i$ are non-empty, for all $t \in \{x, y\}$ and $i \in \{0, ..., k-1\}$. Without loss of generality we have the following three cases (by Lemma 1):

(i) Assume the four non-empty sets are $S_x \cap R_1, S_y \cap R_0, S_y \cap R_1$ and $S_y \cap R_2$ (only possible for $|C| \ge 3$). We have the edges $c_0 c_{k-1}$ and $c_1 c_2$ by triangulation. Lemma 2 implies that $S_x \cap R_1 = \{x_1\}$ and that the vertices of $S_y \cap R_0$ and the vertices of $S_y \cap R_2$ disagree with x_1 on c_0 and c_1 respectively. Hence by Lemma 3, we have $|S_y \cap R_0|, |S_y \cap R_2| \le 3$. For $|C| = 3$, if every vertex from $S_y \cap R_1$ is adjacent to either c_0 or c_1, then $\{c_0, c_1\}$ will be a dominating set with at least four common neighbours $\{x, y, x_1, c_2\}$ contradicting the maximality of D. If not, then triangulation will force x_1 to be adjacent to at least two vertices y_1, y_2 (say) from S_y. But then, $\{x_1, y\}$ will become a dominating set with at least four common neighbours $\{y_1, y_2, c_0, c_1\}$ contradicting the maximality of D. For $|C| = 4$ and 5, Lemma 1 implies that vertices of $S_y \cap R_0$ and vertices of $S_y \cap R_2$ disagree with each other on y. Now by Lemma 2, any vertex of $S_y \cap R_1$ is adjacent to either c_0 (if it agrees with the vertices of $S_y \cap R_0$ on y) or c_1 (if it agrees with the vertices of $S_y \cap R_2$ on y). Also vertices of $S_y \cap R_0$ and $S_y \cap R_2$ are connected to x_1 by a 2-dipath through c_0 and c_1 respectively. Now by (1) we have $|S_y| \geq (16 - 2 - 5 - 1) = 8$. Hence without loss of generality, at least four vertices y_1, y_2, y_3, y_4 of S_y are adjacent to c_0. Hence $\{c_0, y\}$ is a dominating set with at least five common neighbours $\{y_1, y_2, y_3, y_4, c_{k-1}\}$ contradicting the maximality of D for $|C| = 4$. For $|C| = 5$, each vertex of $S_y \cap R_1$ disagree with c_3 by Lemma 1 and hence without loss of generality are all adjacent to c_0. Now $|S_y \cap R_2| \leq 3$ and $|S_y| \geq 8$ implies $|S_y \cap (R_0 \cup R_1)| \geq 5$. But every vertex of $S_y \cap (R_0 \cup R_1)$ and c_4 are adjacent to c_0. Hence $\{c_0, y\}$ is a dominating set with at least six common neighbours, contradicting the maximality of D for $|C| = 5$.

(ii) Assume the four non-empty sets are $S_x \cap R_0, S_x \cap R_1, S_y \cap R_0$ and $S_y \cap R_1$. For $|C| = 2$ every vertex in S is adjacent to either c_0 or c_1 (by Lemma 2). So, $\{c_0, c_1\}$ is a dominating set. Hence no vertex $w \in S$ can be adjacent to both c_0 and c_1 because otherwise $\{c_0, c_1\}$ will be a dominating set with at least three common neighbours $\{x, y, w\}$ contradicting the maximality of D. By (1) we have $|S| \geq 16 - 2 - 2 = 12$ and hence without loss of generality we may assume $|S_x \cap R_0| \geq 3$. Assume $\{x_1, x_2, x_3\} \subseteq S_x \cap R_0$. Now all vertices of $S_x \cap R_0$ must be adjacent to c_0 (or c_1), as otherwise it will force all vertices of $S_y \cap R_1$ to be adjacent to both c_0 and c_1 (by Lemma 2). Without loss of generality assume all vertices of $S_x \cap R_0$ are adjacent to c_0. Then all $w \in S_y$ will be adjacent to c_0, as otherwise $d(w, x_i) > 2$ for some $i \in \{1, 2, 3\}$. But then $\{c_0, x\}$ will be a dominating set with at least three common vertices $\{x_1, x_2, x_3\}$ contradicting the maximality of D. For $|C| = 3, 4$ and 5, every vertex of S will be adjacent to c_0 (by Lemma 2). By (1) we have $|S| \geq (16 - 2 - 5) = 9$. Hence without loss of generality, $|S_x| \geq 5$. Hence $\{c_0, x\}$ is a dominating set with at least five common neighbours $S_x \cup \{y\}$ contradicting the maximality of D for $|C| = 3, 4$. For $|C| = 5$, we may assume $|S_x| = 5$ and, x and y are non-adjacent as otherwise it will create the dominating set $\{c_0, x\}$ with at least six common neighbours $S_x \cup \{y\}$ contradicting the maximality of D. Now every vertex of S_y must be connected to x by a 2-dipath through c_i for some $i \in \{0, 1, 2, 3, 4\}$. By Lemma 1 we know that each vertex of $S_y \cap R_i$ disagree with c_{i+2} on y for $i \in \{0, 1\}$. Also each vertex of $S_y \cap R_i$ disagree with each vertex of $S_x \cap R_{1-i}$ on c_0. Hence, by Lemma 3, we have $|S_y \cap R_0|, |S_y \cap R_1| \leq 3$ and, without loss of generality, we may assume $|S_y \cap R_0| = 3$ and $|S_y \cap R_1| = 2$. Now c_2 and c_3 agree with each other on y, as otherwise both the vertices of $S_y \cap R_1$ must be connected by 2-dipaths to c_3 through c_2, which is not possible to do keeping the graph planar. This implies that vertices of $S_y \cap R_0$ and vertices of $S_y \cap R_1$ agree with each other on y but disagree on c_0. Hence there is a vertex in either $S_y \cap R_0$ or $S_y \cap R_1$ which is neither adjacent nor connected by a 2-dipath to x.

(iii) Assume the four non-empty sets are $S_x \cap R_1, S_x \cap R_2, S_y \cap R_0$ and $S_y \cap R_1$ (only possible for $|C| = 3$). Now Lemma 2 implies that every vertex of $(S_x \cap R_1) \cup (S_y \cap R_0)$ is adjacent to c_0 and every vertex of $(S_x \cap R_2) \cup (S_y \cap R_1)$ is adjacent to c_1. Moreover triangulation forces the edges $c_0 c_2$ and $c_1 c_2$. Triangulation also forces some vertex $v_1 \in S_y \cap R_1$ to be adjacent to c_0. This will create the dominating set $\{c_0, c_1\}$ with at least four common neighbours $\{x, y, v_1, c_2\}$ contradicting the maximality of D.

Hence at most three sets out of the $2k$ sets $S_t \cap R_i$ can be non-empty, where $t \in \{x, y\}$ and $i \in \{0, 1, ..., k-1\}$.

Now assume that exactly three sets out of the sets $S_t \cap R_i$ are non-empty, where $t \in \{x, y\}$ and $i \in \{0, ..., k-1\}$. Without loss of generality we have the following two cases (by Lemma 1):

(i) Assume the three non-empty sets are $S_x \cap R_0, S_y \cap R_0$ and $S_y \cap R_1$. Triangulation implies the edge $c_0 c_1$ inside the region R_1. For $|C| = 2$, there exists $u \in S_y \cup R_1$ such that u is adjacent to both c_0 and c_1 by triangulation. Now if $|S_y \cup R_1| \geq 2$, then some other vertex $v \in S_y \cup R_1$ must be adjacent to either c_0 or c_1. Without loss of generality we may assume that v is adjacent to c_0. Then every $w \in S_x \cap R_0$ will be adjacent to c_0 to have $d(v, w) \leq 2$. But then $\{c_0, y\}$ will be a dominating set with at least three common neighbours $\{c_1, u, v\}$ contradicting the maximality of D. So now let us assume that $S_y \cup R_1 = \{u\}$. Then any $w \in S_x \cap R_0$ is adjacent to either c_0 or c_1. If $|S_x| \geq 5$, then without loss of generality we can assume that at least three vertices of S_x are adjacent to c_0. Now to have at most distance 2 with all those three vertices, every vertex of S_y would be adjacent to c_0. This will create the dominating set $\{c_0, x\}$ with at least three common neighbours contradicting the maximality of D. Also $|S_x| = 1$ clearly creates the dominating set $\{c_0, y\}$ (as x_1 is adjacent to c_0 by triangulation) with at least three common neighbours (a vertex from $S_y \cap R_0$ by triangulation, u and c_1) contradicting the maximality of D. For $2 \leq |S_x| \leq 4$, c_0 (or c_1) can be adjacent to at most two vertices of $S_y \cap R_0$ because otherwise there will be one vertex $v \in S_y \cap R_0$ which will force c_0 (or c_1) to be adjacent to all vertices of $w \in S_x$ in order to satisfy $d(v, w) \leq 2$ and create a dominating set $\{c_0, y\}$ that contradicts the maximality of D. Now, not all vertices of S_x can is adjacent to c_0 (or c_1) as otherwise $\{c_0, y\}$ (or $\{c_1, y\}$) will be a dominating set with at least three common neighbours (u, c_1 (or c_0) and a vertex from $S_y \cap R_0$) contradicting the maximality of D. Note that $|S_y \cap R_0| \geq 11 - S_x$ by (1). Assume $S_x = \{x_1, ..., x_n\}$ with triangulation forcing the edges $c_0 x_1, x_1 x_2, ..., x_{n-1} x_n,$ $x_n c_1$ for $n \in \{2, 3, 4\}$. For $|S_x| = 2$, at most four vertices of $S_y \cap R_0$ can be adjacent to c_0 or c_1. Hence there will be at least five vertices of $S_y \cap R_0$ each connected to x by a 2-dipath through x_1 or x_2. Without loss of generality, x_1 will be adjacent to at least three vertices of S_y and hence $\{x_1, y\}$ will be a dominating set contradicting the maximality of D. For $|S_x| = 3$, without loss of generality assume that x_2 is adjacent to c_0. To satisfy $d(x_1, v) \leq 2$ for all $v \in S_y \cap R_0$, at least four vertices of S_y will be adjacent connected to x_1 by a 2-dipath through x_2. This will create the dominating set $\{x_2, y\}$ contradicting the maximality of D. For $|S_x| = 4$ we have $x_2 c_0$ and $x_3 c_1$ as otherwise at least three vertices of S_x will be adjacent to either c_0 or c_1 which is not possible (because it forces all vertices of S_y to be adjacent to c_0 or c_1). Now each vertex $v \in S_y \cap R_0$ must be adjacent to either c_0 or x_2 (to satisfy $d(v, x_1) \leq 2$) and also to either c_1 or x_3 (to satisfy $d(v, x_4) \leq 2$) which is not possible to do keeping the graph planar. For $|C| = 3, 4, 5$ by Lemma 2, each vertex of S_x disagree with each vertex of $S_y \cap R_1$ on c_0. We also have the edge $x_1 c_2$ for some $x_1 \in S_x$ by triangulation. Now by (1) we have, $|S| \geq (16 - 2 - 5) = 9$. Hence $|S_x| \leq 2$ as otherwise every vertex $u \in S_y$ will be adjacent to c_0 creating a dominating set $\{c_0, t\}$ with at least six common neighbours $S_t \cup \{c_1\}$ for some $t \in \{x, y\}$ contradicting the maximality of D. Now for $|C| = 3$, we can assume x and y are non-adjacent as otherwise $\{c_0, y\}$ will be a dominating set with at least four common neighbours (x, c_1 and, two other vertices each from the sets $S_y \cap R_0$, $S_y \cap R_1$ by triangulation) contradicting the maximality of D. Hence triangulation will imply the edge $c_1 c_2$. Now for $|S_x| \leq 2$, either $\{c_0, c_2\}$ is a dominating set with at least four common neighbours $\{x, y, c_1, x_1\}$ contradicting the maximality of D or x_1 is adjacent to at least two vertices $y_1, y_2 \in S_y \cap R_0$ creating a dominating set $\{x_1, y\}$ (another vertex in S_x must be adjacent to x_1 by triangulation) with at least four common neighbours $\{y_1, y_2, c_0, c_2\}$ contradicting the maximality of D. For $|C| = 4$ we have $|S_y \cap R_1| \leq 2$ as otherwise we will have the dominating set $\{c_0, y\}$ with at least five common neighbours (c_1, vertices of $S_y \cap R_1$ and one vertex of $S_y \cap R_0$ by triangulation) contradicting the maximality of D. Now by (1) we have $|S_y \cap R_0| \geq (16 - |D| - |C| - |S_x| - |S_y \cap R_1|) \geq (16 - 2 - 4 - 2 - 2) = 6$. Now, at most 2 vertices of $S_y \cap R_0$ can be adjacent to c_0 as otherwise $\{c_0, y\}$ will be a dominating set with at least five common neighbours contradicting the maximality of D. Also by triangulation in R_3 we either have the edge xy or have the edge $c_2 c_3$. Also we have either at least four vertices of $S_y \cap R_0$ adjacent to c_3 or at least three vertices of $S_y \cap R_0$ adjacent to x_1 by triangulation. In either of these cases, a dominating set $\{c_3, y\}$ (because every vertex in S_x will be adjacent to c_0 in order to have distance at most 2 with all the four vertices of $S_y \cap R_0$ adjacent to c_3) or $\{x_1, y\}$ with at least five common neighbours is created contradicting the maximality of D. For $|C| = 5$ by Lemma 1, each vertex of $S_y \cap R_i$ must disagree with c_{i+2} on y. If vertices of $S_y \cap R_0$ and vertices of $S_y \cap R_1$ agree with each other on y, then they must disagree with each other on c_0 which implies $|S_y \cap R_i| \leq 3$ for all $i \in \{0, 1\}$. If vertices of $S_y \cap R_0$ and vertices of $S_y \cap R_1$ disagree with each other on y, then vertices of $S_y \cap R_i$ must agree with c_{3-i} on y. Then, by Lemma 2, each vertex of $S_y \cap R_i$ must be connected to c_{3-i} by a 2-dipath through c_{4-3i} which implies $|S_y \cap R_i| \leq 3$ for all $i \in \{0, 1\}$. Hence we have $|S| = |S_x| + |S_y| \leq 2 + (3 + 3) = 8$. Then by (1) we have, $16 \leq |\vec{G}| \leq (2 + 5 + 8) = 15$, a contradiction.

(ii) Assume the three non-empty sets are $S_x \cap R_1, S_y \cap R_0$ and $S_y \cap R_2$ (only possible for $|C| \geq 3$). By Lemma 2, we have $S_x = \{x_1\}$ and the fact that each vertex of $S_y \cap R_i$ disagrees with $c_{i^2/4}$ on x_1. Triangulation implies the edges $x_1 c_0$, $x_1 c_1, c_{k-1} c_0, c_0 c_1$ and $c_1 c_2$. Hence for $|C| = 3$, $\{c_0, c_1\}$ is a dominating set with at least four common neighbours $\{x, y, c_2, x_1\}$ contradicting the maximality of D. For $|C| = 4$ and 5 we have, every vertex of $S_y \cap R_0$ disagree with every vertex of $S_y \cap R_2$ on y. Hence, by Lemma 3, we have $|S_y \cap R_i| \leq 3$ for all $\in \{0, 2\}$. Hence by (1) we have $16 \leq |\vec{G}| = |D| + |C| + |S| \leq [2 + 5 + (1 + 3 + 3)] = 14$, a contradiction.

Hence at most two sets out of the $2k$ sets $S_t \cap R_i$ can be non-empty, where $t \in \{x, y\}$ and $i \in \{0, 1, ..., k - 1\}$.

Now assume that exactly two sets out of the sets $S_t \cap R_i$ are non-empty, where $t \in \{x, y\}$ and $i \in \{0, ..., k - 1\}$, yet there are two non-empty regions.

Without loss of generality assume that the two non-empty sets are $S_x \cap R_0$ and $S_y \cap R_1$. Triangulation would force x and y to have a common neighbour other than c_0 and c_1 for $|C| = 2$ which is a contradiction. For $|C| = 3, 4, 5$ triangulation implies the edges $c_{k-1}c_0$ and c_0c_1. By Lemma 2, we know that each vertex of S is adjacent to c_0. By (1) we have $|S| \geq (16 - 2 - 5) = 9$. Hence without loss of generality we may assume $|S_x| \geq 5$. Then $\{c_0, x\}$ will be a dominating set with at least six common neighbours $S_x \cup \{c_{k-1}, c_1\}$ contradicting the maximality of D.

Hence we are done.   □

Lemma 6. $|C| \geq 6$.

Proof. For $|C| = 2, 3, 4, 5$ without loss of generality by Lemma 5, we may assume R_1 to be the only non-empty region. Then triangulation will force the configuration depicted in Fig 4 as a subgraph of $und(\vec{G})$, where $C = \{c_0, ..., c_{k-1}\}$, $S_x = \{x_1, ..., x_{n_x}\}$ and $S_y = \{y_1, ..., y_{n_y}\}$. Without loss of generality we may assume $|S_y| = n_y \geq n_x = |S_x|$. Then by (1) we have $n_y = |S_y| \geq \lceil (16 - 2 - 5)/2 \rceil = 5$.

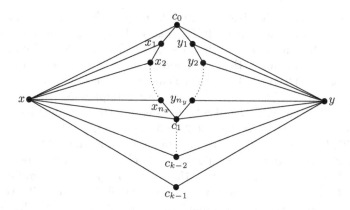

Fig. 4. The only non-empty region is R_1

Claim 1: The Lemma is true when $n_x \leq 2$.

Claim 2: If $n_x \geq 3$, then the edge $t_i t_{i_t + l}$ implies $l = 1$ for $0 \leq i < i + l \leq n_t$.

Claim 3: If $n_x \geq 3$, then the edge $c_i t_j$ implies $j = (n_t - 1)i + 1$ for all $i \in \{0, 1\}$.

Proof of these three claims can be done by using similar ideas used for proving Lemma 4.

Now for $n_x \geq 3$, to satisfy $d(y_{n_y}, x_1) \leq 2$ we must have (because of the above claims) either of the following edges: $y_{n_y - 1} x_1$, $y_{n_y} x_1$ or $y_{n_y} x_2$. If we have $y_{n_y - 1} x_1$ or $y_{n_y} x_1$, then we have a contradiction by forcing $d(y_1, x_{n_x}) > 2$. If we have $y_{n_y} x_2$, then also we have a contradiction by forcing $d(y_1, x_{n_x}) > 2$ when $n_x \geq 4$. For $n_x = 3$, the edge $y_{n_y} x_2$ will force all the edges $y_j x_2$ to satisfy $d(x_3, y_j) \leq 2$ for all $j \in \{1, 2..., n_y\}$. But in this case $n_y \geq 6$ by (1). Hence $\{x_2, y\}$ is a dominating set with at least six common neighbours contradicting the maximality of D. Hence using Lemma 4 we are done.   □

Lemma 7. *If $|C| \geq 6$, then the following holds:*

(a) $|C^{\alpha\beta}| \leq 3$, $|C_t^\alpha| \leq 6$, $|C| \leq 12$. Moreover, if $|C^{\alpha\beta}| = 3$, then $\vec{G}[C^{\alpha\beta}]$ is a 2-dipath.

(b) $|C_t^\alpha| \geq 5$ (respectively $4, 3, 2, 1, 0$) implies $|S_t^\alpha| \leq 0$ (respectively $1, 4, 5, 6, 7$).

Proof. (a) If $|C^{\alpha\beta}| \geq 4$, then there will be two vertices $u, v \in C^{\alpha\beta}$ with $d(u, v) > 2$ which is a contradiction. Hence we have the first enequality which implies the other two. Also if $|C^{\alpha\beta}| = 3$, then the only way to connect the two non-adjacent vertices u, v of $C^{\alpha\beta}$ is to connected them with a 2-dipath through the other vertex (other than u, v) of $C^{\alpha\beta}$.

(b) Lemma 1(b) implies that, if all elements of C_t^α does not belong to the four boundary points of any three consecutive regions (like R, R^1, R^2 in Lemma 2), then $|S_t^\alpha| = 0$. Hence we have $|C_t^\alpha| \geq 5$ implies $|S_t^\alpha| \leq 0$.

By Lemma 2, if all the elements of C_t^α belongs to the four boundary points $c^1, c^2, \overline{c^1}, \overline{c^2}$ of three consecutive regions R, R^1, R^2 (like in Lemma 2) and contains both $\overline{c^1}, \overline{c^2}$, then $|S_t^\alpha| \leq 1$. Also $S_t^\alpha \subseteq R$ by Lemma 2. Hence we have $|C_t^\alpha| \geq 4$ implies $|S_t^\alpha| \leq 1$.

Now assume that all the elements of C_t^α belongs to the three boundary points $c^1, c^2, \overline{c^1}$ of two adjacent regions R, R^1 (like in Lemma 2) and contains both $\overline{c^1}, c^2$. Then by Lemma 1, $v \in S_t^\alpha$ implies v is in R or R^1. Now if both $S_t^\alpha \cap R$ and $S_t^\alpha \cap R^1$ are non-empty then each vertex of $(S_t^\alpha \cap R) \cup \{c^2\}$ disagree with each vertex of $(S_t^\alpha \cap R^1) \cup \{\overline{c^1}\}$ on c^1 (by Lemma 2). Hence by Lemma 3, we have $|(S_t^\alpha \cap R) \cup \{\overline{c^1}\}|, |(S_t^\alpha \cap R^1) \cup \{c^2\}| \leq 3$ which clearly implies $|S_t^\alpha \cap R|, |S_t^\alpha \cap R^1| \leq 2$ and $|S_t^\alpha| \leq 4$. If one of $S_t^\alpha \cap R$ and $S_t^\alpha \cap R^1$ is empty then we must have $|S_t^\alpha| \leq 3$ by Lemma 2 and 3. Hence we have $|C_t^\alpha| \geq 3$ implies $|S_t^\alpha| \leq 4$.

Let $R, R^1, R^2, c^1, c^2, \overline{c^1}, \overline{c^2}$ be like in Lemma 2 and assume $C_t^\alpha = \{c^1, c^2\}$. By Lemma 1, $v \in S_t^\alpha$ implies v is in R, R^1 or R^2 and also that both $S_t^\alpha \cap R^1$ and $S_t^\alpha \cap R^2$ can not be non-empty. Hence without loss of generality assume $S_t^\alpha \cap R^2 = \emptyset$. Then by Lemma 2, vertices of $S_t^\alpha \cap R^1$ disagree with vertices of $(S_t^\alpha \cap R) \cup \{c^2\}$ on c^1. Hence by Lemma 3 we have, $|S_t^\alpha \cap R^1|, |(S_t^\alpha \cap R) \cup \{c^2\}| \leq 3$ which implies $|S_t^\alpha| \leq 5$. Now if $S_t^\alpha \cap R^1 = \emptyset$, then we have $S_t^\alpha = S_t^\alpha \cap R$. Let $|S_t^\alpha \cap R| \geq 6$. Now consider the induced graph $\vec{O} = \vec{G}[(S \cap R) \cup \{c^1, c^2\}]$. In this graph the vertices of $(S_t^\alpha \cap R) \cup \{c^1, c^2\}$ are pairwise at directed distance at most two. Hence $\chi_o(\vec{O}) \geq 8$. But this is a contradiction as $\vec{O}$ is an outerplanar graph and every outerplanar graph has an oriented 7-colouring [1]. Hence we have $|C_t^\alpha| \geq 2$ implies $|S_t^\alpha| \leq 5$.

In general S_t^α is contained in two distinct adjacent regions by Lemma 1. Without loss of generality assume $S_t^\alpha \subseteq R_1 \cup R_2$. If both $S_t^\alpha \cap R_1$ and $S_t^\alpha \cap R_2$ are non-empty, then by Lemma 2 we know that vertices of $S_t^\alpha \cap R_1$ disagree with vertices of $S_t^\alpha \cap R_2$ on c_1. Hence $|S_t^\alpha \cap R_1|, |S_t^\alpha \cap R_2| \leq 3$ which implies $|S_t^\alpha| \leq 6$. In particular, if $|C_t^\alpha| = 1$, then $|S_t^\alpha| = 6$ implies $C_t^\alpha = \{c_1\}$. Now assume only one of the two sets $S_t^\alpha \cap R_1$ and $S_t^\alpha \cap R_2$ is non-empty. Without loss of generality assume $S_t^\alpha \cap R_1 \neq \emptyset$. If $c_0, c_1 \notin C_t^\alpha$ and $|C_t^\alpha| = 1$, then we have $|S_t^\alpha \cap R_1| \leq 3$ by Lemma 2 and 3. In the induced outerplanar graph $\vec{O} = \vec{G}[(S \cap R_1) \cup \{c_1, c_2\}]$

vertices of $S_t^\alpha \cup (c_t^\alpha \cap \{c^1, c^2\})$ are pairwise at directed distance at most two. Hence $7 \geq \chi_o(\vec{O}) \geq |S_t^\alpha \cup (c_t^\alpha \cap \{c^1, c^2\})|$. Therefore, $|C_t^\alpha| \geq 0$ (respectively 1) implies $|S_t^\alpha| \leq 6$ (respectively 7). $\qquad \square$

Lemma 8. $|C| \leq 5$.

Proof. Without loss of generality we can suppose $|C_x^\alpha| \geq |C_y^\beta| \geq |C_y^{\overline{\beta}}| \geq |C_x^{\overline{\alpha}}|$ (the last inequality is forced). We know that $|C| \leq 12$ and $|C_x^\alpha| \leq 6$ (Lemma 7(a)). So it is enough to show that $|S| \leq 13 - |C|$ for all possible values of $(|C|, |C_x^\alpha|, |C_y^\beta|)$ since it contradicts (1).

For $(|C|, |C_x^\alpha|, |C_y^\beta|) = (12, 6, 6), (11, 6, 6), (10, 6, 6), (10, 6, 5), (10, 5, 5),$ $(9, 5, 5), (8, 4, 4)$ we have $|S| \leq 13 - |C|$ using Lemma 7(b). For $(|C|, |C_x^\alpha|, |C_y^\beta|) = (8, 6, 6), (7, 6, 6), (7, 6, 5), (6, 6, 6), (6, 6, 5), (6, 6, 4), (6, 5, 5)$ we are forced to have $|C^{\alpha\beta}| > 3$, which is not possible by Lemma 7(a).

For $(|C|, |C_x^\alpha|, |C_y^\beta|) = (9, 6, 6)$ we are forced to have $|C^{\alpha\beta}| = |C^{\alpha\overline{\beta}}| = |C^{\overline{\alpha}\beta}| = 3$ in order to satisfy the first inequality of Lemma 7(a). So $\vec{G}[C^{\alpha\beta}], \vec{G}[C^{\alpha\overline{\beta}}]$ and $\vec{G}[C^{\overline{\alpha}\beta}]$ are 2-dipaths by Lemma 7(a). Without loss of generality we can assume $C^{\alpha\beta} = \{c_0, c_1, c_2\}$ and $C^{\overline{\alpha}\beta} = \{c_3, c_4, c_5\}$. Hence by Lemma 1 we have $u \in R_1 \cup R_2$ and $v \in R_4 \cup R_5$ for any $(u, v) \in S_y^{\overline{\beta}} \times S_x^{\overline{\alpha}}$. Hence by Lemma 1, either $S_y^{\overline{\beta}}$ or $S_x^{\overline{\alpha}}$ is empty. Without loss of generality assume $S_y^{\overline{\beta}} = \emptyset$. Therefore we have, $|S| = |S_x| = |S_x^{\overline{\alpha}}| \leq 4$ (by Lemma 7(b)). So this case is not possible.

Similarly for $(|C|, |C_x^\alpha|, |C_y^\beta|) = (7, 6, 4)$ without loss of generality we can assume that $\vec{G}[C^{\alpha\beta}]$ and $\vec{G}[C^{\alpha\overline{\beta}}]$ are 2-dipaths and, $C^{\alpha\beta} = \{c_0, c_1, c_2\}, C^{\alpha\overline{\beta}} = \{c_3, c_4, c_5\}$ and $C^{\overline{\alpha}\beta} = \{c_6\}$. By Lemma 7 we have $|S_x| \leq 6$ and $|S_y| \leq 4 + 1 = 5$. So we are done if either $S_x = \emptyset$ or $S_y = \emptyset$. So assume both S_x and S_y are non-empty. First assume that $S_y^\beta \neq \emptyset$. Then by Lemma 1 we have $S_y^\beta \subseteq R_5$, $S_x^{\overline{\alpha}} \subseteq R_5 \cup R_6$ and hence $S_y^{\overline{\beta}} = \emptyset$. By Lemma 2, vertices of S_y^β and vertices of $S_x^{\overline{\alpha}} \cap R_5$ must disagree with c_6 on c_5 while disagreeing with each other on c_5, which is not possible. Hence, $S_x^{\overline{\alpha}} \cap R_5 = \emptyset$. Also $|S_x^{\overline{\alpha}} \cap R_6| \leq 3$ as they all disagree on c_5 with the vertex of S_y^β. So $|S| \leq 4$ when $S_y^\beta \neq \emptyset$. Now assume $S_y^\beta = \emptyset$ hence $S_y^{\overline{\beta}} \neq \emptyset$. Then by Lemma 1 we have $S_y^{\overline{\beta}} \subseteq R_1 \cup R_2, S_x^{\overline{\alpha}} \subseteq R_0 \cup R_1$ and hence $S_y^\beta = \emptyset$. Assume $S_y^{\overline{\beta}} \cap R_2 = \emptyset$ as otherwise vertices of $S_x^{\overline{\alpha}}$ will be adjacent to both c_0 and c_1 (to be connected to c_6 and vertices of $S_y^{\overline{\beta}} \cap R_2$ by a 2-dipath) implying $|S_x^{\overline{\alpha}}| \leq 1$ and hence $|S| \leq 5$. If $S_x^{\overline{\alpha}} \cap R_0 \neq \emptyset$, then $|S_y^{\overline{\beta}} \cap R_1| = 1, |S_x^{\overline{\alpha}} \cap R_1| \leq 1$ and $|S_y^{\overline{\alpha}} \cap R_0| \leq 3$ by Lemma 2 and hence $|S| \leq 5$. If $S_x^{\overline{\alpha}} \cap R_0 = \emptyset$ then we have $|S_y^{\overline{\beta}} \cap R_1| \leq 2, |S_x^{\overline{\alpha}} \cap R_1| \leq 3$ and hence $|S| \leq 5$. So this case is not possible.

In a similar way one can handle the other cases. $\qquad \square$

Proof of Theorem 1. By Lemma 6 and 8 we get that it is not possible to extend $und(\vec{H})$ to a planar oclique of order at least 16. Hence we are done. $\qquad \square$

References

1. Sopena, E.: Oriented graph coloring. Discrete Mathematics 229(1-3), 359–369 (2001)
2. Marshall, T.H.: Homomorphism bounds for oriented planar graphs. J. Graph Theory 55, 175–190 (2007)
3. Raspaud, A., Sopena, E.: Good and semi-strong colorings of oriented planar graphs. Inf. Process. Lett. 51(4), 171–174 (1994)
4. Sopena, E.: There exist oriented planar graphs with oriented chromatic number at least sixteen. Inf. Process. Lett. 81(6), 309–312 (2002)
5. Klostermeyer, W., MacGillivray, G.: Analogs of cliques for oriented coloring. Discussiones Mathematicae Graph Theory 24(3), 373–388 (2004)
6. Goddard, W., Henning, M.A.: Domination in planar graphs with small diameter. J. Graph Theory 40, 1–25 (2002)

Sufficient Condition for $\{C_4,\ C_{2t}\}$ - Decomposition of $K_{2m,2n}$ — An Improved Bound

Shanmugasundaram Jeevadoss and Appu Muthusamy

Department of Mathematics, Periyar University,
Salem, Tamil Nadu, India
{raazdoss,appumuthusamy}@gmail.com

Abstract. In this paper, we have improved the bounds of the sufficient conditions obtained by C.C.Chou and C.M.Fu [J. Comb. Optim. 14, 205-218 (2007)] for the existence of decomposition of complete bipartite graph $K_{2m,2n}$ into cycles of length 4 and $2t$, $t > 2$. Further an algorithm is presented to provide such bound which in turn reduce the number of constructions for the existence of required decomposition.

Keywords: complete bipartite graph, cycle decomposition.

1 Introduction

All the graphs considered here are simple. Let $K_{m,n}$ denotes the complete bipartite graph with part sizes m, n and let C_k denotes the cycle of length k. By a decomposition of a graph G we mean a partition of G into edge-disjoint subgraphs $G_1, \ldots, G_n$ such that $\bigcup_{i=1}^{n} E(G_i) = E(G)$. If each $G_i \cong H$, for all i, then we say that H decomposes G, or G has an $H - decomposition$ and we denote it by $H|G$; If $H \cong C_k$, we say that G has a $C_k - decomposition$. If G can be decomposed into p copies of C_{2t} and q copies of C_4 then we say that G has a $\{C_4,\ C_{2t}\}$ - decomposition and we write $G = p\,C_{2t} \oplus q\,C_4$ where $p, q \in \mathbb{N} \cup \{0\}$, the set of nonnegative integers. For the standard graph-theoretic terminology the reader is referred to [1].

For our convenience, we use some notations as in [3].
Let $D(G) = \{(p,q)\mid G = pC_{2t} \oplus qC_4 \text{ where } p, q \in \mathbb{N} \cup \{0\}\}$ and $S_r = \{(p,q)\mid 2tp + 4q = r \text{ where } p, q \in \mathbb{N} \cup \{0\}\}$. It is easy to see that $D(G) \subseteq S_r$ if G has r edges. For the two sets $A, B \subseteq S_r$ we define $A + B = \{(a_1 + b_1, a_2 + b_2)\mid (a_1, a_2) \in A, (b_1, b_2) \in B\}$ and $rA = A + A + \cdots + A$ (r times). Let U be the set of positive integers and for each $u, v \in U$ and $v \geq u$ we define $K_{u,U} = \bigoplus_{v \in U} K_{u,v}$,
$$D(K_{u,U}) = \bigcup_{v \in U} D(K_{u,v}) \text{ and } S_{uU} = \bigcup_{v \in U} S_{uv}.$$

1.1 Program Code

Program 1

The following MATHEMATICA program provides all posible p, q and its corresponding u, v such that $2t\,p + 4\,q = 4uv$, where t is even and $\frac{t}{2} \leq u, v < t$.

S. Arumugam and B. Smyth (Eds.): IWOCA 2012, LNCS 7643, pp. 143–147, 2012.

```
t = input  even positive integer;
For[u = t/2, u < t, u++,  For[v = u, v < t, v++,
For[p = 0, p <= (4*u*v/2*t), p++, For[q = 0, q <= (u*v), q++,
If[(2*t*p) + (4*q) == (4*u*v),
Print["u=", u,"v=",v, 2*t,"-", p,"4-", q ]
]]]]]
```

Program 2

The following MATHEMATICA program provides required p, q and its corresponding u, v such that $2t\,p + 4\,q = 4uv$, where t is even and $\frac{t}{2} \le u, v < t$

```
t = input  even positive integer; r = 0;
For[u = t/2, u < t, u++,  For[v = u, v < t, v++,
For[p = r, p <= ((4*u*v)/(2*t)), p++,
For[q = 0, q <= (4*u*v - 2*t*p)/4, q++,
If[((2*t*p) + (4*q)) == (4*u*v),
Print["u=", u,"v=", v, 2*t "-", p,"4-", q]; u = v + 1; v = v;
For[x = u, x < v, x++, For[y = x, y < v, y++,
For[s = 0, s < x*y, s++, If[((2*t*p) + (4*s)) == (4*x*y),
Print["v=", x,"v=", y, 2*t "-", p,"4-", s]; Break[]]];
If[((2*t*p) + (4*s)) == (4*x*y), Break[]]];
If[x == y || x + 1 == y, Break[]]];r += 1; Break[]
]]]]]
```

Program 3

The following MATHEMATICA program provides all posible p, q and its corresponding u, v such that $2t\,p + 4\,q = 4uv$, where t is odd and $\frac{t+1}{2} \le u, v \le \frac{3t-1}{2}$.

```
t = input  odd  positive integer;
For[u = ((t + 1)/2), u <= ((3*t - 1)/2), u++,
For[v = u, v <= ((3*t - 1)/2), v++,
For[p = 0, p <= (4*u*v/2*t), p++,
For[q = 0, q <= (u*v), q++,
If[(2*t*p) + (4*q) == (4*u*v),
Print["u=", u,"v=", v,2*t,"-", p,"4-", q ]
]]]]]
```

Program 4

The following MATHEMATICA program provides required p, q and its corresponding u, v such that $2t\,p + 4\,q = 4uv$, where t is odd and $\frac{t+1}{2} \le u, v \le \frac{3t-1}{2}$.

```
t = input   odd   positive integer; r = 0;
For[u = (t + 1)/2, u <= (3*t - 1)/2, u++,
For[v = u, v <= (3*t - 1)/2, v++,
For[p = r, p <= ((4*u*v)/(2*t)), p++,
For[q = 0, q <= (4*u*v - 2*t*p)/4, q++,
If[((2*t*p) + (4*q)) == (4*u*v),
Print["u=", u,"v=", v, 2*t "-", p,"4-", q]; u = u + 1; v = v;
For[x = u, x < v, x++, For[y = x, y < v, y++,
For[s = 0, s < x*y, s++, If[((2*t*p) + (4*s)) == (4*x*y),
Print["u=", x,"v=", y,  2*t "-", p,"4-", s]; Break[]]];
If[((2*t*p) + (4*s)) == (4*x*y), Break[]]];
If[x == y || x + 1 == y, Break[]]];
r += 2; Break[] ]]]]]
```

Let $X_t = \{(p,q)|p,q \in \mathbb{N} \cup \{0\}$ obtained from Program 1 $\}$, when t is even and $Y_t = \{(p,q)|p,q \in \mathbb{N} \cup \{0\}$ obtained from Program 3$\}$, when t is odd .

Let $P_t = \{(p,q)|p,q \in \mathbb{N} \cup \{0\}$ obtained from Program 2 $\}$, when t is even and $Q_t = \{(p,q)|p,q \in \mathbb{N} \cup \{0\}$ obtained from Program 4$\}$, when t is odd .

Sotteau [4] has shown that $K_{m,\,n}$ has a C_{2k}-decomposition if and only if (i) $m, n \geq k$ (ii) m and n are even and (iii) $mn \equiv 0 \pmod{2k}$.

C.C.Chou, C.M.Fu and W.C. Huang [2] have shown that G can be decomposed into p copies of C_4, q copies of C_6 and r copies of C_8 for each triple p, q, r of nonnegative integers such that $4p + 6q + 8r = |E(G)|$, in the following two cases: (a) $G = K_{m,n}$, if $m \geq 4, n \geq 6$, and m, n are even, (b) $G = K_{n,n}$ minus a $1 - factor$, if n is odd.

C.C.Chou and C.M.Fu [3] have shown that the existence of $\{C_4, C_{2t}\}-$ decomposition of $K_{2u,\,2v}$, $\frac{t}{2} \leq u, v < t$ (i.e. for all $(p,q) \in X_t$) when t even (respectively $\frac{t+1}{2} \leq u, v \leq \frac{3t-1}{2}$,(i.e. for all $(p,q) \in Y_t$) when t odd) implies such decomposition in $K_{2m,\,2n}$, $m,\,n \geq t$ (respectively in $K_{2m,\,2n}$, $m,\,n \geq \frac{3t+1}{2}$).

In this paper, we show that the existence of $\{C_4, C_{2t}\}-$ decomposition of $K_{2u,\,2v}$, for all $(p,q) \in P_t$ when t even (respectively $(p,q) \in Q_t$ when t odd) implies such decomposition in $K_{2m,\,2n}$, $m, n \geq t$ (respectively in $K_{2m,\,2n}$, $m, n \geq \frac{3t+1}{2}$). Since $P_t \subseteq X_t$ and $Q_t \subseteq Y_t$, our result reduce the bounds given by C.C.Chou and C.M.Fu [3] which in turn reduce the number of constructions for the existence of such decomposition. Further the existence of $\{C_4, C_{2t}\}-$ decomposition of $K_{2u,\,2v}$ was assured by providing constructions for such decomposition in $K_{2u,\,2v}$.

2 $\{C_4, C_{2t}\}-$ Decompositions of $K_{2m,\,2n}$

Before proving our main results, we require the following properties of S_r.

Lemma 1 ([3]). *Let a, b and t be positive integers.*

(i) *If t is even and one of a, b is a multiple of t then $S_{2a} + S_{2b} = S_{2a+2b}$.*
(ii) *If t is odd and one of a, b is a multiple of t then $S_{4a} + S_{4b} = S_{4a+4b}$.*

Lemma 2. *Let $U = \{u \in \mathbb{Z}^+ | \frac{t}{2} \leq u < t\}$, and $p, q, s \in \mathbb{Z}^+ \cup \{0\}$, the set of nonnegative integers, where t is even. If $P_t \subseteq D(K_{t, 2U})$, then for each pair $(p, s) \in S_{2tU} \setminus P_t$, there exists a pair $(p, q) \in P_t$, $q < s$ such that $(p, s) \in D(K_{t, 2U})$.*

Proof. Let $(p, q) \in P_t$ and $P_t \subseteq D(K_{t, 2U})$. Then $K_{t, 2u} = p\, C_{2t} \oplus q\, C_4$ for a positive integer $u \in U$ and hence $(p, q) \in S_{2tu}$. Suppose $(p, s) \in S_{2tU} \setminus P_t$ and $q < s$, i.e. $(p, s) \in S_{2tv}$, for a positive integer $v \neq u \in U$ then $s - q = \frac{t(v-u)}{2}$. We decompose $K_{t, 2v}$ as follows $K_{t, 2v} \cong K_{t, 2u} \oplus K_{t, 2(v-u)} \cong K_{t, 2u} \oplus \frac{t(v-u)}{2} K_{2,2} \cong p\, C_{2t} \oplus s\, C_4$. Thus $(p, s) \in D(K_{t, 2v})$, therefore $S_{2tU} \setminus P_t \subseteq D(K_{t, 2U})$. Hence $D(K_{t, 2U}) = S_{2tU}$. $\qquad\square$

Lemma 3. *Let p be positive integer and let U be as defined in Lemma 2. If t is even and $P_t \subseteq D(K_{t, 2U})$, then $D(K_{t, 2p}) = S_{2tp}$ for all $p \geq \frac{3t+1}{2}$.*

Proof. Since t is even and $2p \geq 3t + 1$, there is a nonnegative integer r such that $2p = rt + 2u$, $\frac{t}{2} \leq u < t$. Therefore we can decompose $K_{t, 2p}$ into $r\, K_{t, t}$ and $K_{t, 2u}$ i.e. $K_{t, 2p} \cong r K_{t, t} \oplus K_{t, 2u}$. By the hypothesis, $P_t \subseteq D(K_{t, 2U})$. Then by Lemmas 1 and 2, we have $D(K_{t, 2p}) \supseteq r\, D(K_{t, t}) + D(K_{t, 2u}) = r\, S_{t^2} + S_{2tu} = S_{2tp}$. Therefore $D(K_{t, 2p}) = S_{2tp}$. $\qquad\square$

Theorem 1. *Let m, n, u and v be positive integers and let U be defined as in Lemma 2. If t is even and $P_t \subseteq \bigcup\limits_{u, v \in U} D(K_{2u, 2v})$ then $D(K_{2m, 2n}) = S_{4mn}$ for all $m, n \geq t$.*

Proof. For $2m, 2n \geq t$, we can decompose $K_{2m, 2n}$ as follows: $K_{2m, 2n} \cong K_{2m-t, 2n-t} \oplus K_{2m-t, t} \oplus K_{t, 2n}$. $D(K_{2m, 2n}) \supseteq D(K_{2m-t, 2n-t}) + D(K_{2m-t, t}) + D(K_{t, 2n})$. By the hypothesis, $D(K_{2m-t, 2n-t}) = S_{(2m-t)(2n-t)}$. By Lemmas 2 and 3 we have $D(K_{2m-t, t}) = S_{t(2m-t)}$ and $D(K_{t, 2n}) = S_{2nt}$. By Lemma 1 and the hypothesis, we have $D(K_{2m, 2n}) \supseteq S_{(2m-t)(2n-t)} + S_{(2m-t)t} + S_{2nt} = S_{4mn}$. Thus $D(K_{2m, 2n}) = S_{4mn}$. $\qquad\square$

Lemma 4. *Let $V = \{u \in \mathbb{Z}^+ | \frac{t+1}{2} \leq u \leq \frac{3t-1}{2}\}$, and $p, q, s \in \mathbb{Z}^+ \cup \{0\}$, the set of nonnegative integers where t is odd. If $Q_t \subseteq D(K_{2t, 2V})$, then for each pair $(p, s) \in S_{4tV} \setminus Q_t$, there exists a pair $(p, q) \in Q_t$, $q < s$ such that $(p, s) \in D(K_{2t, 2V})$.*

Proof. Let $(p, q) \in Q_t$ and $Q_t \subseteq D(K_{2t, 2V})$. Then $K_{2t, 2u} = p\, C_{2t} \oplus q\, C_4$ for a positive integer $u \in V$ and hence $(p, q) \in S_{4tu}$. Suppose $(p, s) \in S_{4tV} \setminus Q_t$ and $q < s$, i.e. $(p, s) \in S_{4tv}$, for a positive integer $v \neq u \in V$ then $s - q = t(v - u)$. We decompose $K_{2t, 2v}$ as follows $K_{2t, 2v} \cong K_{2t, 2u} \oplus K_{2t, 2(v-u)} \cong K_{2t, 2u} \oplus t(v - u) K_{2, 2} \cong p\, C_{2t} \oplus s\, C_4$. Thus $(p, s) \in D(K_{2t, 2v})$, therefore $S_{4tV} \setminus Q_t \subseteq D(K_{2t, 2V})$. Hence $D(K_{2t, 2V}) = S_{4tV}$. $\qquad\square$

Lemma 5. *Let p be positive integer and let V be defined as in Lemma 4. If t is odd and $Q_t \subseteq D(K_{2t, 2V})$, then $D(K_{2t, 2p}) = S_{4tp}$ for all $p \geq \frac{3t+1}{2}$.*

Proof. Since t is odd and $2p \geq 3t+1$, there is a nonnegative integer r such that $2p = 2rt+2u$, $\frac{t+1}{2} \leq u \leq \frac{3t-1}{2}$. Therefore we can decompose $K_{2t,\,2p}$ into r $K_{2t,\,2t}$ and $K_{2t,\,2u}$ i.e. $K_{2t,\,2p} \cong rK_{2t,\,2t} \oplus K_{2t,\,2u}$. By the hypothesis, $Q_t \subseteq D(K_{2t,\,2U})$. Then by Lemmas 1 and 2, we have $D(K_{2t,\,2p}) \supseteq r\,D(K_{2t,\,2t}) + D(K_{2t,\,2u}) = r\,S_{4t^2} + S_{4tu} = S_{4tp}$. Therefore $D(K_{2t,\,2p}) = S_{4tp}$. $\qquad\square$

Theorem 2. *Let m, n, u and v be positive integers and let V be defined as in Lemma 4. If t is odd and $Q_t \subseteq \bigcup\limits_{u,\,v \in V} D(K_{2u,\,2v})$, then $D(K_{2m,2n}) = S_{4mn}$ for all $m, n \geq \frac{3t+1}{2}$.*

Proof. For $2m, 2n \geq 3t+1$, we can decompose $K_{2m,\,2n}$ as follows: $K_{2m,\,2n} \cong K_{2m-2t,\,2n-2t} \oplus K_{2m-2t,\,2t} \oplus K_{2t,\,2n-2t} \oplus K_{2t,\,2t}$. $D(K_{2m,\,2n}) \supseteq D(K_{2m-2t,\,2n-2t}) + D(K_{2m-2t,\,2t}) + D(K_{2t,\,2n-2t}) + D(K_{2t,\,2t})$. By the hypothesis, $D(K_{2m-2t,\,2n-2t}) = S_{(2m-2t)(2n-2t)}$. By Lemmas 4 and 5 we have $D(K_{2m-2t,2t}) = S_{2t(2m-2t)}$, $D(K_{2t,2n-2t}) = S_{2t(2n-2t)}$ and $D(K_{2t,2t}) = S_{4t^2}$. By Lemma 1 and the hypothesis, we have $D(K_{2m,2n}) \supseteq S_{(2m-2t)(2n-2t)} + S_{(2m-2t)2t} + S_{2t(2n-2t)} + S_{4t^2} = S_{4mn}$. Thus $D(K_{2m,2n}) = S_{4mn}$. $\qquad\square$

References

1. Bondy, J.A., Murty, U.S.R.: Graph theory with applications. The Macmillan Press Ltd., New York (1976)
2. Chou, C.C., Fu, C.M., Huang, W.C.: Decomposition of $K_{m,n}$ into short cycles. Discrete Math. 197/198, 195–203 (1999)
3. Chou, C.C., Fu, C.M.: Decomposition of $K_{m,n}$ into 4-cycles and 2t-cycles. J. Comb. Optim. 14, 205–218 (2007)
4. Sotteau, D.: Decomposition of $K_{m,n}(K^*_{m,n})$ into cycles (circuits) of length $2k$. J. Combin. Theory Ser. B 30, 75–81 (1981)

Incomparability Graphs of Lattices II

Meenakshi Wasadikar and Pradnya Survase[*]

Department of Mathematics, Dr. B.A.M. University, Aurangabad 431004, India
wasadikar@yahoo.com, survase.pradnya5@gmail.com

Abstract. In this paper, we study some graphs which are realizable and some which are not realizable as the incomparability graph (denoted by $\Gamma'(L)$) of a lattice L with at least two atoms. We prove that for $n \geq 4$, the complete graph K_n with two horns is realizable as $\Gamma'(L)$. We also show that the complete graph K_3 with three horns emanating from each of the three vertices is not realizable as $\Gamma'(L)$, however it is realizable as the zero-divisor graph of L. Also we give a necessary and sufficient condition for a complete bipartite graph with two horns to be realizable as $\Gamma'(L)$ for some lattice L.

Keywords: Incomparability graph, bipartite graph, horn, double star graph, zero-divisor graph.

1 Introduction

Filipov [5] discuses the comparability graphs of partially ordered sets by defining the adjacency between two elements of a poset by using the comparability relation, that is a, b are adjacent if either $a \leq b$ or $b \leq a$. Duffus and Rival [4] discuss the covering graph of a poset. The papers of Gadenova [6], Bollobas and Rival [2] discuss the properties of covering graphs derived from lattices. Nimbhorkar, Wasadikar and Pawar [10] defined the zero-divisor graphs of a lattice L with 0, by defining the adjacency of two elements $x, y \in L$ by $x \wedge y = 0$.

Also, the concept of the cozero divisor graph of a commutative ring was introduced by M. Afkhami and K. Khashyarmanesh in [1]. Let R be a commutative ring with identity and let $W(R)^*$ be the set of all nonzero and nonunit elements of R. Two distinct vertices a and b in $W(R)^*$ are adjacent if and only if $a \notin bR$ and $b \notin aR$.

Recently, Bresar et al. [3] introduced the cover incomparability graphs of posets and called these as $C - I$ graphs of P. They defined the graph in which the edge set is the union of the edge sets of the corresponding covering graph and the corresponding incomparability graph.

In a lattice L, if a, b are incomparable then we write $a \parallel b$. Let L be a finite lattice and let $W(L) = \{x \mid \text{there exists } y \in L \text{ such that } x \parallel y\}$. The incomparability graph of L, denoted by $\Gamma'(L)$, is a graph with the vertex set

[*] The second author gratefully acknowledges the financial assistance in the form of Rajiv Gandhi National Senior Research Fellowship from UGC, New Delhi.

S. Arumugam and B. Smyth (Eds.): IWOCA 2012, LNCS 7643, pp. 148–161, 2012.
© Springer-Verlag Berlin Heidelberg 2012

$W(L)$ and two distinct vertices $a, b \in W(L)$ are adjacent if and only if they are incomparable. Note that $\Gamma'(L)$ does not contain any isolated vertex.

Wasadikar and Survase [12] introduced the incomparability graph of a lattice. Throughout this paper, L is a finite lattice with at least two atoms.

In this paper, we study some more properties of $\Gamma'(L)$. In section 2 we show that, if G is a graph on five vertices without any isolated vertex then G is realizable as $\Gamma'(L)$ for some lattice L if and only if G is not isomorphic to a member of a set of four graphs. Also we show when the zero-divisor graph and the incomparability graph of a lattice L are isomorphic. In section 3 we show that, the complete graph K_3 with exactly one pendant emanating from all the three vertices is not realizable as the incomparability graph of a lattice. However it is realizable as the zero-divisor graph of a lattice L.

The undefined terms are from West [14], Harary [8] and Gratzer [7].

A graph G is *connected* if there exists a path between any two distinct vertices. A graph G is *complete* if each pair of distinct vertices is joined by an edge. For a positive integer n, we use K_n to denote the complete graph with n vertices. A *complete bipartite graph* is a simple bipartite graph such that two vertices are adjacent if and only if they belong to different partite sets. The complete bipartite graph is denoted by $K_{m,n}$. A graph in which one vertex is adjacent to every other vertex and no other adjacencies is called a *star graph*. A vertex of a graph G is called a *pendant vertex* if its degree is 1. A graph which is the union of two star graphs whose centers a and b are connected by a single edge is called a *double star* graph.

2   Some Realizable and Non Realizable Graphs

Nimbhorkar, Wasadikar and Pawar [10] associated a zero-divisor graph to a lattice L with 0, whose vertices are the elements of L and two distinct elements are *adjacent* if and only if their meet is 0. Similarly in [11] we define a graph of a lattice L with 0. We say that an element $x \in L$ is a *zero-divisor* if there exists a non zero $y \in L$ such that $x \wedge y = 0$. We denote by $Z(L)$ the set of zero-divisors of L. We associate a graph $\Gamma(L)$ to L with the vertex set $Z^*(L) = Z(L) - \{0\}$, the set of all nonzero zero-divisors of L and distinct $a, b \in Z^*(L)$ are adjacent if and only if $a \wedge b = 0$. We call this graph as the zero-divisor graph of L.

Wasadikar and Survase [12] have shown that all connected graphs with at most four vertices can be realized as $\Gamma'(L)$.

In this section we discuss graphs with five vertices. There are 34 graphs with five vertices (see [8] Appendix 1) out of which 19 are realizable as $\Gamma'(L)$.

Definition 1. *In a lattice L with 0, a nonzero element $a \in L$ is called an atom if there is no $x \in L$ such that $0 < x < a$.*

Lemma 1. If a lattice L contains n atoms, then these atoms induce a K_n in the incomparability graph.

We denote by $A^l = \{x \in L \mid x \leq y \text{ for all } y \in A\}$.

The next theorem characterizes which graphs are realizable as the incomparability graph of a lattice.

Theorem 2. Let G be a graph on five vertices without any isolated vertex. Then G is realizable as $\Gamma'(L)$ for some L if and only if G is not isomorphic to any of the four graphs shown in Figures 1 to 4 given below.

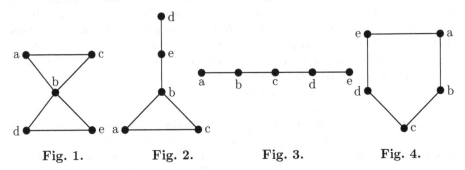

Fig. 1.   Fig. 2.   Fig. 3.   Fig. 4.

Proof. We know that, in a lattice the greatest lower bound of any nonempty finite subset of L exists. Here we show that, the greatest lower bound of some nonempty finite subset of L does not exist.

Consider the Figure 1. Suppose that $G = \Gamma'(L)$ for some lattice L. Since $\Gamma'(L)$ contains a 3 - cycle, L can contain two or three atoms and any two atoms are adjacent in $\Gamma'(L)$. we have the following cases.

Case (i) Suppose, without loss of generality, L has two atoms d, b. We show that $a \wedge c$ does not exist. Since from Figure 1, d and a are comparable and d is an atom hence $d \leq a$. Similarly $d \leq c$. Also a, e are comparable. If $a \leq e$, then $d \leq a$ implies $d \leq e$, a contradiction since d and e are adjacent. Hence $e \leq a$. Similarly $e \leq c$. Thus $\{a, c\}^l = \{0, d, e\}$ but $d \parallel e$ hence $a \wedge c$ does not exist.

Now suppose d, e are the two atoms in L then in a similar manner $\{a, c\}^l = \{0, d, e\}$ but $d \parallel e$. Thus $a \wedge c$ does not exist.

Case (ii) Suppose L has three atoms a, b and c. We show that $d \wedge e$ does not exist. We note that, $\{d, e\}^l = \{0, a, c\}$ but $a \parallel c$ hence $d \wedge e$ does not exist. So Figure 1 cannot be realizable as $\Gamma'(L)$.

Now for Figure 2 Suppose that $G = \Gamma'(L)$ for some lattice L. We have the following cases.

Case (i) Suppose L has two atoms a, b. We show that $d \wedge e$ does not exist. Since a is an atom we have $a \leq d$ and $a \leq e$. From Figure 2, c and d are comparable. If $d \leq c$, then $a \leq d$ implies $a \leq c$, a contradiction since a and c are adjacent. Hence $c \leq d$. Similarly $c \leq e$. Thus $\{d, e\}^l = \{0, a, c\}$ but $a \parallel c$. Hence $d \wedge e$ does not exist.

Suppose e, d are the two atoms in L then $e \leq a$, $e \leq c$ and $d \leq a$, $d \leq c$ that is $\{a, c\}^l = \{0, d, e\}$ but $e \parallel d$. Hence $a \wedge c$ does not exist.

Case (ii) Suppose L has three atoms a, b and c. Then by similar arguments as in the case (ii) of Figure 1, $d \wedge e$ does not exist. So Figure 2 is not realizable as $\Gamma'(L)$.

Consider the Figure 3. Suppose $P_5 = \Gamma'(L)$ for some lattice L. Then by Lemma 1 L has exactly two atoms.

Let b and c be the two atoms. Then we have $b \leq d$, $b \leq e$ and $c \leq a$, $c \leq e$.

Also we have $a \leq d$ or $d \leq a$.

If $a \leq d$, then $c \leq a$ implies $c \leq d$, a contradiction since c and d are adjacent.

If $d \leq a$, then $b \leq d$ implies $b \leq a$, a contradiction since a and b are adjacent.

Hence neither $a \leq d$ nor $d \leq a$, a contradiction since a and d are not adjacent.

Now let d and e be the two atoms in L. We show that $a \wedge b$ does not exist. We note that $\{a, b\}^l = \{0, d, e\}$ but $d \parallel e$ hence $a \wedge b$ does not exist. So the path P_5 cannot be realized as $\Gamma'(L)$.

Consider the Figure 4. Suppose that $G = \Gamma'(L)$ for some lattice L. By Lemma 1 L has exactly two atoms. Let, without any loss of generality, a and b be the two atoms. Then $a \leq c$, $a \leq d$ and $b \leq e$, $b \leq d$.

Also we have $c \leq e$ or $e \leq c$. If $c \leq e$, then $a \leq c$ implies $a \leq e$, a contradiction since a and e are adjacent.

If $e \leq c$, then $b \leq e$ implies $b \leq c$, a contradiction since b and c are adjacent. Neither $c \leq e$ nor $e \leq c$, a contradiction since c and e are nonadjacent. Hence $\Gamma'(L)$ cannot be a 5 - gon.

To show the converse, as mentioned earlier, $\Gamma'(L)$ cannot have any isolated vertex. There are 23 graphs on five vertices without isolated vertices. Hence there are 19 graphs other than the graphs shown in Figures 1 to 4. Each of these 19 graphs is realizable as the incomparability graph of a lattice. These graphs are shown in Figure 5 to Figure 23.

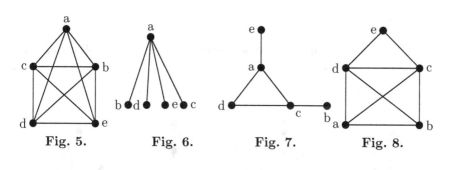

Fig. 5.   Fig. 6.   Fig. 7.   Fig. 8.

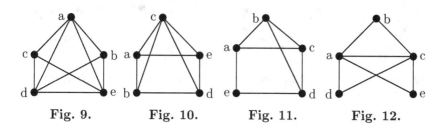

Fig. 9.   Fig. 10.   Fig. 11.   Fig. 12.

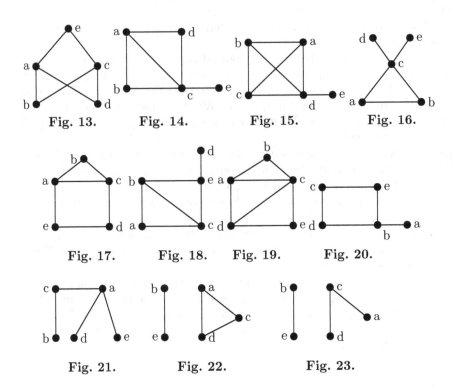

Fig. 13.   Fig. 14.   Fig. 15.   Fig. 16.

Fig. 17.   Fig. 18.   Fig. 19.   Fig. 20.

Fig. 21.   Fig. 22.   Fig. 23.

The following are examples of lattices corresponding to the above graphs respectively.

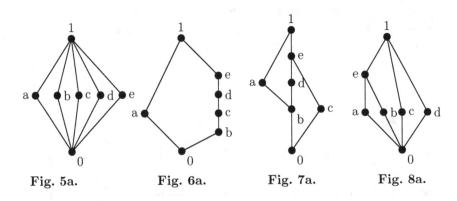

Fig. 5a.   Fig. 6a.   Fig. 7a.   Fig. 8a.

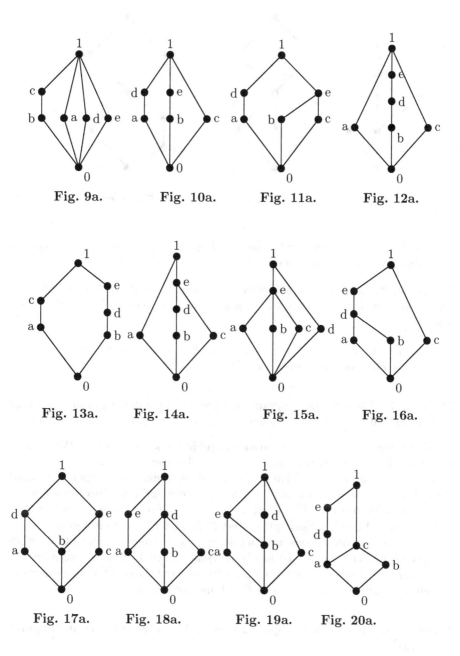

Fig. 9a.   Fig. 10a.   Fig. 11a.   Fig. 12a.

Fig. 13a.   Fig. 14a.   Fig. 15a.   Fig. 16a.

Fig. 17a.   Fig. 18a.   Fig. 19a.   Fig. 20a.

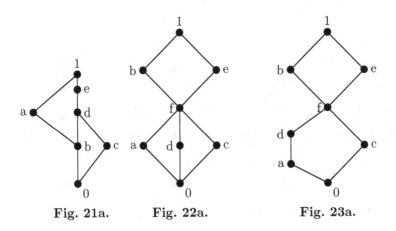

Fig. 21a.   Fig. 22a.   Fig. 23a.

Remark 1. However each graph shown in Figure 1 to Figure 4 can be realized as a subgraph of $\Gamma'(L)$ for some lattice L.

Definition 2. *Let L be a lattice then a non-zero element $a \in L$ is called meet-irreducible if $a = b \wedge c$ implies $a = b$ or $a = c$. Otherwise it is called meet-reducible.*

For example, in Figure 17(a), the elements a, c, d and e are meet-irreducible whereas the element b is meet-reducible.

Theorem 3. *The zero-divisor graph and the incomparability graph of a lattice L are isomorphic if and only if L does not contain any meet-reducible element.*

Proof. Suppose $\Gamma(L)$ and $\Gamma'(L)$ are isomorphic for some lattice L. We want to show that, L does not contain any meet-reducible element.

Suppose on the contrary L has a meet-reducible element say b then there exist $a, c \in L$ and $a, c \neq b$ such that $b = a \wedge c$. Hence a and c are incomparable. So there is an edge $a - c$ in $\Gamma'(L)$ but $a \wedge c \neq 0$. So a and c are not adjacent in $\Gamma(L)$, a contradiction to assumption that $\Gamma(L)$ and $\Gamma'(L)$ are isomorphic.

Conversely suppose L does not contain any meet-reducible element. We want to show that, $\Gamma(L)$ and $\Gamma'(L)$ are isomorphic. Since L does not contain any meet-reducible element the set of all zero-divisors and the set of all incomparable elements are equal hence $\Gamma(L)$ and $\Gamma'(L)$ are isomorphic.

Theorem 4. *The complete graph K_n is realizable as the incomparability graph of a lattice.*

Proof. Consider the complete graph K_n. Let a_i, $i = 1, 2, \ldots, n$ be the vertices of K_n. The corresponding lattice is as shown in Figure 24.

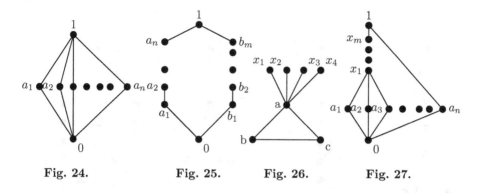

Fig. 24.   **Fig. 25.**   **Fig. 26.**   **Fig. 27.**

Theorem 5. Any complete bipartite graph $K_{m,n}$ is realizable as the incomparability graph of a lattice.

Proof. Consider the complete bipartite graph $K_{m,n}$. Let $V_1 = \{a_1, a_2, \ldots, a_n\}$ and $V_2 = \{b_1, b_2, \ldots, b_m\}$ be the two partitions. The corresponding lattice is as shown in Figure 25.

3   Graphs with Horns

Let G be a graph. All pendant vertices which are adjacent to the same vertex of G together with edges is called a *horn*.

For example, in Figure 26, $X = \{x_1, x_2, x_3, x_4\}$ together with the edges $a - x_1, a - x_2, a - x_3, a - x_4$ is a horn at a, and is denoted as $a - X$.

We denote the complete graph K_n together with m horns $X_1, X_2, \ldots, X_m$ by $K_n(m)$ where $a_1 - X_1, a_2 - X_2, \ldots, a_m - X_m$, $a_i \in V(K_n)$ and $0 \le m \le n$.

We note that $K_1(1)$, $K_2(1)$ and $K_2(0)$ are star graphs, $K_2(2)$ is a double star graph.

Theorem 6. The complete graph $K_n(1)$, $n \ge 3$ is realizable as the incomparability graph of a lattice.

Proof. Consider the complete graph K_n. Let X be a horn in K_n at the vertex a_n where $X = \{x_1, x_2, \ldots, x_m\}$ and let a_i, $i = 1, 2, \ldots, n$ be the vertices of K_n. The corresponding lattice is as shown in Figure 27.

Corollary 7. The complete graph $K_3(1)$ is realizable as the incomparability graph of a lattice.

Proof. Consider the complete graph K_3. Let a, b and c be the three vertices of K_3 and let X be horn at c. Let $X = \{x_1, x_2, \ldots, x_n\}$. The corresponding lattice is as shown in Figure 28.

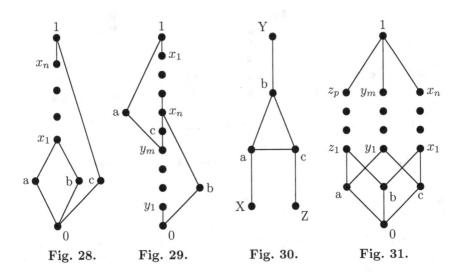

Fig. 28.   Fig. 29.   Fig. 30.   Fig. 31.

Lemma 8. The complete graph $K_3(2)$ is realizable as the incomparability graph of a lattice.

Proof. Consider the complete graph K_3. Let a, b and c be the three vertices of K_3 and let X and Y be horns at a and b respectively. Let $X = \{x_1, x_2, \ldots, x_n\}$ and $Y = \{y_1, y_2, \ldots, y_m\}$. The corresponding lattice is as shown in Figure 29.

Theorem 9. The complete graph $K_3(3)$ is not realizable as the incomparability graph of a lattice. However it is realizable as the zero-divisor graph of a lattice L.

Proof. Consider the complete graph K_3. Let a, b and c be the three vertices of K_3 and let X, Y and Z be horns at a, b and c respectively. Let $X = \{x_1, x_2, \ldots, x_n\}$, $Y = \{y_1, y_2, \ldots, y_m\}$ and $Z = \{z_1, z_2, \ldots, z_p\}$.

Case (i) Suppose L has two atoms a and b. Then $a \leq z_j$ for $j = 1, 2, \ldots, p$, $a \leq y_k$ for $k = 1, 2, \ldots, m$ and $b \leq x_i$ for $i = 1, 2, \ldots, n$, $b \leq z_j$ for $j = 1, 2, \ldots, p$.

Also we have $y_k \leq x_i$ or $x_i \leq y_k$.

If $y_k \leq x_i$, then $a \leq y_k$ implies $a \leq x_i$, a contradiction since a and x_i are adjacent.

If $x_i \leq y_k$, then $b \leq x_i$ implies $b \leq y_k$, a contradiction since b and y_k are adjacent.

Hence neither $y_k \leq x_i$ nor $x_i \leq y_k$, a contradiction since x_i and y_k are not adjacent.

Suppose a and x_1 are the two atoms in L. We have $a \leq z_j$ for $j = 1, 2, \ldots, p$, $a \leq y_k$ for $k = 1, 2, \ldots, m$ and $x_1 \leq x_i$ for $i = 2, \ldots, n$, $x_1 \leq z_j$ for $j = 1, 2, \ldots, p$, $x_1 \leq b$, $x_1 \leq c$, $x_1 \leq y_k$ for $k = 1, 2, \ldots, m$.

Also we have

(i) $y_k \leq c$ or $c \leq y_k$ (ii) $y_k \leq z_j$ or $z_j \leq y_k$ (iii) $b \leq z_j$ or $z_j \leq b$.

If $y_k \leq c$, then $a \leq y_k$ implies $a \leq c$, a contradiction since a and c are adjacent. Hence $c \leq y_k$.

If $y_k \leq z_j$, then $c \leq y_k$ implies $c \leq z_j$, a contradiction since c and z_j are adjacent. Hence $z_j \leq y_k$.

We have $b \leq z_j$ or $z_j \leq b$.

If $b \leq z_j$, then $z_j \leq y_k$ implies $b \leq y_k$, a contradiction since b and y_k are adjacent.

If $z_j \leq b$, then $a \leq z_j$ implies $a \leq b$, a contradiction since a and b are adjacent.

Hence neither $b \leq z_j$ nor $z_j \leq b$, a contradiction since b and z_j are not adjacent.

Case (ii) Suppose L has three atoms a, b and c. Then we have $a \leq z_j$, $a \leq y_k$, $b \leq x_i$, $b \leq z_j$, $c \leq x_i$ and $c \leq y_k$.

Also we have $y_k \leq x_i$ or $x_i \leq y_k$.

If $y_k \leq x_i$, then $a \leq y_k$ implies $a \leq x_i$, a contradiction since a and x_i are adjacent.

If $x_i \leq y_k$, then $b \leq x_i$ implies $b \leq y_k$, a contradiction since b and y_k are adjacent.

Hence neither $y_k \leq x_i$ nor $x_i \leq y_k$ since x_i and y_k are not adjacent. Hence $K_3(3)$ cannot be realized as $\Gamma'(L)$.

However it is realizable as the zero-divisor graph of a lattice L see Figure 31.

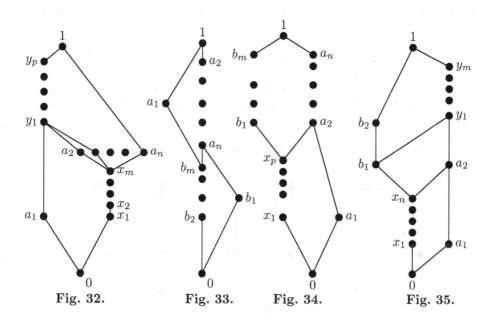

Fig. 32.   Fig. 33.   Fig. 34.   Fig. 35.

Theorem 10. The complete graph $K_n(2)$, $n \geq 4$ is realizable as the incomparability graph of a lattice.

Proof. Consider the complete graph K_n. Let a_i, $i = 1, 2, \ldots, n$ be the vertices of K_n and let X and Y be horns at a_1 and a_n respectively. Let $X = \{x_1, x_2, \ldots, x_m\}$ and $Y = \{y_1, y_2, \ldots, y_p\}$. The corresponding lattice is as shown in Figure 32.

Theorem 11. A double star graph is realizable as the incomparability graph of a lattice.

Proof. Let $G = \Gamma'(L)$ be a double star graph with centers a_1, b_1 and end vertices b_j, $j = 2, 3, \ldots, m$ and a_i, $i = 2, \ldots, n$. The corresponding lattice is as shown in Figure 33.

Next we discuss some Theorems for complete bipartite graphs with a horn. We denote the complete bipartite graph $K_{m,n}$ together with P horns by $K_{m,n}(P)$.

Remark 2. Let $K_{m,n}$ be the complete bipartite graph with partitions $V_1 = \{a_1, a_2, \ldots, a_n\}$ and $V_2 = \{b_1, b_2, \ldots, b_m\}$. Then by Theorem 5, $K_{m,n}$ is realizable as $\Gamma'(L)$. Since the a_i are non-adjacent in $\Gamma'(L)$, they are comparable in L. So we can arrange them as $a_1 < a_2 < a_3 < \ldots < a_n$. Similarly, we can arrange b_j as $b_1 < b_2 < \ldots < b_m$.

Using this Remark we have the following Theorems.

Theorem 13. $K_{2,2}(2)$ is realizable as $\Gamma'(L)$ if and only if both the horns are at vertices a_1 and b_2.

Proof. Consider the complete bipartite graph $K_{2,2}$. Let $V_1 = \{a_1, a_2\}$ and $V_2 = \{b_1, b_2\}$ be the two partitions. Let $X = \{x_1, x_2, \cdots, x_n\}$ and $Y = \{y_1, y_2, \cdots, y_m\}$ be the two horns. If the two horns are at a_1 and b_2 respectively as shown in Figure (i), then the corresponding lattice is as shown in Figure 35.

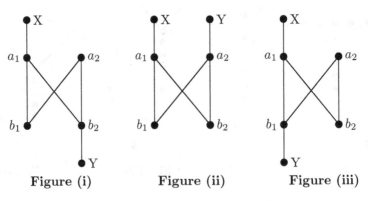

Figure (i)   Figure (ii)   Figure (iii)

Conversely, we consider the two cases.

Case (i) Suppose the horns X and Y are at a_1 and a_2 respectively, see Figure (ii) and let this graph be realizable as $\Gamma'(L)$ for some lattice L. Clearly L does not contain three atoms as $K_{2,2}$ does not contain a 3 - cycle.

Subcase (i) Suppose a_1 and b_1 are the two atoms.

Then $a_1 \leq a_2$, $a_1 \leq y_k$, $b_1 \leq y_k$ for each k and $b_1 \leq x_i$ for each i.

Also we have $a_2 \leq x_1$ or $x_1 \leq a_2$.

If $a_2 \leq x_1$, then $a_1 \leq a_2$ implies $a_1 \leq x_1$, a contradiction since a_1 and x_1 are adjacent.

If $x_1 \leq a_2$, then $b_1 \leq x_1$ implies $b_1 \leq a_2$, a contradiction since a_2 and b_1 are adjacent.

Hence neither $a_2 \leq x_1$ nor $x_1 \leq a_2$, a contradiction since x_1 and a_2 are not adjacent.

Subcase (ii) Suppose a_1 and x_1 are the two atoms. Then $a_1 \leq a_2$, $x_1 \leq b_1$, $x_1 \leq b_2$, $x_1 \leq a_2$, $a_1 \leq y_k$ and $x_1 \leq y_k$ for each k .

We know that, in a lattice the greatest lower bound of any nonempty finite subset of L exists. We now show that the greatest lower bound of $A = \{a_2, y_1, y_2, \ldots, y_m\}$ does not exist. The possible set of lower bounds of A is $\{0, a_1, x_1, \ldots, x_n\}$. If a_1 is the greatest lower bound, then $x_i \leq a_1$, a contradiction since a_1 is an atom.

If any x_i is the greatest lower bound then $a_1 \leq x_i$, a contradiction since $a_1 \parallel x_i$. Hence the greatest lower bound of A does not exist. So $K_{2,2}$ is not realizable as $\Gamma'(L)$ if both the horns are at vertices a_1 and a_2 respectively.

Case (ii) Suppose both the horns are at vertices a_1 and b_1 respectively see Figure (iii).

Subcase (i) Suppose a_1 and b_1 are the two atoms. Then by similar manner as in case (i) we get a contradiction.

Subcase (ii) a_1 and x_1 are the two atoms. Then $a_1 \leq a_2$, $x_1 \leq b_1$, $x_1 \leq b_2$, $x_1 \leq a_2$, $a_1 \leq y_k$ and $x_1 \leq y_k$ for each k .

By Remark 2 we have $b_1 \leq b_2$.

Also we have $b_2 \leq y_k$ or $y_k \leq b_2$.

If $b_2 \leq y_k$, then $b_1 \leq b_2$ implies $b_1 \leq y_k$, a contradiction since b_1 and y_k are adjacent.

If $y_k \leq b_2$, then $a_1 \leq y_k$ implies $a_1 \leq b_2$, a contradiction since a_1 and b_2 are adjacent.

Hence neither $b_2 \leq y_k$ nor $y_k \leq b_2$, a contradiction since b_2 and y_k are not adjacent.

Hence $K_{2,2}(2)$ is not realizable as $\Gamma'(L)$ if both the horns are at vertices a_1 and b_1 respectively.

Theorem 14. A complete bipartite graph with two horns, that is $K_{m,n}(2)$, $m > 2$ or $n > 2$ is realizable as $\Gamma'(L)$ for some lattice L if and only if the two horns are at vertices a_1, a_n or at vertices a_1, b_m.

Proof. Consider the complete bipartite graph $K_{m,n}$. Suppose, without loss of generality, $n > 2$. Let $V_1 = \{a_1, a_2, \ldots, a_n\}$ and $V_2 = \{b_1, b_2, \ldots, b_m\}$ be the two partitions. Let $X = \{x_1, x_2, \ldots, x_p\}$ and $Y = \{y_1, y_2, \ldots, y_r\}$ be the two horns. If the horns are at a_1 and a_n respectively then the corresponding lattice is shown in Figure 36. If the horns are at a_1 and b_m respectively then the corresponding lattice is shown in Figure 37.

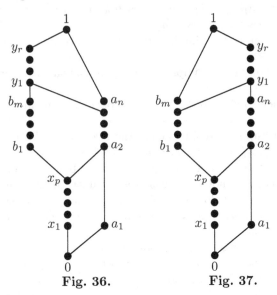

Fig. 36.   Fig. 37.

Conversely consider the complete bipartite graph $K_{m,n}$ and let both the horns be at vertices from the same partite set say V_1.

We have $V_1 = \{a_1, a_2, \ldots, a_n\}$ and $V_2 = \{b_1, b_2, \ldots, b_m\}$. Let X and Y be the two horns at a_1 and a_i, $i \neq n$ respectively where $X = \{x_1, x_2, \ldots, x_p\}$ and $Y = \{y_1, y_2, \ldots, y_r\}$. Let this graph be realizable as $\Gamma'(L)$ for some lattice L. Clearly L does not contain three atoms as $K_{m,n}$ does not contain a 3 - cycle.

Case (i) Suppose a_1 and b_1 are the two atoms. We have $a_1 \leq y_j$ for $j = 1, 2, \ldots, r$, $b_1 \leq y_j$ for $j = 1, 2, \ldots, r$ and $b_1 \leq x_l$, $l = 1, 2, \ldots, p$.

Also we have $a_2 \leq x_1$ or $x_1 \leq a_2$.

If $a_2 \leq x_1$, then $a_1 \leq a_2$ implies $a_1 \leq x_1$, a contradiction since a_1 and x_1 are adjacent.

If $x_1 \leq a_2$, then $b_1 \leq x_1$ implies $b_1 \leq a_2$, a contradiction since b_1 and a_2 are adjacent.

Hence neither $a_2 \leq x_1$ nor $x_1 \leq a_2$, a contradiction since a_2 and x_1 are not adjacent.

Case (ii) Suppose that a_1 and x_1 are the two atoms. Since $x_1, x_2, \ldots, x_p$ are comparable we can arrange them as $x_1 < x_2 < \ldots < x_p$. Similarly we have $y_1 < y_2 < \ldots < y_r, a_1 < a_2 < \ldots < a_n$ and $b_1 < b_2 < \ldots < b_m$. Now $x_k \leq y_1$ or $y_1 \leq x_k$ for each k. $y_1 \leq x_k$ then $a_1 \leq y_1$ implies $a_1 \leq x_k$, a contradiction. Hence $x_k \leq y_1$ for each k. Thus we have $x_1 < x_2 < \ldots < x_p < y_1 < y_2 < \ldots < y_r$.

Now $y_r \leq a_{i+1}$ or $a_{i+1} \leq y_r$.

If $a_{i+1} \leq y_r$ then $a_i \leq y_r$, a contradiction. Hence $y_r \leq a_{i+1}$. Thus we have the chain $x_1 < x_2 < \ldots < x_p < y_1 < y_2 < \ldots < y_r < a_{i+1} < \ldots < a_n$.

Now for $k \leq i-1$, either $y_j \leq a_k$ or $a_k \leq y_j$ for each j. If $y_j \leq a_k$ then $a_k \leq a_i$ implies $y_j \leq a_i$, a contradiction. Hence $a_k \leq y_j$.

Now since y_1, b_1 are not adjacent, we have $y_1 \leq b_1$ or $b_1 \leq y_1$. If $y_1 \leq b_1$ then $a_2 \leq y_1$ implies $a_2 \leq b_1$, a contradiction since a_2 and b_1 are adjacent.

If $b_1 \leq y_1$ then $y_1 \leq a_{i+1}$ implies $b_1 \leq a_{i+1}$, a contradiction since b_1 and a_{i+1} are adjacent. Hence neither $y_1 \leq b_1$ nor $b_1 \leq y_1$ and y_1 and b_1 are not adjacent, a contradiction.

Acknowledgment. The authors are thankful to the referees for many fruitful suggestions for improvement of the paper.

References

1. Afkhami, M., Khashyarmanesh, K.: The cozero divisor graph of a commutative ring. Southeast Asian Bull. Math. 35, 753–762 (2011)
2. Bollobas, B., Rival, I.: The maximal size of the covering graph of a lattice. Algebra Universalis 9, 371–373 (1979)
3. Bresar, B., Changat, M., Klavzar, S., Kovse, M., Mathews, J., Mathews, A.: Cover - incomparability graphs of posets. Order 25, 335–347 (2008)
4. Duffus, D., Rival, I.: Path lengths in the covering graph. Discrete Math. 19, 139–158 (1977)
5. Filipov, N.D.: Comparability graphs of partially ordered sets of different types. Collq. Maths. Soc. Janos Bolyai 33, 373–380 (1980)
6. Gedenova, E.: Lattices, whose covering graphs are s- graphs. Colloq. Math. Soc. Janos Bolyai 33, 407–435 (1980)
7. Grätzer, G.: General Lattice Theory. Birkhauser, Basel (1998)
8. Harary, F.: Graph Theory, Narosa, New Delhi (1988)
9. Nimbhorkar, S.K., Wasadikar, M.P., Demeyer, L.: Coloring of meet semilattices. Ars Combin. 84, 97–104 (2007)
10. Nimbhorkar, S.K., Wasadikar, M.P., Pawar, M.M.: Coloring of lattices. Math. Slovaca 60, 419–434 (2010)
11. Wasadikar, M., Survase, P.: Some properties of graphs derived from lattices. Bull. Calcutta Math. Soc. 104, 125–138 (2012)
12. Wasadikar, M., Survase, P.: Incomparability Graphs of Lattices. In: Balasubramaniam, P., Uthayakumar, R. (eds.) ICMMSC 2012. CCIS, vol. 283, pp. 78–85. Springer, Heidelberg (2012)
13. Wasadikar, M., Survase, P.: The zero-divisor graph of a meet-semilattice. J. Combinatorial Math. and Combinatorial Computing (accepted)
14. West, D.B.: Introduction to Graph Theory. Prentice-Hall, New Delhi (1996)

On Antimagic Labeling of Odd Regular Graphs

Tao-Ming Wang and Guang-Hui Zhang

Department of Applied Mathematics
Tunghai University
Taichung, Taiwan 40704, R.O.C
wang@go.thu.edu.tw

Abstract. An antimagic labeling of a finite simple undirected graph with q edges is a bijection from the set of edges to the set of integers $\{1, 2, \cdots, q\}$ such that the vertex sums are pairwise distinct, where the vertex sum at vertex u is the sum of labels of all edges incident to such vertex. A graph is called antimagic if it admits an antimagic labeling. It was conjectured by N. Hartsfield and G. Ringel in 1990 that all connected graphs besides K_2 are antimagic. Another weaker version of the conjecture is every regular graph is antimagic except K_2. Both conjectures remain unsettled so far. In this article, certain classes of regular graphs of odd degree with particular type of perfect matchings are shown to be antimagic. As a byproduct, all generalized Petersen graphs and some subclass of Cayley graphs of $\mathbb{Z}_n$ are antimagic.

Keywords: antimagic labeling, regular graph, perfect matching, 2-factor, generalized Petersen graph, Cayley graph, circulant graph.

1 Introduction

All graphs in this paper are finite simple, undirected, and without loops unless otherwise stated. In 1990, N. Hartsfield and G. Ringel [9] introduced the concepts called antimagic labeling and antimagic graphs.

Definition 1. *For a graph $G = (V, E)$ with q edges and without any isolated vertex, an antimagic edge labeling is a bijection $f : E \to \{1, 2, \cdots, q\}$, such that the induced vertex sum $f^+ : V \to \mathbb{Z}^+$ given by $f^+(u) = \sum\{f(uv) : uv \in E\}$ is injective. A graph is called antimagic if it admits an antimagic labeling.*

N. Hartsfield and G. Ringel showed that paths, cycles, complete graphs K_n ($n \geq 3$) are antimagic. They conjectured that all connected graphs besides K_2 are antimagic, which remains unsettled. In 2004 N. Alon et al [1] showed that the last conjecture is true for dense graphs. They showed that all graphs with $n(\geq 4)$ vertices and minimum degree $\Omega(\log n)$ are antimagic. They also proved that if G is a graph with $n(\geq 4)$ vertices and the maximum degree $\Delta(G) \geq n-2$, then G is antimagic and all complete partite graphs except K_2 are antimagic. In 2005, T.-M. Wang [15] studied antimagic labeling of sparse graphs, and showed that the toroidal grid graphs are antimagic. In 2008, T.-M. Wang et al. [16]

S. Arumugam and B. Smyth (Eds.): IWOCA 2012, LNCS 7643, pp. 162–168, 2012.

showed various types of graph products are antimagic. In 2009, D. Cranston [7] proved that all regular bipartite graphs are antimagic. While many various types of graphs have been shown to be antimagic [2,3,4,5,6,10,11,17,18], the question of antimagic-ness of regular graphs still remains open. In this paper, we consider the antimagic labeling of certain classes of regular graphs with perfect matchings. For more conjectures and open problems on antimagic graphs and related type of graph labeling problems, please see the dynamic survey article of J. Gallian [8].

2   Antimagic Labeling of 3-Regular Graphs

In 2000, M. Miller and M. Bača studied antimagic labelings of arithmetic type for generalized Petersen graphs [2], which are referred as (a, d)-antimagic labelings. Note that (a, d)-antimagic labelings are requiring all vertex sums form an arithmetic progression, hence also antimagic. M. Miller and M. Bača showed (a, d)-antimagic-ness of $GP(n, 2)$ for certain n, and also listed conjectures for other generalized Petersen graphs.

In this section we show the generalized Petersen graphs are antimagic by proving a more general theorem regarding 3-regular graphs with a particular type of perfect matchings, which contain generalized Petersen graphs as special cases. A r-factor of a graph is a r-regular spanning subgraph, and a 1-factor is a perfect matching. A factorization of a graph is a decomposition of the graph into union of factors so that the edge set is partitioned.

Theorem 2. *Let G be 3-regular with $2n$ vertices $\{u_1, u_2, \cdots, u_n, v_1, v_2, \cdots, v_n\}$ and $M = \{u_i v_i \mid 1 \leq i \leq n\}$ be a perfect matching of G. Assume additionally that $\{u_1, u_2, \cdots, u_n\}$ and $\{v_1, v_2, \cdots, v_n\}$ induce two 2-regular subgraphs of G respectively. Then G is antimagic.*

Proof. Let $G = M \bigoplus F = M \bigoplus (F_1 \cup F_2)$, where the 2-factor F is a disjoint union of two 2-regular subgraphs F_1 and F_2, each with n vertices. Let $V(F_1) = \{u_1, u_2, \cdots, u_n\}$ and $V(F_2) = \{v_1, v_2, \cdots, v_n\}$. Now we give an edge labeling f by the following steps. First we label the edges of M via $f(u_i v_i) = 3i$ for each $1 \leq i \leq n$. Then labeling over edges of $F = F_1 \cup F_2$ as follows. Since F_1 and F_2 are 2-regular graphs, we assign an orientation so that over each connected component (connected 2-cycle) the flow is either clockwise or counter-clockwise. We labeling over F by setting $f^{out}(w)$ and $f^{in}(w)$ respectively to be the outgoing edge label from the vertex w and the incoming edge label to the vertex w, according to the given orientation. Precisely we give the labeling as follows:

$$f^{out}(u_i) = 3n + 1 - 3i, \, f^{out}(v_i) = 3n + 2 - 3i$$

for each $1 \leq i \leq n$. We claim this labeling f is antimagic. Note that the vertex sum $f^+(u_i)$ for each vertex u_i is $f^+(u_i) = f^{out}(u_i) + f(u_i v_i) + f^{in}(u_i)$, which is $(3n + 1 - 3i) + (3i) + f^{in}(u_i) = 3n + 1 + f^{in}(u_i)$ for each $1 \leq i \leq n$. Also note that $f^{in}(u_i) = f^{out}(u_{k_i})$ for a unique k_i, where $1 \leq k_i \neq i \leq n$, and $\{1, \cdots, n\} = \{k_1, \cdots, k_n\}$. Therefore $f^+(u_i) = 3n + 1 + f^{in}(u_i) = 6n + 2 - 3k_i$

are pairwise distinct for $1 \leq i \leq n$. Similarly we obtain $f^+(v_i) = 3n + 2 + f^{in}(v_i)$ which are pairwise distinct for $1 \leq i \leq n$. Then f is antimagic since for each $1 \leq i \leq n$, we see $f^+(u_i) \equiv 2 \ (mod \ 3)$ and $f^+(v_i) \equiv 1 \ (mod \ 3)$.   ∎

Definition 3. *Let n, k be integers such that $n \geq 3$ and $1 \leq k \leq \lfloor \frac{n-1}{2} \rfloor$. The generalized Petersen graph $GP(n, k)$ is defined by $V(GP(n, k)) = \{u_i, v_i | 1 \leq i \leq n\}$, and $E(GP(n, k)) = \{u_i u_{i+1}, u_i v_i, v_i v_{i+k} | 1 \leq i \leq n\}$ where the subscripts are taken modulo n. (See Figure 1.) We call $u_1, u_2, \cdots, u_n$ an outer cycle, and $v_1, v_2, \cdots, v_n$ an inner cycle.*

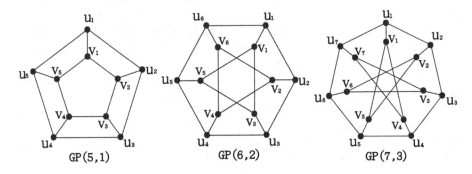

Fig. 1. Examples of generalized Petersen graphs

Note that all generalized Petersen graphs are 3-regular with $2n$ vertices, $3n$ edges, and admitting perfect matchings $\{u_i v_i | 1 \leq i \leq n\}$. Obviously $\{u_1, u_2, \cdots, u_n\}$ and $\{v_1, v_2, \cdots, v_n\}$ induce two 2-regular subgraphs respectively. Therefore, as a byproduct of the above Theorem 2:

Corollary 4. *Every generalized Petersen graph $GP(n, k)$ is antimagic.*

Example 5. *In the following Figure 2 antimagic labelings of $GP(5, 2)$ and $GP(6, 2)$ are given.*

3   Antimagic Labeling of Odd Regular Graphs

In this section, we extend previous Theorem 2 to a more general situation for regular graphs of odd degree. First we state a result we need here and also in later sections:

Theorem 6. (J. Petersen, 1891) *Let G be a $2r$-regular graph. Then there exists a 2-factor in G.*

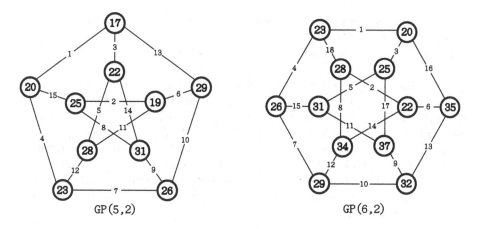

Fig. 2. GP(5,2) and GP(6,2) are antimagic

Notice that after removing edges of the 2-factor by the Petersen Theorem, we will get an even regular graph again and again. Thus an even regular graph has a 2-factorization.

Theorem 7. *Let $r \geq 1$ and let G be a $(2r + 1)$-regular graph with $2n$ vertices $\{u_1, u_2, \cdots, u_n, v_1, v_2, \cdots, v_n\}$ and $M = \{u_i v_i \mid 1 \leq i \leq n\}$ be a perfect matching of G. Assume additionally that $\{u_1, u_2, \cdots, u_n\}$ and $\{v_1, v_2, \cdots, v_n\}$ induce two $2r$-regular subgraphs of G respectively. Then G is antimagic.*

Proof. Let $G = M \bigoplus (F_1 \cup F_2)$, where F_1 and F_2 are two $2r$-regular subgraphs F_1 and F_2, each induced by n vertices, $\{u_1, u_2, \cdots, u_n\}$ and $\{v_1, v_2, \cdots, v_n\}$ respectively. Note that by Petersen's Theorem 6, F_1 and F_2 can be factored as unions of 2-factors, say $F_1 = F_1^1 \bigoplus F_1^2 \bigoplus \cdots \bigoplus F_1^r$ and $F_2 = F_2^1 \bigoplus F_2^2 \bigoplus \cdots \bigoplus F_2^r$ respectively, where F_1^j and F_2^k are 2-factors for each $1 \leq j \leq r$ and each $1 \leq k \leq r$ respectively.

Now we give an antimagic labeling f by the following steps. Note that G has $(2r+1)n$ edges. First we split all edge labels $1, 2, \cdots, (2r+1)n$ into $2r+1$ groups as follows: $\{1, 2, \cdots, n\}, \{n+1, n+2, \cdots, 2n\}, \cdots \cdots \{2rn+1, 2rn+2, \cdots, (2r+1)n\}$. Then we will put these groups of labels in order over the edges of $F_1^1, F_1^2, \cdots, F_1^r, F_2^1, F_2^2, \cdots, F_2^r$, and M respectively in below.

We define recursively that $G_k = M \bigoplus (F_1^1 \cup F_2^1) \bigoplus \cdots \bigoplus (F_1^k \cup F_2^k)$ for $1 \leq k \leq r$, and it is not hard to see $G = G_r$. Therefore $G_1 = M \bigoplus (F_1^1 \cup F_2^1)$, $G_2 = G_1 \bigoplus (F_1^2 \cup F_2^2), \cdots$, till $G_r = G_{r-1} \bigoplus (F_1^r \cup F_2^r) = G$. Since F_1^j and F_2^k are 2-factors for each $1 \leq j \leq r$ and each $1 \leq k \leq r$ respectively, as before we assign an orientation so that over each connected component (connected 2-cycle) the flow direction is either clockwise or counter-clockwise. We set $f_k^{out}(w)$ and $f_k^{in}(w)$ respectively, for each $1 \leq k \leq r$, to be the outgoing edge label over the 2-factor $(F_1^k \cup F_2^k)$ from the vertex w and the incoming edge label to the vertex w according to the given orientation. On the other hand, we denote $f^+(w)$ to be the induced vertex sum at the vertex w, and we use $f_k^+(w)$ to stand for the partial vertex sum at w for G_k for each $1 \le k \le r$. Then we may start labeling recursively over $G_1, G_2, \cdots, G_r = G$.

Precisely we give the labeling in the following steps:

Step 1: For $G_1 = M \bigoplus (F_1^1 \cup F_2^1)$: first for the edges of the perfect matching M we set $f(u_i v_i) = 2rn + i$ for each $1 \le i \le n$. Then over $(F_1^1 \cup F_2^1)$ we set $f_1^{out}(u_i) = 1 + (2r+1)n - f(u_i v_i)$ and $f_1^{out}(v_i) = rn + 1 + (2r+1)n - f(u_i v_i)$ respectively for each $1 \le i \le n$. Therefore $f_1^+(u_i) = f_1^{in}(u_i) + f(u_i v_i) + f_1^{out}(u_i) = 1 + (2r+1)n + f_1^{in}(u_i)$. Also note that $f_1^{out}(u_i) = f_1^{in}(u_j)$ for a unique j, where $1 \le j \ne i \le n$. Therefore $f_1^+(u_i) = 1 + (2r+1)n + f_1^{in}(u_i) = i + 1 + (2r+1)n$, for $1 \le i \le n$, which form a sequence of consecutive integers. Similarly $f_1^+(v_i) = 1 + (3r+1)n + f_1^{in}(v_i) = i + 1 + (3r+1)n$, for $1 \le i \le n$, which form a sequence of consecutive integers.

Step 2: For $G_2, G_3, ..., G_r$ we proceed recursively as follows: For $2 \le k \le r$, over $(F_1^k \cup F_2^k)$ we set $f_k^{out}(u_i) = (2r + 1 + k^2 - k)n + k - f_{k-1}^+(u_i v_i)$ and $f_k^{out}(v_i) = (2kr + r + k^2 - k + 1)n + k - f_{k-1}^+(v_i)$ respectively for each $1 \le i \le n$. Therefore $f_k^+(u_i) = f_k^{in}(u_i) + f_{k-1}^+(u_i) + f_k^{out}(u_i) = (2r + 1 + k^2 - k)n + k + f_k^{in}(u_i)$. Also note that $f_k^{out}(u_i) = f_k^{in}(u_j)$ for a unique j, where $1 \le j \ne i \le n$. Therefore $f_k^+(u_i) = i + k + (2r + k^2)n$ for $1 \le i \le n$, which form a sequence of consecutive integers. Similarly $f_k^+(v_i) = i + k + (2kr + 2r + k^2)n$ for $1 \le i \le n$, which form a sequence of consecutive integers.

Then this labeling f is antimagic, since the vertex sum at the vertex u_i is $f^+(u_i) = i + r + (2r + r^2)n$ for $1 \le i \le n$, and similarly $f^+(v_i) = i + r + (2r + 3r^2)n$ for $1 \le i \le n$, which shows that the vertex sums form a strictly monotone sequence $f^+(u_1) < f^+(u_2) < \cdots < f^+(u_n) < f^+(v_1) < f^+(v_2) < \cdots < f^+(v_n)$. ∎

To obtain more examples, we consider the Cayley graphs of $\mathbb{Z}_n$, which are also known as circulant graphs as follows:

Definition 8. *A circulant graph* $CIR_n(S)$ *with n vertices, with respect to $S \subset \{1, 2, \cdots, \lfloor \frac{n}{2} \rfloor\}$, is a graph with the vertex set $V(CIR_n(S)) = \{0, 1, 2, \cdots, n-1\}$, and the edge set is formed by the following rule:*

$$E(CIR_n(S)) = \{ij : i - j \equiv \pm s \pmod{n}, s \in S\}.$$

Note that the circulant graph $CIR_n(S)$ *is also called a Cayley graph of the finite cyclic group $\mathbb{Z}_n$ generated by S.*

Example 9. *Note that for $n \ge 5$, the circulant graphs $CIR_{2n}(\{a, b, n\})$ (where $0 < a \ne b < n$, n odd, and $\gcd(2n, a) = \gcd(2n, b) = 2$) are 5-regular graphs with perfect matchings, which satisfy the assumption in Theorem 7. Therefore $CIR_{2n}(\{a, b, n\})$ are antimagic. See Figure 3 for the example $CIR_{14}(\{4, 6, 7\})$.*

In a similar fashion, we may construct an infinite class of circulant graphs which represent the class of odd $(2r + 1)$-regular graphs, for each $r \ge 2$, with perfect matchings, as stated in Theorem 7.

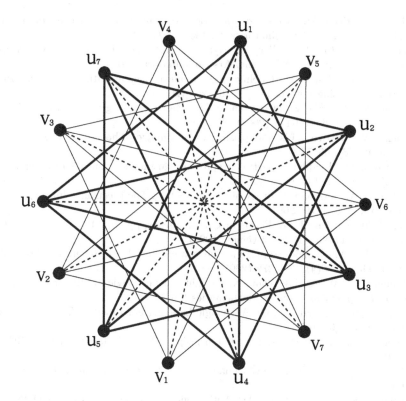

Fig. 3. Circulant graph $CIR_{14}(\{4,6,7\})$ (Cayley graph of $\mathbb{Z}_{14}$ generated by $\{4,6,7\}$)

4 Concluding Remark

In this article, we obtain antimagic labelings of a class of odd regular graphs with particular types of 1-factors, which contain the generalized Petersen graphs and certain circulant graphs as subclasses. Hopefully these results may be helpful to resolve more general situations regrading the conjecture that every regular graph except K_2 is antimagic, or helpful to resolve the Hartsfields-Ringel conjecture that every connected graph except K_2 is antimagic.

Acknowledgement. This research is partially supported by the National Center of Theoretical Sciences (NCTS) of Taiwan, via the CTS Fellowship. The first author wishes to express his most sincere thanks to the organization NCTS, which makes the presentation of the paper possible in IWOCA conference held on July 19-24, 2012, in Krishnankoil, Tamil Nadu, India.

References

1. Alon, N., Kaplan, G., Lev, A., Roditty, Y., Yuster, R.: Dense graphs are antimagic. Journal of Graph Theory 47(4), 297–309 (2004)

2. Miller, M., Bača, M.: Antimagic valuations of generalized Petersen graphs. Australasian Journal of Combinatorics 22, 135–139 (2000)
3. Bača, M., Jendrol, S., Miller, M., Ryan, J.: Antimagic Labelings of Generalized Petersen Graphs That Are Plane. Ars Combinatoria (2004)
4. Barrus, M.D.: Antimagic labeling and canonical decomposition of graphs. Information Processing Letters 110(7), 261–263 (2010)
5. Cheng, Y.: A new class of antimagic Cartesian product graphs. Discrete Mathematics 308(24), 6441–6448 (2008)
6. Cheng, Y.: Lattice grids and prisms are antimagic. Theoretical Computer Science 374(1V3), 66–73 (2007)
7. Cranston, D.: Regular bipartite graphs are antimagic. Journal of Graph Theory 60(3), 173–182 (2009)
8. Gallian, J.A.: A dynamic survey of graph labeling. The Electronic Journal of Combinatorics DS6, 1–79 (2001)
9. Hartsfield, N., Ringel, G.: Pearls in Graph Theory, pp. 108–109. Academic Press, Inc., Boston (1990) (Revised version 1994)
10. Hefetz, D.: Anti-magic graphs via the combinatorial nullstellensatz. Journal of Graph Theory 50(4), 263–272 (2005)
11. Huang, P.Y., Wong, T.L., Zhu, X.: Weighted-1-antimagic graphs of prime power order. To appear in Discrete Mathematics (2011)
12. Lee, M., Lin, C., Tsai, W.: On Antimagic Labeling For Power of Cycles. Ars Combinatoria 98, 161–165 (2011)
13. Petersen, J.: Die Theorie der regularen graphs. Acta Mathematica (15), 193–220 (1891)
14. Stewart, B.M.: Magic graphs. Canadian Journal of Mathematics 18, 1031–1059 (1966)
15. Wang, T.-M.: Toroidal Grids Are Anti-magic. In: Wang, L. (ed.) COCOON 2005. LNCS, vol. 3595, pp. 671–679. Springer, Heidelberg (2005)
16. Wang, T.M., Hsiao, C.C.: On Antimagic Labeling for Graph Products. Discrete Mathematics 308, 3624–3633 (2008)
17. Wong, T.L., Zhu, X.: Antimagic labelling of vertex weighted graphs. To appear in Journal of Graph Theory (2011)
18. Zhang, Y., Sun, X.: The antimagicness of the Cartesian product of graphs. Theoretical Computer Science 410, 727–735 (2009)

A Graph Theoretic Model to Solve the Approximate String Matching Problem Allowing for Translocations

Pritom Ahmed, A.S.M. Shohidull Islam, and M. Sohel Rahman

AℓEDA Group,
Department of CSE, BUET, Dhaka 1000, Bangladesh
{pritom.11,sohansayed}@gmail.com, msrahman@cse.buet.ac.bd

Abstract. In this paper, we visit the problem of approximate string matching allowing for translocations. We study the graph theoretic model proposed by [5] and extending the model, devise an efficient algorithm to solve the approximate string matching allowing for translocations. The resulting algorithm is an adaptation of the classic shift-and algorithm. For patterns having length similar to the word-size of the target machine, the algorithm runs in $O(n + mk^2)$ time for fixed length translocation where n, m and k are the length of the text, pattern and the translocation respectively.

1 Introduction

In text processing, approximate string matching is a fundamental problem which consists of finding inexact (as opposed to exact) matches of a pattern in a string. The accuracy of a match is measured in terms of the sum of the costs of the edit operations necessary to convert the string into an exact match.

Most biological string matching methods are based on *Levenshtein distance* [6] or on the *Damerau edit distance* [4]. The edit operations in the Levenshtein distance are deletion, insertion and substitution of characters, while the Damerau edit distance allows swaps of characters, i.e., transpositions of two adjacent characters. These distances assume that changes between strings occur locally, i.e., only a small portion of the string is involved in the *mutation event*. However, large scale changes are also possible. For example, large pieces of DNA can be moved from one location to an adjacent location (Translocations) where more than two characters may be involved.

1.1 Our Contribution

In this paper, we investigate the approximate string matching problem under a string distance whose edit operations are translocations of equal length factors. Our algorithm is based on a graph-theoretic approach which is an extension of

S. Arumugam and B. Smyth (Eds.): IWOCA 2012, LNCS 7643, pp. 169–181, 2012.

the work of Rahman & Illiopoulos [5]. Using the model, we devise an efficient algorithm to solve the approximate string matching problem allowing fixed length translocation. The resulting algorithm is an adaptation of the classic shift-and algorithm [3] and runs in $O(mk^2 + n)$ time if the pattern is similar in size to the size of the word in the target machine where n, m and k are the length of text, pattern and allowed translocation respectively. We also extend our work to handle the version when all length translocations are allowed. Notably, to the best of our knowledge the only attempt to solve the problem can be found in [2]. And our result is comparable to the result of [2].

1.2   RoadMap

The rest of the paper is organized as follows. In Section 2, we present some preliminary definitions. Section 3 presents our Graph-Theoretic model to solve the pattern matching problem allowing fixed length translocations. In Section 4, we present our algorithm to solve the problem. In Section 5, we analyse the time complexity of our algorithm. Finally, we briefly conclude in Section 7.

2   Preliminaries

A *string* is a sequence of zero or more symbols from an alphabet Σ. A string X of length n is denoted by $X[1..n] = X_1 X_2 \ldots X_n$, where $X_i \in \Sigma$ for $1 \leq i \leq n$. The *length* of X is denoted by $|X| = n$. A string w is called a *factor* of X if $X = uwv$ for $u, v \in \Sigma^*$; in this case, the string w occurs at position $|u| + 1$ in X. The factor w is denoted by $X[|u| + 1..|u| + |w|]$. A *k-factor* is a factor of length k. A *prefix (or suffix)* of X is a factor $X[x..y]$ such that $x = 1$ $(y = n)$, $1 \leq y \leq n$ $(1 \leq x \leq n)$. We define i-th prefix to be the prefix ending at position i i.e. $X[1..i], 1 \leq i \leq n$. On the other hand, i-th suffix is the suffix starting at position i i.e. $X[i..n], 1 \leq i \leq n$.

Definition 1. *Given two strings X and Y, the mutation distance, $md(X,Y)$, is based on the following operation:*

1. **Translocation:** *a factor of the form ZW is transformed into WZ, provided that $|Z| = |W| > 0$. The translocation size in this case is said to be $|W| = |Z|$.*

Each translocation operation is assigned unit cost. Mutation distance, $md(X,Y)$ gives the number of Translocation operation required to convert X into Y.

Notice that, by definition, the maximum length of the factors involved in a translocation is $\lfloor |X|/2 \rfloor$, where X is the string under consideration. Moreover, there are strings X, Y such that X can not be converted into Y by any sequence of translocations. In these cases, $md(X,Y) = \infty$. When $md(X,Y) < \infty$, we say that X *md*-matches Y and vice-versa, i.e., $X \Leftrightarrow_{md} Y$. If X *md*-matches with a factor $Y[i - |X| + 1 \ldots i]$ of Y, then we say X has an *md*-match at position i

of Y or alternatively, X has an md-occurrence at i of Y^1. If the translocation size is fixed (say, k) and $X \Leftrightarrow_{md} Y$ then we say that X k-md-matches Y, i.e., $X \Leftrightarrow_{k-md} Y$.

Problem "FLT" (Pattern Matching with Fixed Length Translocations).
Given a text $T = T_1 T_2 \ldots T_n$ where $|T| = n$ and a pattern $P = P_1 P_2 \ldots P_m$ where $|P| = m$, we want to find each location i such that P has a k-md-match in T at position i where, $m \leq i \leq n$ and k is the given size of translocation.

Problem "ALT" (Pattern Matching with All Length Translocations).
Given a text $T = T_1 T_2 \ldots T_n$ where $|T| = n$ and a pattern $P = P_1 P_2 \ldots P_m$ where $|P| = m$, we want to find each location i such that P has a k-md-match in T at location i where, $m \leq i \leq n$ and $k = 1, 2, \ldots, \lfloor n/2 \rfloor$ is the size of translocation.

Problem "MLT" (Pattern Matching with Multiple Length Translocations). *Given a text $T = T_1 T_2 \ldots T_n$ where $|T| = n$ and a pattern $P = P_1 P_2 \ldots P_m$ where $|P| = m$, we want to find each location i such that P has a k_j-md-match in T at location i where, $m \leq i \leq n$ and $k_j \in \{k_1, k_2, \ldots, k_q\}$ are the desired translocation sizes such that $k_j \leq \lfloor n/2 \rfloor$ for all $j \in \{1, \ldots, q\}$.*

We conclude this section with a definition of the degenerate string with an example.

Definition 2. *A string X is said to be degenerate, if it is built over the potential $2^{|\Sigma|} - 1$ non-empty sets of letters belonging to Σ.*

Example 1. Suppose we are considering DNA alphabet i.e. $\Sigma = \Sigma_{DNA} = \{A, C, T, G\}$. Then we have 15 non-empty sets of letters belonging to Σ_{DNA}. In what follows, the set containing A and T will be denoted by $[AT]$ and the singleton $[C]$ will be simply denoted by C for ease of reading.

3   A Graph-Theoretic Model for Pattern Matching with Fixed Length Translocation

In this section, we present a new graph-theoretic model to solve Problem FLT. Our model is inspired by the work of [5]. In our model, we view the text and the pattern as two separate graphs. We start with the following definition of a $\mathcal{T}$-graph borrowed from [5].

Definition 3. *Given a text $T = T_1 \ldots T_n$, a $\mathcal{T}$-graph, denoted by $T^G = (V^T, E^T)$, is a directed graph with n vertices and $n - 1$ edges such that $V^T = \{1, 2, \ldots n\}$ and $E^T = \{(i, i + 1) | 1 \leq i < n\}$. For each $i \in V^T$ we define $label(i) = T_i$ and for each edge $e \equiv (i, j) \in E_T$ we define $label(e) \equiv label((i, j)) \equiv (label(i), label(j)) = (T_i, T_j)$.*

---

[1] Note that, contrary to the usual practice of reporting the starting position as the occurrence, we have used the end point of a match as the occurrence.

Note that the labels in the above definition may not be unique. Also, we normally use the labels of the vertices and the edges to refer to them.

Now we define a $\mathcal{P}$-graph which is extended significantly from the model of [5] to handle our problems.

Definition 4. *Given a text* $P = P_1 \ldots P_m$, *and a translocation size* k, *a* $\mathcal{P}$-*graph, denoted by* $P^G = (V^P, E^P)$, *is a directed graph. The vertex set* V^P *can be partitioned into three disjoint vertex sets namely* $V^P_{down}, V^P_{middle}, V^P_{up}$ *such that* $-k \leq up \leq -1$, $1 \leq down \leq k$, *and* $middle = 0$. *The partition is defined in a* $(2k+1) \times m$ *matrix* M *as follows. For the sake of notational symmetry, we use* $M[up]$, $M[middle]$ *and* $M[down]$ *to denote respectively the rows* $M[-k] \ldots M[-1]$, $M[0]$ *and* $M[1] \ldots M[k]$ *of the matrix* M, *respectively (please see Figure 1). The following terms are required for the rest of the definitions:*

$$x = m - \left\lfloor \frac{m}{k} \right\rfloor \times k$$

$$y_{up} = \begin{cases} (m - x - up - 1) & \text{if } (m - x - up - 1) \geq m \text{ for } -k \leq up \leq -1, \\ (m - x - up - 1 - k) & \text{Otherwise.} \end{cases}$$

$$z_{down} = \begin{cases} (m - x + down - 1 - k) & \text{if } (m - x + down - 1) \geq m \text{ for } 1 \leq down \leq k, \\ (m - x + down - 1 - 2k) & \text{Otherwise.} \end{cases}$$

The vertex partitions are defined as follows:

1. $V^P_{up} = \{M[up, k-up], M[up, k-up+1], \ldots M[up, y_{up}]\}$ *where,* $-k \leq up \leq -1$.
2. $V^P_{middle} = \{M[middle, 0], M[middle, 1], \ldots M[middle, m]\}$ *where,* $middle = 0$.
3. $V^P_{down} = \{M[down, down], M[down, down + 1], \ldots M[down, z_{down}]\}$ *where,* $1 \leq down \leq k$.

The labels of the vertices are derived from P *as follows:*

1. *For each vertex* $M[up, i] \in V^P_{up}$ *where,* $-k \leq up \leq -1$ *and* $k - up \leq i \leq y_{up}$:

$$label(M[up, i]) = P_{i-k} \tag{1}$$

2. *For each vertex* $M[middle, i] \in V^P_{middle}$ *where,* $1 \leq i \leq m$ *and* $middle = 0$:

$$label(M[middle, i]) = P_i \tag{2}$$

3. *For each vertex* $M[down, i] \in V^P_{down}$ *where,* $1 \leq down \leq k$ *and* $down \leq i < z_{down}$:

$$label(M[down, i]) = P_{i+k} \tag{3}$$

The edge set E^P *is defined as the union of the sets* E^P_{up}, E^P_{middle} *and* E^P_{down} *as follows:*

1. $E^P_{up} = \{(M[up, i], M[-up, i + 1]) \mid (i + 1 + up) = kq\} \cup \{(M[up, i], M[middle, i+1]) \mid (i+1+up) = kq\} \cup \{(M[up, i], M[up, i+1]) \mid (i+1+up) \neq kq\}$, *where* $k - up \leq i \leq y_{up} - 1$, $q = 2, 3, \ldots$.

2. $E^P_{middle} = \{(M[0,i], M[0,i+1]) \mid 1 \le i \le m-1\} \cup \{((M[0,i], M[j,i+1]) \mid 1 \le i \le m-2k , j = (i \bmod k)+1\}$

3. $E^P_{down} = \{(M[down,i], M[-down,i+1]) \mid (i+1-down) = kq\} \cup \{(M[down,i]), M[down,i+1]) \mid (i+1-down) = kq\}$, where $down \le i \le z_{down}-1$, $q = 1, 2....$

The labels of the edges are derived from using the labels of the vertices in an obvious way.

Example 2. Suppose, $P = abbcdbcddacbbbaac$ and translocation size, $k = 3$. The corresponding $\mathcal{P}$-graph is shown in Figure 1. The row and column numbers are shown on the left and above respectively. V^P_{up} consists of the vertices of rows $-1, -2$ and -3, V^P_{middle} consists of the vertices of rows 0 and V^P_{down} consists of the vertices of row $1, 2$ and 3. The edges starting from vertices in V^P_{up}, V^P_{middle} and V^P_{down} are called E^P_{up}, E^P_{middle} and E^P_{down} respectively.

Lemma 1. *The graph P^G has at most $2mk + 2k + m - 4k^2$ vertices and at most $2mk + 3m - 4k - 4k^2 - 1$ edges. This will happen only when $m + 1 = kq$, where q (≥ 3) is a natural number.*

Proof. Due to space constraint the proof will be given in the journal version.

Definition 5. *Given a $\mathcal{P}$-graph P^G, a path $Q = u_1 \rightsquigarrow u_\ell = u_1 u_2 \ldots u_\ell$ is given by a sequence of consecutive directed edges $\langle (u_1, u_2), (u_2, u_3), \ldots (u_{\ell-1}, u_\ell) \rangle$ in P^G starting at node u_1 and ending at node u_ℓ. The label of the path $u_1 \rightsquigarrow u_\ell$ is $label(u_1 \rightsquigarrow u_\ell) = label(u_1)label(u_2) \ldots label(u_\ell)$. The length of the path Q, denoted by $len(Q)$, is the number of edges on the path and hence is $\ell - 1$ in this case. It is easy to note that the length of a longest path in P^G is $m - 1$.*

Definition 6. *If we have two strings X, Y and $X \leftrightarrow_{md} Y$. Then a factor of Y is referred to as a PSfactor of X. A k-**PSfactor** is a PSfactor of length $k + 1$ where, k is the size of translocation. The first edge of a path, $u_0 \rightsquigarrow u_k$ (in $\mathcal{P}$-graph) corresponding to a PSfactor is called an $\mathcal{F}$-**edge**.*

Definition 7. *If we have a path $u_1 \rightsquigarrow u_k$, where k is the size of the translocation and all of the vertices lie in the same row either in V^P_{up} or V^P_{down} (not in V^P_{middle}), then we call those k vertices a **Block**. There are $O(\frac{m}{k})$ blocks in each row of $M[up]$ and $M[down]$. Furthermore, consider a block represented by the path $u_1 \rightsquigarrow u_k$ with $col(u_k) = j$. Then, if $M[0, j + k - 1]$ has two outgoing edges, the Block is called a $\mathcal{C}$-**block**.*

Example 3. In Figure 1, on row -3, $\{b, c, d\}$ of column $6, 7, 8$ respectively forms a block. It is also a $\mathcal{C}$-block as $M[0, 8 + 3 - 1] = M[0, 10]$ has two outgoing edges one to $M[0, 11]$ another to $M[2, 11]$. On the other hand, $\{b, c, d\}$ on row -3, column $9, 10, 11$, is a block but not a $\mathcal{C}$-block.

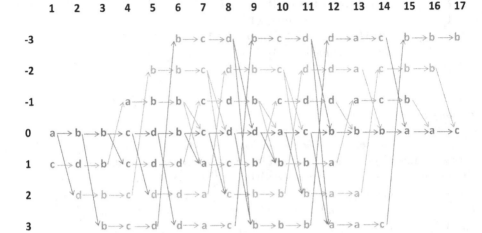

Fig. 1. $\mathcal{P}$-graph for pattern *abbcdbcddacbbbaac* and $k = 3$

Definition 8. *Given a $\mathcal{P}$-graph P^G and a $\mathcal{T}$-graph T^G for translocation size k, we say that P^G matches T^G at position $i \in [1..n]$ if and only if there exists a path $Q = u_1 u_2 \ldots u_m$ in P^G having $u_1 \in \{M[0,1], M[+1,1]\}$ and $u_m \in \{M[-k, m], M[0, m]\}$ such that for $j \in [1..m]$ we have $label(u_j) = T_{i-m+j}$.*

The above definitions set up our model to solve Problem FLT. The following Lemma presents the idea for the solution. Due to space constraint the proof of the lemma will be provided in the journal version.

Lemma 2. *Given a pattern P of length m, a text T of length n and an integer k as the translocation size, suppose P^G and T^G are the $\mathcal{P}$-graph and $\mathcal{T}$-graph of P and T, respectively. Then, P has k-md-occurrence at location $i \in [m, m+1 \ldots n]$ of T if and only if P^G matches T^G at position $i \in [m, m+1 \ldots n]$ of T^G.*

It is clear that the number of possible paths of length $m-1$ in P^G is exponential in m. So spelling all the paths and then performing a pattern matching against, possibly, an index of T is very time consuming unless m is a constant. We on the other hand exploit the above model in a different way and apply a modified version of the classic shift-and algorithm to solve the problem FLT.

4   Algorithms for Problem FLT

In this section, we use the model proposed in Section 3 to devise a novel and efficient algorithm for the approximate string matching allowing for fixed length translocations. We are using a modified version of Shift-And algorithm [3]. Due to space constraints we are not giving any description on the Shift-And algorithm. The reader unfamiliar with the Shift-And algorithm is referred to [3].

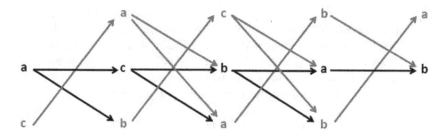

Fig. 2. $\mathcal{P}$-graph for pattern *acbab* with $k = 1$

The idea is described as follows. First of all, the shift-and algorithm can be extended easily for the degenerate patterns [1]. In our model for solving problem FLT, the pattern can be thought of having a set of letters at each position i as follows: $\tilde{P} = [M[-k, i], M[-k + 1, i], M[-k + 2, i] \ldots, M[0, i], \ldots, M[k - 2, i], M[k - 1, i], M[k, i]]$. Note that we have used $\tilde{P}$ instead of P above because, in our case, the sets of characters in the consecutive positions in the pattern P do not have the same relation as in a usual degenerate pattern. Particularly in our case, a match at position $i + 1$ of P will depend on the previous match of position i as the following example shows.

Example 4. Suppose, $P = acbab$ and $T = bcbaaabcba$ and $k = 1$. The $\mathcal{P}$-graph of P is shown in Figure 2. So, in line of above discussion, we can say that $\tilde{P} = [ac][acb][cba][ba][ab]$. Now, as can be easily seen, if we consider degenerate match, then $\tilde{P}$ matches T at position 2 and 6. However, $P \Leftrightarrow_{1-md} T$ only at position 6; not at position 2. To elaborate, note that at position 2, the match is due to c. So, according to the graph P^G the next match has to be an a and hence at position 2 we can not have a match.

In what follows, we present a novel technique to adapt the shift-and algorithm to tackle the above situation. We use the $\mathcal{P}$-graph as follows. For the sake of convenience, in the discussion that follows, we refer to both $\tilde{P}$ and the pattern P as though they were equivalent; but it will be clear from the context what we really mean. Suppose, we have a match up to position $i < m$ of $\tilde{P}$ in $T[j-i+1..j]$. Now we have to check whether there is a 'match' between T_{j+1} and P_{i+1}. For a simple degenerate match, we only need to check whether $T_{j+1} \in P_{i+1}$ or not. However, as the Example 4 shows, for our case we need to do more than that. What we do is as follows. Suppose that $T_j = c = M[z, i], -k \le z \le k$. Now, from the $\mathcal{P}$-graph we know which of the $M[r, i + 1], -k \le r \le k$ can follow $M[z, i]$. So, for example, even if $M[q, i + 1] = T_{j+1}$ we can't continue if there is no edge from $M[z, i]$ to $M[q, i + 1]$ in the $\mathcal{P}$-graph.

In what follows, we will use the following concept. Two edges (u, v), (x, y) of the $\mathcal{P}$-graph are said to be the 'same' if $label(u) = label(x)$ and $label(v) = label(y)$, i.e., if the two edges have the same labels. Also, given an edge $(u_0, u_1) \equiv (M[i_1, j_1], M[i_2, j_2])$ we say that edge (u_0, u_1) 'belongs to' column j_2, i.e., where the edge ends and we say $col((u_0, u_1)) = j_2$. Now we traverse all the edges and construct a set of k-PSfactors $\mathcal{S} = \{S_1 \ldots S_\ell\}$. We use the label of the path $u_1 \rightsquigarrow u_k$ which corresponds to a k-PSfactor S_i as the name of S_i. Now, we construct P-masks $P_{S_j}, 1 \leq j \leq \ell$ such that $P_{S_j}[g] = 1$, $1 \leq g \leq m$, if and only if, there is a k-PSfactor, S_j having $col(\mathcal{F}\text{-edge}) = g$.

With the P-masks at our hand, we now compute R_{j+1} as follows:

$$R_{j+1} = SHIFT(R_j) \ AND \ D_{T_{j+1}} \ AND \ P_{S_j}$$

Note that, to locate the appropriate P-mask, we need to perform a look up in the database using the factor $T_j T_{j+1} \ldots T_{j+k}$. However, we still have a problem to take care of as discussed in the following subsection.

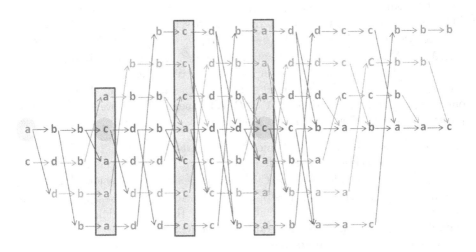

Fig. 3. The problem described in Section 4.1

4.1   The Problem

The problem occurs when the pattern has the following form [As shown in Figure 3]:

$\langle (P[i] = P[i+2k]) \wedge (P[i+k] = P[i+3k]) \rangle \bigvee \langle (P[i+2k] = P[i+4k]) \wedge (P[i+3k] = P[i+5k]) \rangle \bigvee \ldots \ldots$

In such cases, if $j = i \mod k$ then we can not distinguish between $((j+1) \mod k)$th row and $((j+2) \mod k)$th row of the $\mathcal{P}$-graph thus yielding wrong result which is shown in the following example.

Example 5. The algorithm described in Section 4 accepts '*adbabbccbaddabaac*' as a *3-md*-match string of '*abbcdbaddccbabaac*' which is shown in Figure 4.

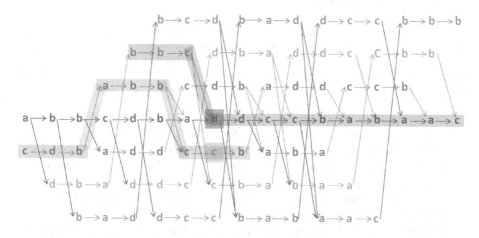

Fig. 4. The wrong result described in Section 4.1

Algorithm 1. Algorithm for Approximate String Matching Allowing for Fixed Length Translocation

---

1: $result \leftarrow 2^{pattern-size} - 1$
2: $checkup \leftarrow checkdown \leftarrow 0$
3: find **D-Mask** for T_0
4: $result \leftarrow result$ & **D-Mask**$_{value}$
5: $result \leftarrow result >> 1$
6: **for** $j = 0$ to $(n - k)$ **do**
7:     find **P-Mask, Up-mask, Down-Mask, Middle-Mask** for snippet s_j
8:     find **D-Mask** for T_{j+1}
9:     $result \leftarrow result$ & **P-Mask** & **D-Mask**
10:     $temp \leftarrow prevcheckup[j \mod k] >> k$
11:     $checkup \leftarrow checkup \mid upmask$
12:     $checkup \leftarrow checkup$ & $\sim downmask$ & $\sim middlemask$
13:     $prevcheckup[j \mod k] \leftarrow checkup$
14:     $result \leftarrow \sim (temp$ & $checkup)$ & $result$
15:     $temp \leftarrow prevcheckdown[j \mod k] >> k$
16:     $checkdown \leftarrow checkdown \mid downmask$
17:     $checkdown \leftarrow checkdown$ & $\sim upmask$
18:     $prevcheckdown[j \mod k] \leftarrow checkdown$
19:     $result \leftarrow \sim (temp$ & $checkdown)$ & $result$
20:     $x \leftarrow 2^{k-1}$
21:     **if** $(check$ & $x) = x$ && $((checkup$ & $x) < x)$ && $((checkdown$ & $x) < x)$ **then**
22:         Match found ending at position $(j + k - 1)$
23:     **end if**
24:     $result \leftarrow result >> 1$
25: **end for**

4.2 Fixing the problem

We start this section with the following definition that will be useful in the rest of this paper with k as the size of translocation.

Definition 9. *A* level change *means change of row in the Matrix M for one of the following cases :*

- *Ongoing Translocation, i.e., going up from a position (i_1, j) to $(i_2, j+1)$ where $(i_1 > middle)$ AND $(i_2 < middle)$ AND $(i_1 = -i_2)$.*
- *End of Previous & Beginning of new Translocation, i.e., going down from a position (i_1, j) to $(i_2, j+1)$ where $(i_1 < middle)$ AND $(i_2 > middle)$ AND $(i_1 = -i_2)$.*
- *A End of Translocation, i.e., going from a position (i_1, j) to $(i_2, j+1)$ where $(i_1 < middle)$ AND $(i_2 = middle)$.*
- *A Beginning of Translocation, i.e., going down from a position (i_1, j) to $(i_2, j+1)$ where $(i_1 = middle)$ AND $(i_2 > middle)$.*

In order to fix the problem mentioned in Section 4.1, we will introduce *three* new masks namely *up*, *down* and *middle* in our algorithm as follows.

1. This time we construct **up-masks**, $ups_j, 1 \leq j \leq \ell$ such that $ups_j[g] = 1$ if and only if, there is a k-PSfactor S_i with $\mathcal{F}$-edge $\equiv (M[i_1, j_1], M[i_2, g])$ having $i_1 > 0$ and $i_2 = -i_1$.
2. We construct **down-masks**, $downs_j, 1 \leq j \leq \ell$ such that $downs_j[g] = 1$ if and only if, there is a k-PSfactor S_i with $\mathcal{F}$-edge $\equiv (M[i_1, j_1], M[i_2, g])$ having $\langle (i_1 < 0$ and $i_2 = -i_1)$ or $(i_1 = 0$ and $i_2 > 0)\rangle$.
3. We construct **middle-masks**, $middles_j, 1 \leq j \leq \ell$ such that $middles_j[g] = 1$ if and only if, there is a k-PSfactor S_i with $\mathcal{F}$-edge $\equiv (M[i_1, j_1], M[i_2, g])$ having $\langle (i_1 < 0$ and $i_2 = 0)$ or $(i_1 = 0$ and $i_2 = 0)\rangle$.

We have to ensure that after a *level change* at a particular position (i, j), another *level change* must follow just after k positions, i.e., at the $(i, j + k)$th position in the Matrix M otherwise there can be no match. That is the reason why we have used these *three* masks. The detailed explanation is as follows:

1. If a *level change* has occurred and it is a **Beginning of Translocation** then we have to check whether another *level change* occurs after k positions and it has to be an **Ongoing Translocation**, otherwise there can be no match. So we do the following operations.
   At first, we perform the *not* operation on the *up-mask* (ups_{j+k}) then perform the *and* operation on the negated value and *down-mask* ($downs_j$). Then, we perform *not* operation on the resulting value. Finally, we perform *and* operation on the negated value and $result_{j+k}$.
2. If a *level change* has occurred and it is **Ongoing Translocation** then we have to check whether another *level change* occurs after k positions and it has to be an **End of Translocation** or **End of Previous & Beginning**

of new **Translocation**, otherwise there can be no match. So we do the following operations.

At first, we perform *not* operations on the *down-mask* ($down_{S_{j+k}}$) and *middle-mask* ($middle_{S_{j+k}}$) then perform *and* operation on the negated values and *up-mask* (up_{S_j}). Then, we perform *not* operation on the resulting value. Finally, we perform *and* operation on the negated value and $result_{j+k}$.

3. This process continues repeatedly until an **End of Translocation** occurs or end of pattern is encountered. So we do the following operations.

At first, we perform *not* operation on the *middle-mask* ($middle_{S_{j+k}}$) then perform *and* operation on the negated value and *up-mask* (up_{S_j}). Then, we perform *not* operation on the resulting value. Finally, we perform *and* operation on the negated value and $result_{j+k}$.

The algorithm is formally presented in Algorithm 1.

5   Algorithm Analysis

We use the following lemma for the analysis of the algorithm.

Lemma 3. *A Hash table can be used for the database so that insertion and searching of each PSfactor in database can be done with constant amortized cost for alphabet of fixed size.*

We analyse the algorithm in two parts.

Preprocessing Phase: We have to calculate the total number of k-PSfactors. Note that, the total number of k-PSfactors beginning at $V_{down}^{\mathcal{P}}$ is $= O(\frac{m}{k}) \times k \times k$ $= O(mk)$. The total number of k-PSfactors beginning at $V_{middle}^{\mathcal{P}}$ is $= O(m) \times O(k)$ $= O(mk)$. If $k \leq \frac{m}{6}$, then there is at least one $\mathcal{C}$-block in every $M[up]$ where, $-1 \leq up \leq -k$. In that case, the total number of k-PSfactors that are needed to be computed is $\frac{1}{2}(mk^2 - 5k^3)$. Thus total number of k-PSfactors is $O(mk^2)$ when $k \leq \frac{m}{6}$. Otherwise, The number is same as the for the case of $V_{up}^{\mathcal{P}}$, that is $O(mk)$. So we get the following lemma.

Lemma 4. *The time complexity of the preprocessing phase is $O((mk^2)(m/w))$ when $k \leq \frac{m}{6}$, otherwise $O((mk)(m/w))$. Here w is the target machine word size.*

Matching Phase: The retrieval of the mask for each factor of T from the database and other operations can be done in $O(1)$ i.e. constant in time, according to Lemma 3. Since there can be at most $O(n)$ number of factors of T having length $k + 1$, so the runtime complexity of the algorithm is $O(nm/w)$.

So, the worst case time complexity of the algorithm is $= O((mk^2 + n)(m/w))$. When $k > \frac{m}{6}$, we get a better time complexity $O((mk+n)(m/w))$. If the pattern size is similar to machine word size then we get $O(mk^2 + n)$ and when $k > \frac{m}{6}$, we get $O(mk + n)$.

6   Solution for Problem ALT and Problem FLT

We can solve Problem ALT and Problem MLT by extending the solution to Problem FLT. Due to space constraint, we could not give the full algorithm in this paper. Hence, we are presenting the results in the following theorems.

Theorem 1. *Problem ALT can be solved in $O((nk+mk^3)(m/w))$. If the pattern size is similar to the word size of the target machine then the time complexity will be $O(nk + mk^3)$.*

Theorem 2. *Problem MLT can be solved in $O((nq + \sum_{i=k_1}^{k_q} mi^2)(m/w))$. If the pattern size is similar to the word size of the target machine then the time complexity will be $O(nq + \sum_{i=k_1}^{k_q} mi^2)$.*

For this problem, for short patterns, the algorithm of [2] will run in $O(nr)$. Here $r = r_2 - r_1$ where r_2 and r_1 are respectively the *maximum* and *minimum* of the given translocation sizes i.e., $k_1, k_2, \ldots k_q$. As $q \leq r$, there is a clear improvement in our result.

7   Conclusion

In this paper, we have revisited the Pattern Matching with fixed length translocation, a variant of the classic pattern matching problem. We have extended the graph-theoretic approach [5] to model the problem. Then, using the model, we have devised an efficient algorithm to solve the problem. The resulting algorithm is an adaptation of the classic shift-and algorithm and runs in $O(n + mk^2)$ for fixed length translocation if the pattern-length is similar to the word-size in the target machine. Notably, the best known algorithm for pattern matching allowing for all length translocation runs in $O(nk)$ and uses dynamic programming approach [2]. This seems to be the first attempt to provide an efficient solution to the pattern matching problem for fixed length translocation without using dynamic programming approach. Moreover, the techniques used in our algorithm are quite simple and easy to implement. We believe that this new variant of the graph theoretic model could be used to devise more efficient algorithms and a similar approach can be taken to model similar other variants allowing operations like inversion, swap matching, of the classic pattern matching problem.

References

1. Baeza-Yates, R.A., Gonnet, G.H.: A new approach to text searching. Commun. ACM 35(10), 74–82 (1992)
2. Cantone, D., Faro, S., Giaquinta, E.: Approximate string matching allowing for inversions and translocations. In: Holub, J., Žďárek, J. (eds.), pp. 37–51 (2010)
3. Charras, C., Lecroq, T.: Handbook of Exact String Matching Algorithms. College Publications (2004)

4. Damerau, F.: A technique for computer detection and correction of spelling errors. Commun. ACM 7(3), 171–176 (1964)
5. Iliopoulos, C.S., Rahman, M.S.: A New Model to Solve the Swap Matching Problem and Efficient Algorithms for Short Patterns. In: Geffert, V., Karhumäki, J., Bertoni, A., Preneel, B., Návrat, P., Bieliková, M. (eds.) SOFSEM 2008. LNCS, vol. 4910, pp. 316–327. Springer, Heidelberg (2008)
6. Levenshtein, V.: Binary Codes Capable of Correcting Deletions, Insertions and Reversals. Soviet Physics Doklady 10, 707 (1966)

Deterministic Fuzzy Automata on Fuzzy Regular ω-Languages

R. Arulprakasam[1], V.R. Dare[2], and S. Gnanasekaran[3]

[1] Department of Mathematics, CK College of Engineering and Technology,
Cuddalore-607 003
[2] Department of Mathematics, Madras Christian College, Chennai-600 059
[3] Department of Mathematics , Periyar Arts College, Cuddalore-607 001
r.arulprakasam@yahoo.com,
rajkumardare@yahoo.com,
sargunam.g.sekaran@gmail.com

Abstract. In this paper, the concept of fuzzy local ω-language, Büchi fuzzy local ω-language are studied and we give some closure properties of fuzzy local ω-languages. We also establish relationship between deterministic fuzzy local automaton and fuzzy local ω-language. Further we show that every fuzzy regular ω-language is a projection of a Büchi fuzzy local ω-language.

Keywords: Local automaton, Local ω-language, Fuzzy set, Deterministic fuzzy automaton, Fuzzy regular ω-language.

1 Introduction

The notion of a fuzzy set was introduced by Zadeh [9] and it has application in many fields of sciences. Fuzzy regular language have many important applications including learning systems, pattern recognition, database theory, lexical analysis in programming language compilations and user-interface translations. Roughly speaking, in recent years their application have been further extended to include parallel processing, image generation, compression type theory for object-oriented languages and DNA computing, etc. Fuzzy automata was introduced by Wee in [8]. Fuzzy automaton provide a reliable formal basis for the theory of computing with words. Formal languages are precise while natural languages are quite imprecise. To reduce a gap between these two constructs, it becomes advantageous to introduce fuzziness into the structures of formal languages. This leads to the concept of fuzzy languages. In [8] Wee started the studies of fuzzy languages accepted by fuzzy automata. More recent development of algebraic theory of fuzzy automata and fuzzy languages can be found in book by Mordeson and Malik [6]. Berstel and Pin [2] have defined local automata and shown that a language is local if and only if it is recognized by a local automaton. Béal [1] gave a more general definition of local automata. Caron [3] has made use of equivalent definition in order to generalize the result stated by Berstel

S. Arumugam and B. Smyth (Eds.): IWOCA 2012, LNCS 7643, pp. 182–188, 2012.

and Pin [7]. D.S.Malik et al [5] and S.Gnanasekaran [4] studied the closure properties of fuzzy regular languages and fuzzy local languages on finitary case. The present paper is mainly concerned with the languages of infinite words (that is, ω-words) rather than finite words. The basic definitions of this paper are given in Section 2. In Section 3 we introduce two subclasses of fuzzy regular ω-languages that are fuzzy local ω-languages and Büchi fuzzy local ω-languages. Further, we give some closure properties of these classes of languages under intersection and union. In Section 4 deterministic fuzzy automaton acceptance condition on infinite words, fuzzy local automaton, are introduced and we characterize fuzzy local ω-languages by fuzzy automata. Finally, we show that every fuzzy regular ω-language is a projection of a Büchi fuzzy local ω-language.

2 Preliminaries

In this section, we recall some basic concepts on fuzzy set, local ω-languages, deterministic fuzzy automata.

2.1 Fuzzy Subset

Suppose that X is a universal set. A fuzzy set A, or rather a fuzzy subset A of X, is defined by a function assigning to each element x of X a value $A(x)$ in the real unit closed interval $[0, 1]$. Such a function is called a membership function, which is a generalization of the characteristic function associated to a crisp subset of X. The value $A(x)$ characterizes the degree of membership of x in A.

We denote by $\mathcal{F}(X)$ the set of all fuzzy subsets of X and by $\mathcal{P}(X)$ the power set of X. For any $A, B \in \mathcal{F}(X)$, we say that A is contained in B (or B contains A), denoted by $A \subseteq B$, if $A(x) \leq B(x)$ for all $x \in X$. We say that $A = B$ if and only if $A \subseteq B$ and $B \subseteq A$. A fuzzy set is said to be empty if its membership function is identically zero on X. We use ϕ to denote the empty fuzzy set.

For any family λ_i, $i \in I$, of elements of $[0, 1]$, we write $\vee_{i \in I} \lambda_i$ or $\vee \{\lambda_i \ | i \in I\}$ for the supremum of $\{\lambda_i \ | i \in I\}$, and $\wedge_{i \in I} \lambda_i$ or $\wedge \{\lambda_i \ | i \in I\}$ for the infimum. In particular, if I is finite, then $\vee_{i \in I} \lambda_i$ and $\wedge_{i \in I} \lambda_i$ are the greatest element and the least element of $\{\lambda_i \ | i \in I\}$ respectively. Given $A, B \in \mathcal{F}(X)$, the union of A and B, denoted $A \cup B$, is defined by the membership function $(A \cup B)(x) = A(x) \vee B(x)$ for all $x \in X$, the intersection of A and B, denoted $A \cap B$, is given by the membership function $(A \cap B)(x) = A(x) \wedge B(x)$ for all $x \in X$. Let $\lambda \in [0, 1]$ and $A \in \mathcal{F}(X)$. The scale product λA of λ and A is defined by $(\lambda A)(x) = \lambda \wedge A(x)$ for every $x \in X$, which is again a fuzzy subset of X.

2.2 Local ω-Languages

Let Σ be a finite alphabet and Σ^* be the set of all finite words over Σ. For each $u \in \Sigma^*$, we denote by $P_1(u)$, the prefix of u of length 1 and by $F_2(u)$, the set of all factors of u of length 2. We denote by $S_1(u)$, the suffix of u of length 1. An infinite word α over Σ is a function $\alpha : N \to \Sigma$ from the set N of all positive integers to Σ. We represent the infinite word α as $\alpha = a_1 a_2 \cdots$ where $\alpha(i) = a_i \in \Sigma$, for all i. We denote by Σ^ω, the set of all infinite words over Σ. For $\alpha \in \Sigma^\omega$, $inf_2(\alpha)$ denotes the set of all elements of $F_2(\alpha)$, each of which repeats infinite number of times in α. A language $L \subseteq \Sigma^\omega$ is called local if there exists a pair (I, C), where $I \subseteq \Sigma$ and $C \subseteq \Sigma^2$ such that $L = \{\alpha \in \Sigma^\omega : P_1(\alpha) \in I, F_2(\alpha) \subseteq C\}$.

2.3   Deterministic Fuzzy Automaton

A deterministic fuzzy automaton is a tuple $M = (Q, \Sigma, \delta, q_0, F)$ where Q is a finite non-empty set of states, Σ is a finite alphabet, $\delta : Q \times \Sigma \to Q$ is a transition function, $q_0 \in Q$ is the initial state and F is a fuzzy final subset of Q (that is, $F : Q \to [0, 1]$). The language accepted by M is the fuzzy subset $L(M)$ of Σ^*, defined by $L(M)(u) = F(\delta(q_0, u))$. A fuzzy language L is said to be a fuzzy regular language if there exists a deterministic fuzzy automaton M such that $L = L(M)$.

3   Fuzzy Local ω-Language

In this section we define fuzzy local ω-languages, Büchi fuzzy local ω-languages and study their properties.

Definition 31. *The pair $S = (\lambda_1, \lambda_2)$ is called a fuzzy local system if λ_1 is a fuzzy subset of Σ and λ_2 is a fuzzy subset of Σ^2. The fuzzy ω-language L over Σ whose membership function defined by $L(\alpha) = \lambda_1(P_1(\alpha)) \bigwedge (\bigwedge_{x \in F_2(\alpha)} \lambda_2(x))$, $\forall \alpha \in \Sigma^\omega$ is called the fuzzy ω-language generated by S and we write $L = L^\omega(S)$.*

Definition 32. *The fuzzy ω-language L over Σ is called a fuzzy local ω-language if $L = L_L^\omega(S)$ for some fuzzy local system S.*

Example 33. *Consider the fuzzy ω-language whose membership function is given by*

$$L(\alpha) = \begin{cases} 0.5 & \text{if } \alpha = (ab)^\omega, \\ 0 & \text{otherwise.} \end{cases}$$

Let us consider $S = (\lambda_1, \lambda_2)$, where

$$\lambda_1(x) = \begin{cases} 0.5 & \text{if } x = a, \\ 0 & \text{otherwise} \end{cases}$$

and

$$\lambda_2(x) = \begin{cases} 0.6 & \text{if } x = ab, \\ 0.5 & \text{if } x = ba, \\ 0 & \text{otherwise.} \end{cases}$$

Then $L = L^\omega(S)$ and therefore L is a fuzzy local ω-language.

Remark. The class of all local ω-languages is a proper subset of the class of all fuzzy local ω-languages.

Theorem 34. *If L_1 and L_2 are fuzzy local ω-languages over Σ, then $L_1 \cap L_2$ is a fuzzy local ω-language over Σ.*

Proof: If L_1 and L_2 are fuzzy local ω-languages, then $L_1 = L^{\omega}(S_1)$ for some fuzzy local system $S_1 = (\lambda'_1, \lambda'_2)$ and $L_2 = L^{\omega}(S_2)$ for some fuzzy local system $S_2 = (\lambda''_1, \lambda''_2)$. Consider the local system $S = (\lambda_1, \lambda_2)$ where $\lambda_1 = \lambda'_1 \wedge \lambda''_1$ and $\lambda_2 = \lambda'_2 \wedge \lambda''_2$. We show that $L^{\omega}(S) = L^{\omega}(S_1) \cap L^{\omega}(S_2) = L_1 \cap L_2$.
For $\alpha \in \Sigma^{\omega}$,

$$
\begin{aligned}
L^{\omega}(S)(\alpha) &= \lambda_1(P_1(\alpha)) \wedge \left(\wedge_{x \in F_2(\alpha)} \lambda_2(x) \right) \\
&= \left((\lambda'_1 \wedge \lambda''_1)(P_1(\alpha)) \right) \wedge \left(\wedge_{x \in F_2(\alpha)} (\lambda'_2 \wedge \lambda''_2)(x) \right) \\
&= \left(\lambda'_1(P_1(\alpha)) \wedge \lambda''_1(P_1(\alpha)) \right) \wedge \left(\wedge_{x \in F_2(\alpha)} (\lambda'_2(x) \wedge \lambda''_2(x)) \right) \\
&= \left(\lambda'_1(P_1(\alpha)) \wedge (\wedge_{x \in F_2(\alpha)} \lambda'_2(x)) \right) \wedge \left(\lambda''_2(P_1(\alpha)) \wedge (\wedge_{x \in F_2(\alpha)} \lambda''_2(x)) \right) \\
&= L^{\omega}(S_1)(\alpha) \wedge L^{\omega}(S_2)(\alpha) \\
&= L_1(\alpha) \wedge L_2(\alpha) \\
&= (L_1 \cap L_2)(\alpha)
\end{aligned}
$$

Thus $L^{\omega}(S) = L_1 \cap L_2$.

Therefore $L_1 \cap L_2$ is a fuzzy local ω - language.

Note that union of two fuzzy local ω-languages over Σ needs not be a fuzzy local ω-language.

Example 35. *Consider the fuzzy ω-languages L_1 and L_2 over $\Sigma = \{a, b\}$ whose membership functions are defined by*

$$
L_1(\alpha) = \begin{cases} 0.3 & \text{if } \alpha = a(bc)^{\omega}, \\ 0 & \text{otherwise.} \end{cases}
$$

$$
L_2(\alpha) = \begin{cases} 0.4 & \text{if } \alpha = a^{\omega}, \\ 0 & \text{otherwise.} \end{cases}
$$

and

$$
(L_1 \cup L_2)(\alpha) = \begin{cases} 0.4 & \text{if } \alpha = a^{\omega}, \\ 0.3 & \text{if } \alpha = a(bc)^{\omega}, \\ 0 & \text{otherwise.} \end{cases}
$$

If $L_1 \cup L_2$ is fuzzy local ω-language, then there exists a fuzzy local system $S = (\lambda_1, \lambda_2)$ such that $L_1 \cup L_2 = L^{\omega}(S)$. Here $\lambda_1(a), \lambda_2(aa), \lambda_2(ab), \lambda_2(bc), \lambda_2(cb)$ are all greater than zero and therefore $L_1 \cup L_2(a^n(bc)^{\omega}) \neq \phi$, $n \geq 1$. But $L_1(a^n(bc)^{\omega}) = 0$ and $L_2(a^n(bc)^{\omega}) = 0$ which is a contradiction.

Theorem 36. *If Σ_1 and Σ_2 are two disjoint subsets of the alphabet Σ whose union is Σ and if $L_1 \subseteq \Sigma_1^{\omega}$ and $L_2 \subseteq \Sigma_2^{\omega}$ are fuzzy local ω-languages, then $L_1 \cup L_2$ is also a fuzzy local ω-language.*

Proof: Since L_1 and L_2 are fuzzy local ω-languages over Σ_1^ω and Σ_2^ω, we have $L_1 = L^\omega(S_1)$ for some fuzzy local system $S_1 = (\lambda_1', \lambda_2')$ and $L_2 = L^\omega(S_2)$ for some fuzzy local system $S_2 = (\lambda_1'', \lambda_2'')$. Consider the local system $S = (\lambda_1, \lambda_2)$ where $\lambda_1 = \lambda_1' \vee \lambda_1''$ and $\lambda_2 = \lambda_2' \vee \lambda_2''$. Here L_1 and L_2 are defined on disjoint domains Σ_1^ω and Σ_2^ω, respectively. We can view them as having same domain Σ^ω by defining $L_1(\alpha) = 0$ for every $\alpha \in \Sigma^\omega - \Sigma_1^\omega$ and $L_2(\alpha) = 0$ for every $\alpha \in \Sigma^\omega - \Sigma_2^\omega$. We have to show that $L^\omega(S) = L^\omega(S_1) \cup L^\omega(S_2) = L_1 \cup L_2$. For $\alpha \in \Sigma^\omega$,

$$
\begin{aligned}
L^\omega(S)(\alpha) &= \lambda_1(P_1(\alpha)) \wedge (\wedge_{x \in F_2(\alpha)} \lambda_2(x)) \\
&= \left((\lambda_1' \vee \lambda_1'')(P_1(\alpha)) \right) \wedge \left(\wedge_{x \in F_2(\alpha)} (\lambda_2' \vee \lambda_2'')(x) \right) \\
&= \left(\lambda_1'(P_1(\alpha)) \vee \lambda_1''(P_1(\alpha)) \right) \wedge \left(\wedge_{x \in F_2(\alpha)} (\lambda_2'(x) \vee \lambda_2''(x)) \right) \\
&= \left(\lambda_1'(P_1(\alpha)) \wedge (\wedge_{x \in F_2(\alpha)} \lambda_2'(x)) \right) \vee \left(\lambda_1''(P_1(\alpha)) \wedge (\wedge_{x \in F_2(\alpha)} \lambda_2''(x)) \right) \\
&= L^\omega(S_1)(\alpha) \vee L^\omega(S_2)(\alpha) \\
&= L_1(\alpha) \vee L_2(\alpha) \\
&= (L_1 \cup L_2)(\alpha)
\end{aligned}
$$

Thus $L^\omega(S) = L_1 \cup L_2$.

Therefore $L_1 \cup L_2$ is a fuzzy local ω-language.

Definition 37. *A fuzzy ω-language L over Σ is called a Büchi fuzzy local ω-language if there exists a triple $S = (\lambda_1, \lambda_2, \lambda_3)$ where λ_1 is a fuzzy subset of Σ and λ_2, λ_3 are fuzzy subsets of Σ^2 such that $\lambda_3 \leq \lambda_2$ and whose membership function is $L(\alpha) = \lambda_1(P_1(\alpha)) \bigwedge (\wedge_{x \in F_2(\alpha)} \lambda_2(x)) \bigwedge (\vee_{x \in inf_2(\alpha)} \lambda_3(x)), \forall \alpha \in \Sigma^\omega$ and we write $L = L_B^\omega(S)$.*

Example 38. *Consider a fuzzy ω-language L whose membership function is given by*

$$
L(\alpha) = \begin{cases} 0.5 & \text{if } \alpha = a^+ b^\omega, \\ 0 & \text{otherwise.} \end{cases}
$$

Let us consider the fuzzy local system $S = (\lambda_1, \lambda_2, \lambda_3)$, where

$$
\lambda_1(x) = \begin{cases} 0.5 & \text{if } x = a, \\ 0 & \text{otherwise.} \end{cases}
$$

$$
\lambda_2(x) = \begin{cases} 0.6 & \text{if } x = ab, \\ 0.5 & \text{if } x \in \{aa, bb\}, \\ 0 & \text{otherwise.} \end{cases}
$$

and

$$
\lambda_3(x) = \begin{cases} 0.5 & \text{if } x = bb, \\ 0 & \text{otherwise.} \end{cases}
$$

Then $L = L^\omega(S)$ and therefore L is a Büchi fuzzy local ω-language.

Remark. The class of all fuzzy local ω-languages $\mathcal{L}_L^\omega$ is a subset of the class of all Büchi fuzzy local ω-languages $\mathcal{L}_B^\omega$.

Example 39. *The language L in example 3.8 is a Büchi fuzzy local ω-language. But L is not a fuzzy local ω-language, otherwise, $a^\omega \in L$. Therefore $\mathcal{L}_L^\omega \subset \mathcal{L}_B^\omega$.*

4   Fuzzy Local Automaton

In this section we define deterministic fuzzy automaton on infinite words, fuzzy local automaton and we characterize the fuzzy local ω-languages by fuzzy automaton.

Definition 41. *Let* $M = (Q, \Sigma, \delta, q_0, F)$ *be a deterministic fuzzy automaton. If* $\alpha = a_1 a_2 \cdots \in \Sigma^\omega$, *the sequence* $\rho = q_0 q_1 q_2 \cdots$ *of states from* Q *is called a run of* M *for* α, *if for each* $n \geq 1$, $q_n = \delta(q_{n-1}, a_n)$. *The range of* α, *denoted by* $ran(\rho)$, *is the set* $ran(\rho) = \{q_0, q_1, \cdots\}$. *The acceptance value of* ρ *on* α *is* $acc(\rho, \alpha)$ *where* $acc(\rho, \alpha) = \bigwedge_{q \in ran(\rho)} F(q)$. *The fuzzy* ω-*language* $L(M)$ *accepted by* M *is a fuzzy subset of* Σ^ω *with membership function defined by* $L^\omega(M)(\alpha) = acc(\rho, \alpha)$. *A fuzzy* ω-*language* L *is said to be a fuzzy regular* ω-*language if there exists a deterministic fuzzy automaton* M *such that* $L = L^\omega(M)$.

Definition 42. *A deterministic fuzzy automaton* $M = (Q, \Sigma, \delta, q_0, F)$ *is said to be local if for every* $a \in \Sigma$, *the set* $\{\delta(q, a) : q \in Q\}$ *contains at most one element.*

Theorem 43. $L \subseteq \Sigma^\omega$ *is a fuzzy local* ω-*language if and only if* L *is recognized by a fuzzy local automaton.*

Proof: Let L be a fuzzy local ω-languages, then there exists a fuzzy local system $S = (\lambda_1, \lambda_2)$ where λ_1 is a fuzzy subset of Σ and λ_2 is a fuzzy subset of Σ^2 such that $L^\omega(S) = \lambda_1(P_1(\alpha)) \bigwedge (\bigwedge_{x \in F_2(\alpha)} \lambda_2(x)), \forall \alpha \in \Sigma^\omega$. Consider the fuzzy finite automaton $M = (Q, \Sigma, \delta, q_0, F)$ where $Q = \{\{[\lambda]\} \cup \{[a] : a \in \Sigma\} \cup \{[u] : u \in \Sigma^2\}\}$, $q_0 = \{[\lambda]\}$, and δ is defined as follows: For all $a, b \in \Sigma$,

$$\delta([\lambda], a) = [a] \quad if \ \lambda_1(a) \neq 0,$$
$$\delta([a], b) = [ab] \quad if \ \lambda_2(ab) \neq 0 \ \text{ and}$$

For all $u = ab \in Q$ and $c \in \Sigma$,

$$\delta([ab], c) = [bc] \quad if \ \lambda_2(bc) \neq 0.$$

The fuzzy final state F is defined by

$$F([p]) = \begin{cases} \lambda_1(p) & \text{if } p \in \Sigma, \\ \lambda_2(p) & \text{if } p \in \Sigma^2. \end{cases}$$

Then $L = L^\omega(M)$.

Conversely, assume that $L \subset \Sigma^\omega$ is recognized by fuzzy local automaton $M = (Q, \Sigma, \delta, q_0, F)$. Consider the fuzzy local system $S = (\lambda_1, \lambda_2)$ where, for each $a \in \Sigma$,

$$\lambda_1(x) = \begin{cases} 1 & \text{if } \delta(q_0, a) \in Q, \\ 0 & \text{otherwise.} \end{cases}$$

and for each $u \in \Sigma^2$,

$$\lambda_2(u) = \begin{cases} F(\delta(q, u)) & \text{if } \delta(q, u) \in Q, \\ 0 & \text{otherwise.} \end{cases}$$

Then $L^\omega(M) = L^\omega(S)$.   •

Theorem 44. *Every fuzzy regular ω-language is a projection of a Büchi fuzzy local ω-language.*

Proof: Let L be a fuzzy regular ω-language. Let L be recognized by the fuzzy automaton $M = (Q, \Sigma, \delta, q_0, F)$. Let $\Gamma = Q \times \Sigma \times Q$. Let $\lambda_1 : \Gamma \longrightarrow [0, 1]$ be defined by

$$\lambda_1(q_1, a, q_2) = \begin{cases} 1 & \text{if } q_1 = q_0, \\ 0 & \text{otherwise.} \end{cases}$$

Let $\lambda_2 : \Gamma^2 \longrightarrow [0, 1]$ be defined by

$$\lambda_2((q_1, a, q_2)(q_3, a, q_4)) = \begin{cases} 1 & \text{if } q_2 = q_3, \\ 0 & \text{otherwise.} \end{cases}$$

Let $\lambda_3 : \Gamma^2 \longrightarrow [0, 1]$ be defined by

$$\lambda_3((q_1, a, q_2)(q_3, a, q_4)) = \begin{cases} F(q_4) & \text{if } q_2 = q_3, \\ 0 & \text{otherwise} \end{cases}.$$

Let $L_1 = L^{\omega}(S)$, where S is the fuzzy local system $S = (\lambda_1, \lambda_2, \lambda_3)$ over Σ. Then L_1 is a Büchi fuzzy local ω-language. Let the projection map $f : \Gamma \longrightarrow \Sigma$ be defined by $f(q_1, a, q_2) = a$. This map can be extended to Γ^{ω} as $f((q_1, a, q_2)(q_3, b, q_4) \ldots) = ab \ldots$. Then $f(L_1) = L$.

References

1. Béal, M.P.: Codes circulaires, automates locaux et entropie. Theoretical Compute Science 57, 283–302 (1988)
2. Berstel, J., Pin, J.-E.: Local languages and the Berry - Sethi algorithm. Theoretical Computer Science 155, 439–446 (1996)
3. Caron, P.: Families of locally testable languages. Theoretical Computer Science 242, 361–376 (2000)
4. Gnanasekaran, S.: Fuzzy local languages. International Mathematical Forum 5(44), 2149–2155 (2010)
5. Malik, D.S., Mordeson, J.N.: On Fuzzy regular languages. Information Sciences 88, 263–273 (1996)
6. Mordeson, J.N., Malik, D.S.: Fuzzy automata and languages. Chapman and Hall, CRC (2002)
7. Perrin, D., Pin, J.-E.: Infinite Words, Automata, Pure and Applied Mathematics, vol. 141. Elsevier (2004)
8. Wee, W.G., Fu, K.S.: A Formation of Fuzzy Automata and its application as a model of learning system. IEEE Transactions on in Systems Science and Cybernetics 5(3), 215–223 (1969)
9. Zadeh, L.A.: Fuzzy sets. Information and Control 8, 338–353 (1965)

Border Array for Structural Strings

Richard Beal and Donald Adjeroh

West Virginia University
Lane Department of Computer Science and Electrical Engineering,
Morgantown, WV 26506
r.beal@computer.org, don@csee.wvu.edu

Abstract. The *border* and parameterized border (*p-border*) arrays are
data structures used in pattern matching applications for traditional
strings from the constant alphabet Σ and parameterized strings (p-
strings) from the constant alphabet Σ and the parameter alphabet Π. In
this work, we introduce the structural border (*s-border*) array as defined
for an n-length structural string (s-string) T. The s-string is a p-string
with the existence of symbol complements in some alphabet Γ. These different alphabets add to both the intricacies and capabilities of pattern
matching. Initially, we provide a construction that executes in $O(n^2)$
time to build the *s-border* array. The paper establishes theory to improve the result to $O(n)$ by proving particular properties of the *s-border*
data structure. This result is significant because of the generalization of
the s-string, which is a step beyond the p-string. Using the same construction algorithm, we show how to modify the s-string alphabets to
also construct the *p-border* and the traditional *border* arrays in linear
time.

Keywords: structural matching, parameterized matching, structural
string, parameterized string, parameterized border, s-match, p-match,
s-string, p-string, p-border, border.

1 Introduction

The *border* array is a fundamental data structure in string theory used for pattern matching, classifying strings, etc. [1]. A parameterized string (p-string), as
identified by Baker [2], is a generalized string from the constant alphabet Σ and
the parameter alphabet Π. The parameterized border array (*p-border*) is the traditional *border* array problem observed in terms of p-strings [3,4,5]. Similarly,
p-border is also useful in parameterized pattern matching (p-matching), which
is a type of pattern matching where constant symbols $\sigma \in \Sigma$ match and there
exists a bijection between the parameter symbols $\pi \in \Pi$. Consider the example
p-strings that represent program statements $z=y*f/++y$; and $a=b*f/++b$;
over the alphabet sets $\Sigma = \{*, /, +, =, ;\}$ and $\Pi = \{a, b, f, y, z\}$. Here, a p-
match exists because constant symbols $\sigma \in \Sigma$ match and parameter symbols
$\pi \in \Pi$ properly align, namely in the first statement z, y, and f are consistently
substituted by a, b, and f respectively in the second statement. The p-match

S. Arumugam and B. Smyth (Eds.): IWOCA 2012, LNCS 7643, pp. 189–205, 2012.

problem offers a new way to address pattern matching in significant applications regarding the identification of plagiarism in academia and industry [6] and also, detecting unauthorized use of source code [7].

A variation of the p-match problem is known as structural matching (s-matching) between structural strings (s-strings) [8]. The s-string adds the notion of complementary pairs of parameter symbols in some alphabet Γ. Detecting an s-match requires identifying a p-match and ensuring that the parameter complements are consistent. For instance, consider the alphabets $\Sigma = \emptyset$, $\Pi = \{A, U, C, G\}$, and $\Gamma = \{(A, U), (C, G)\}$ and consider the sequences $S = UAUAU$ and $T = GCGCG$. Notice that where parameters U and A exist in S, there exist substitutions of parameters G and C respectively in T. Also, notice that where the complements U and A exist in S, the complement symbols G and C align in T. These observations identify that S and T s-match. This type of matching is relevant for analyzing biological data such as RNA sequences or secondary structures, since the complementary base pairing can be analyzed using the s-match [8]. Currently, the s-match problem is handled via structural suffix trees (s-suffix trees) [8]. In many situations, the huge practical space of an s-suffix tree poses a significant problem, which led to the development of the structural suffix array (s-suffix array) [9]. In this paper, we are motivated to introduce yet another significant data structure for the s-match problem: the structural border array.

Main Contributions: We introduce the structural border array ($s\text{-}border$) as defined for an n-length structural string (s-string) T. Initially, we provide constructions that execute in time $O(n^3)$ and $O(n^2)$ to build the $s\text{-}border$ array. The paper establishes theory to improve the result to $O(n)$ by proving particular properties of the $s\text{-}border$ data structure. Using the same construction algorithm, we show how to modify the s-string alphabets to also construct both the parameterized border ($p\text{-}border$) and the traditional $border$ array in linear time. Our solution to the $p\text{-}border$ problem is a symbol-based approach different from the automaton-oriented solution presented in [3]. The following formalizes our main results.

Theorem 2. *Given an n-length s-string T, there is an algorithm that constructs the s-border array $\mathcal{B}_s$ in $O(n)$ time.*

Theorem 3. *Given an n-length s-string T, the algorithm* construct$\mathcal{B}_s$ *constructs the p-border array $\mathcal{B}_p$ and the traditional border array $\mathcal{B}$ each in $O(n)$ time.*

2 Background

Baker [6] identifies three types of pattern matching: (1) exact matching, (2) parameterized matching (p-match), and (3) matching with modifications. The first p-match breakthroughs, namely, the `prev` encoding and the parameterized suffix tree (p-suffix tree), were introduced by Baker [2]. Additional improvements to the p-suffix tree are given in [10,11,12]. Like the traditional suffix tree [1,13,14], the p-suffix tree [2] implementation suffers from a large practical memory footprint.

One p-matching solution to address the space problem is the parameterized suffix array (p-suffix array) in [15,16]. An expected linear time p-suffix array construction is given in [17]. The work of [18] proves the existence of sub-quadratic and near-linear time worst case p-suffix array constructions. Other solutions that address the p-match problem without the space limitations of the p-suffix tree include the parameterized-KMP [19] and parameterized-BM [20], variants of traditional pattern matching approaches. These particular approaches use a variety of heuristics for shifting the matches to p-match efficiently. Further, the p-match problem is addressed via the Shift-OR mechanism in [21]. Idury et al. [3] studied a heuristic known as the `pfail` function to address the multiple p-match problem using the traditional Aho-Corasick automata. This `pfail` function is now known as the parameterized border array (*p-border*), analogous to the traditional *border* array [1], and has been studied in a variety of combinatorial problems in [4,5]. Other p-match data structures are studied in [22]. A closely related variant of the p-match problem is the structural pattern matching (s-match) problem, introduced by Shibuya [8]. The s-match is used in [8] for RNA analysis by a structural suffix tree (s-suffix tree). An s-suffix tree is similar in nature to the p-suffix tree [2] and constructed in similar time. The practical space used by the s-suffix tree was the motivation to introduce a more lightweight data structure known as the structural suffix array (s-suffix array) [9]. In this paper, we introduce the structural border array (*s-border*) for the s-match problem and provide a linear time construction. We show how to use our algorithm to also construct, in linear time, the *p-border* and the traditional *border* arrays.

3   Preliminaries

A string on an alphabet Σ is a production $T = T[1]T[2]...T[n]$ from Σ^n with $n = |T|$ the length of T. We will use the following string notations: $T[i]$ refers to the ith symbol of string T, $T[i...j]$ refers to the substring $T[i]T[i+1]...T[j]$, and $T[i...n]$ refers to the ith suffix of T: $T[i]T[i+1]...T[n]$. The m-length prefix of a suffix is the substring with the first m symbols of the suffix. The notation $S \circ T$ denotes the concatenation between the strings S and T. We divide the remainder of the preliminaries into parameterized string (p-string) theory, structural string (s-string) theory, and current *border* array definitions.

3.1   Parameterized Strings

The area of parameterized pattern matching defines the finite alphabets Σ and Π. Alphabet Σ denotes the set of constant symbols while Π represents the set of parameter symbols. We assume the use of indexed alphabets. Alphabets are defined such that $\Sigma \cap \Pi = \emptyset$. A terminal symbol $\$ \notin (\Sigma \cup \Pi)$ may be appended to a string for completeness or omitted for brevity.

Definition 1. Parameterized string (p-string): *A p-string is a production* T *of length* n *from* $(\Sigma \cup \Pi)^*\$$.

Consider the alphabet arrangements $\Sigma = \{A, B\}$ and $\Pi = \{w, x, y, z\}$. Example p-strings include $S = AxByABxy\$$, $T = AwBzABwz\$$, and $U = AyByAByy\$$.

Definition 2. ([15,16]) Parameterized matching (p-match): *A pair of p-strings S and T are p-matches with $n = |S|$ iff $|S| = |T|$ and each $1 \leq i \leq n$ corresponds to one of the following:*

1. $S[i], T[i] \in (\Sigma \cup \{\$\}) \wedge S[i] = T[i]$
2. $S[i], T[i] \in \Pi \wedge ((a) \vee (b))$ /* parameter bijection */
   (a) $S[i] \neq S[j], T[i] \neq T[j]$ *for any* $1 \leq j < i$
   (b) $S[i] = S[i - q]$ *iff* $T[i] = T[i - q]$ *for any* $1 \leq q < i$

In our example, we have a p-match between the p-strings S and T since every constant/terminal symbol matches and there exists a bijection of parameter symbols between S and T. U does not satisfy the parameter bijection to p-match with S or T. The process of p-matching leads to defining the prev encoding.

Definition 3. ([15,16]) Previous (prev) encoding: *Given $\mathbb{Z}$ as the set of non-negative integers, the function* prev $: (\Sigma \cup \Pi)^*\$ \rightarrow (\Sigma \cup \mathbb{Z})^*\$$ *accepts a p-string T of length n and produces a string Q of length n that (1) encodes constant/terminal symbols with the same symbol and (2) encodes parameters to point to **previous** like-parameters. More formally, Q is constructed of individual $Q[i]$ with $1 \leq i \leq n$ where:*

$$Q[i] = \begin{cases} T[i], \text{ if } T[i] \in (\Sigma \cup \{\$\}) \\ 0, \text{ if } T[i] \in \Pi \wedge T[i] \neq T[j] \text{ for any } 1 \leq j < i \\ i - k, \text{ if } T[i] \in \Pi \wedge k = \max\{j | T[i] = T[j], 1 \leq j < i\} \end{cases}$$

For a p-string T of length n, the above $O(n)$ space prev encoding demands the worst case construction time $O(n \log(\min\{n, |\Pi|\}))$, which follows from the discussions of Baker [2,20] and Amir et al. [19] on the dependency of alphabet Π in p-match applications. Note that with indexed alphabets and an auxiliary $O(|\Pi|)$ mapping structure, prev is constructed in linear time.

Lemma 1. *With an indexed alphabet, the construction of* prev(T) *requires $O(n)$ time where $n = |T|$.*

For a general n-length p-string T, we identify that while $T[i...n]$ is a traditional suffix, these $T[i...n]$ under the prev encoding, i.e. prev$(T[i...n])$, are known as parameterized suffixes or p-suffixes. In practice, prev may be represented by an array of characters or even integers, i.e., perhaps representing distances with negative integers, representing constants with positive integers, and representing the terminal with zero. Note that the convenient notation prev$(T)[j]$ is equivalent to $P[j]$ where $P = $ prev(T) and is similar for other like functions. It is often convenient to work with prev encodings as strings. Using Definition 3, our examples evaluate to prev$(S) = A0B0AB54\$$, prev$(T) = A0B0AB54\$$, prev$(U) = A0B2AB31\$$. The following proposition is fundamental to the p-matching problem.

Proposition 1. ([2]) *Two p-strings S and T p-match when* prev(S) = prev(T).

The example prev encodings show a p-match between S and T since prev(S) = $A0B0AB54\$$ and prev(T) = $A0B0AB54\$$. The encoding prev is supplemented by the encoding forw.

Definition 4. Forward (forw) encoding: *Let the function* rev(T) *reverse the p-string T and* repl(T, x, y) *replace all occurrences in T of the symbol x with y. We define the function* forw *for the p-string T of length n as* forw(T) = rev(repl(prev(rev(T)), 0, n)).

Our definition of the forw encoding generates output mirroring the fw encoding used by Deguchi et al. [15,16]. The forw encodings in our example with $n = 9$ are forw(S) = $A5B4AB99\$$, forw(T) = $A5B4AB99\$$, forw(U) = $A2B3AB19\$$.

3.2   Structural Strings

As an addendum to the p-string preliminaries, we present the following to formalize the theory of structural strings (s-strings). An s-string is an n-length p-string $T = T[1]T[2]...T[n]$ production from the constant symbol alphabet Σ and the parameter alphabet Π with $\Sigma \cap \Pi = \emptyset$. We terminate the string with a terminal $\$ \notin \Sigma \cup \Pi$. An s-string is a p-string with the added notion of *complementary* symbols, by which two symbols may uniquely correspond to one another. The notion that the s-string is a p-string allows us to apply the prev encoding, forw encoding, and the remaining p-string theory presented in this work. The s-string definition follows.

Definition 5. ([8]) **Structural string (s-string):** *An s-string is a p-string T of length n from $(\Sigma \cup \Pi)^*\$$. Pairs of the parameter symbols, say $(\pi_j, \pi_k) \subseteq \Pi = \{\pi_1, \pi_2, ..., \pi_{|\Pi|}\}$, may uniquely correspond to one another and behave as complements. We further define the alphabet $\Gamma = \{p_1, p_2, ..., p_{|\Gamma|}\}$ to represent the complements within Π as a set of pairs $p_i = (\pi_j, \pi_k)$ for some j and k such that only* complement(π_j) = π_k *and* complement(π_k) = π_j.

Consider the alphabet arrangements $\Sigma = \{A, B\}$, $\Pi = \{v, w, x, y, z\}$, and $\Gamma = \{(w, x), (y, z)\}$. Example s-strings include $S = AxBzzywv\$$, $T = AwByyzxv\$$, and $U = AwByyxzv\$$. Analyzing the complement symbols between two s-strings is the additional work in the structural match (s-match) beyond the parameter bijection observed in the p-match problem.

Definition 6. ([8]) **Structural matching (s-match):** *A pair of s-strings S and T are s-matches with $n = |S|$ iff $|S| = |T|$ and each $1 \le i \le n$ corresponds to one of the following:*

1. $S[i], T[i] \in (\Sigma \cup \{\$\}) \wedge S[i] = T[i]$
2. $S[i], T[i] \in \Pi \wedge ((a) \vee (b)) \wedge ((c) \vee (d))$ /*parameter bijection AND complement mapping*/
   (a) $S[i] \neq S[j], T[i] \neq T[j]$ for any $1 \leq j < i$
   (b) $S[i] = S[i - q]$ iff $T[i] = T[i - q]$ for any $1 \leq q < i$
   (c) $S[i] \neq \texttt{complement}(S[j]), T[i] \neq \texttt{complement}(T[j])$ for any $1 \leq j < i$
   (d) $S[i] = \texttt{complement}(S[i - q])$ iff $T[i] = \texttt{complement}(T[i - q])$ for any $1 \leq q < i$

In our working example, S and T s-match. The s-string U does not s-match with either S or T. The act of verifying Definition 6 between a pair of s-strings is quite involved. Shibuya [8] shows that we can use the p-string $\texttt{prev}$ encoding in Definition 3 and the $\texttt{compl}$ encoding in Definition 7 to assist in identifying an s-match.

Definition 7. ([8]) **Complement ($\texttt{compl}$) encoding:** *Given $\mathbb{Z}$ as the set of non-negative integers, the function $\texttt{compl} : (\Sigma \cup \Pi)^* \$ \to (\Sigma \cup \mathbb{Z})^* \$$ accepts an s-string T of length n and produces a string Q of length n that (1) encodes constant/terminal symbols with the same symbol and (2) encodes parameters to point to their* **previous** *complementary parameters. More formally, Q is constructed of individual $Q[i]$ with $1 \leq i \leq n$ where:*

$$Q[i] = \begin{cases} T[i], \text{ if } T[i] \in (\Sigma \cup \{\$\}) \\ 0, \text{ if } T[i] \in \Pi \wedge T[i] \neq \texttt{complement}(T[j]) \text{ for any } 1 \leq j < i \\ i - k, \text{ if } T[i] \in \Pi \wedge k = \max\{j \mid T[i] = \texttt{complement}(T[j]), 1 \leq j < i\} \end{cases}$$

We observe the similarity between $\texttt{compl}$ (of Definition 7) and $\texttt{prev}$ (of Definition 3) where $\texttt{compl}(T) = \texttt{prev}(T)$ with the mapping structure $(\pi, \pi) \in \Gamma \; \forall \; \pi \in \Pi$. In the worst case, since each $\pi \in \Pi$ is, by Definition 5, the complement of only one other symbol, then $|\Gamma| = |\Pi|$. Thus, the $\texttt{compl}$ encoding can be constructed using similar considerations as in $\texttt{prev}$ construction.

Lemma 2. *With an indexed alphabet, the construction of $\texttt{compl}(T)$ requires $O(n)$ time where $n = |T|$.*

In our working example, $\texttt{compl}(S) = \texttt{compl}(T) = A0B00150\$$ and $\texttt{compl}(U) = A0B00420\$$. Moreover, $\texttt{prev}(S) = \texttt{prev}(T) = \texttt{prev}(U) = A0B01000\$$. It is proven in [8] that combining the $\texttt{prev}$ and $\texttt{compl}$ encoding into the encoding $\texttt{sencode}$, as formalized in Definition 8, leads to identifying an s-match. This s-match scheme is presented in Proposition 2.

Definition 8. ([8]) **Structural encoding ($\texttt{sencode}$):** *Given $\mathbb{Z}$ as the set of non-negative integers, the function $\texttt{sencode} : (\Sigma \cup \Pi)^* \$ \to (\Sigma \cup \mathbb{Z} \cup \overline{\mathbb{Z}})^* \$$ accepts an s-string T of length n and produces a string Q of length n that (1) encodes constant/terminal symbols with the same symbol and either (2a) encodes parameters to point to an existing* **previous** *parameter or (2b) encodes remaining parameters to point to* **previous** *complementary parameter symbols. More formally, Q is constructed of individual $Q[i]$ with $1 \leq i \leq n$ where:*

$$Q[i] = \begin{cases} T[i], \text{ if } T[i] \in (\Sigma \cup \{\$\}) \\ \text{prev}(T)[i], \text{ if } \text{prev}(T)[i] > 0 \\ \overline{\text{compl}(T)[i]}, \text{ if } \text{compl}(T)[i] > 0 \ \wedge \ \text{prev}(T)[i] = 0 \\ 0, \textbf{ otherwise} \end{cases}$$

Proposition 2. ([8]) *Two s-strings S and T s-match when* $\text{sencode}(S) = \text{sencode}(T)$.

It follows from the construction of `prev` and `compl` in Lemma 1 and Lemma 2 that the construction of `sencode` (in Definition 8) is accomplished similarly.

Lemma 3. *With an indexed alphabet, the construction of* $\text{prev}(T)$, $\text{compl}(T)$, *and* $\text{sencode}(T)$ *requires* $O(n)$ *time where* $n = |T|$.

In the working example, $\text{sencode}(U) = A0B01\overline{42}0\$$ and $\text{sencode}(S) = \text{sencode}(T) = A0B01\overline{15}0\$$. Thus, S and T are confirmed to s-match.

For a general n-length s-string T, we identify that while $T[i...n]$ is a traditional suffix, these $T[i...n]$ under the `sencode` encoding, i.e. $\text{sencode}(T[i...n])$, are known as structural suffixes or s-suffixes. It is identified in [8] that the following function is used to obtain the symbol at j from the s-suffix at i.

Definition 9. ([8]) **structural-suffix (s-suffix) symbol retrieval:** *Given an n-length s-string T, let $prevT = \text{prev}(T)$, $complT = \text{compl}(T)$, and $\mathbb{Z}$ represent the set of non-negative integers. Further, let $i, j \in \mathbb{Z}$ such that $1 \le i \le n$ and $1 \le j \le (n-i+1)$. The function $\text{sencode} : (T, i, j) \to (\Sigma \cup \mathbb{Z} \cup \overline{\mathbb{Z}} \cup \{\$\})$ retrieves the symbol at j of the s-suffix $\text{sencode}(T[i...n])$, i.e. $\text{sencode}(T[i...n])[j]$.*

$$\text{sencode}(T, i, j) = \begin{cases} T[j + i - 1], \text{ if } T[j + i - 1] \in (\Sigma \cup \{\$\}) \\ prevT[j + i - 1], \text{ if } 0 < prevT[j + i - 1] < j \\ \overline{complT[j + i - 1]}, \text{ if } 0 < complT[j + i - 1] < j \ \wedge \ prevT[j + i - 1] = 0 \\ 0, \textbf{ otherwise} \end{cases}$$

The significance of $\text{sencode}(T, i, j)$ is that we can retrieve s-suffix symbols effortlessly in an algorithmic environment. However, this function still requires alphabet membership questions $x \in X$ even though the preprocessed $prevT = \text{prev}(T)$ and $complT = \text{compl}(T)$ have already answered these questions. Assuming that the symbol alphabets are logged during preprocessing, each call to $\text{sencode}(T, i, j)$ function executes in constant time.

Lemma 4. *Each call to* $\text{sencode}(T, i, j)$ *requires* $O(1)$ *time.*

3.3   Traditional *border* and Parameterized Border (*p-border*) Arrays

The *border* array, used for exact matching, maintains a list of the length of the longest border between a prefix of a traditional string W and a complete proper suffix of the prefix. More formally, the *border* array is defined as:

Definition 10. ([1]) **border array** (*border* or $\mathcal{B}$): *For an n-length traditional string W, the border array is defined for each index $1 \leq i \leq n$ such that $\mathcal{B}[1] = 0$ and otherwise $\mathcal{B}[i] = \max(\{0\} \cup \{k \mid W[1...k] = W[i-k+1...i], k \geq 1 \wedge i-k+1 > 1\})$.*

From the definition, we refer to the substrings $W[1...k]$ and $W[i-k+1...i]$ as borders. For the working example $W = AAABABAB\$$, $\mathcal{B} = \{0, 1, 2, 0, 1, 0, 1, 0, 0\}$.

The parameterized border (*p-border*) array, which was originally defined as the `pfail` function in [3], redefines the traditional *border* array for p-strings and the p-matching problem.

Definition 11. ([3]) **parameterized border array** (*p-border* or $\mathcal{B}_p$): *For an n-length p-string T, the p-border array is defined for each index $1 \leq i \leq n$ such that $\mathcal{B}_p[1] = 0$ and otherwise $\mathcal{B}_p[i] = \max(\{0\} \cup \{k \mid \text{prev}(T[1...k]) = \text{prev}(T[i-k+1...i]), k \geq 1 \wedge i-k+1 > 1\})$.*

We refer to the substrings $T[1...k]$ and $T[i-k+1...i]$ in the definition as borders. The borders under the encodings, i.e. $\text{prev}(T[1...k])$ and $\text{prev}(T[i-k+1...i])$, are referred to as *parameterized borders* or p-borders. The p-string $T = wzwz\$$ yields the array $\mathcal{B}_p = \{0, 1, 2, 3, 0\}$.

4   Structural Border Array

The traditional *border* array as defined in Definition 10 for traditional strings compares prefixes of a string, say W, with suffixes of those prefixes. Working with the individual symbols of the prefixes and suffixes of W is trivial because $W[j...n]$ is always a suffix of $W[1...n]$ for any j, $1 \leq j \leq n$. This trivial use of symbols is not the case with the parameterized border (*p-border*) of Definition 11. In the case of a p-string, the *p-border* identifies the maximum p-match between borders, in which these borders are under the `prev` encoding by Proposition 1. The challenge of working with the *p-border* is the dynamic nature of `prev`, which is fundamentally different from the suffixes in traditional *border* construction. Further, the *way* in which symbols are handled in the traditional *border* construction is not correct for *p-border* because of the following lemma, which is proven in [17,22].

Lemma 5. *Given a p-string T of length n, the suffixes of $\text{prev}(T)$ are not necessarily the p-suffixes of T, i.e. $\text{prev}(T[i...n]), 1 \leq i \leq n$. More formally, if $\pi \in \Pi$ occurs more than once in T, then $\exists i$, s.t. $\text{prev}(T[i...n]) \neq \text{prev}(T)[i...n], 1 \leq i \leq n$.*

A simple intuition of the previous lemma is that the encodings of a p-suffix under `prev` may change depending on *where* a parameter begins. We define the *s-border* array using the encoding `sencode`, which is the encoding that identifies a structural match (s-match) by Proposition 2.

Definition 12. structural border array (*s-border* or $\mathcal{B}_s$): *For an n-length s-string T, the s-border array is defined for each index $1 \leq i \leq n$ such that $\mathcal{B}_s[1] = 0$ and otherwise $\mathcal{B}_s[i] = \max(\{0\} \cup \{k \mid \text{sencode}(T[1...k]) = \text{sencode}(T[i-k+1...i]), k \geq 1 \wedge i-k+1 > 1\})$.*

The substrings $T[1...k]$ and $T[i - k + 1...i]$ in the definition are referred to as borders. When these borders are under the encoding sencode, they are known as *structural borders* or *s-borders*. Since *s-border* is defined on sencode, which from Definition 8 is a combination of symbols from the text T, prev(T), and compl(T), we encounter the same difficulties as *p-border*. The difference this time is that the *pair* of encodings prev and compl dynamically change depending on the locations of parameters in an s-suffix.

Lemma 6. *Given an s-string T of length n, the suffixes of* sencode(T) *are not necessarily the s-suffixes of T, i.e.* sencode($T[i...n]$).

Proof. Consider some s-string $T = \pi_1\pi_2\pi_2\pi_2\$$ over the alphabet of constants $\Sigma = \{\sigma_1\}$, the alphabet of parameters $\Pi = \{\pi_1, \pi_2, \pi_3\}$, and the alphabet of complements $\Gamma = \{(\pi_1, \pi_3)\}$. By Definition 8, it is the case that sencode(T) = prev(T) and hence, this lemma holds by Lemma 5.   □

4.1   Naïve Algorithm

Without investigating properties of our defined *s-border* array, we can still compute the data structure in a naïve way as shown in Algorithm 1. For an n-length s-string T, the algorithm computes *s-border* in roughly $O(n^3)$ time, since the $O(n)$ sencode construction by Lemma 3 is nested within a **while** loop that is bounded by n iterations and this $O(n^2)$ computation is nested within a **for** loop with n iterations. This particular algorithm makes the case for just how combinatorially difficult it can be to compute *s-border*.

Algorithm 1. Naïve construction of $\mathcal{B}_s$

```
1  int [] construct_Bs_naive(char T[n]){
2      int i,k,m,Bs[n]={0,0,...,0}
3      for i=2 to n {
4          k=1, m=0
5          while(i−k+1>1) {
6              if(sencode(T[1...k])=sencode(T[i−k+1...i]))  m=k
7              k++
8          }Bs[i]=m
9      }return Bs }
```

By Lemma 6, the notion that suffixes of sencode($T[1...n]$) are not the s-suffixes of T creates a challenge to correctly retrieve the symbol at j from the s-suffix at i. Shibuya [8] identifies the function sencode(T, i, j) to accomplish this, which we display in Definition 9. However, the problem of *implementing* the function is the added time to answer many symbol alphabet membership questions, i.e. $T[q] \in \Sigma$, $T[q] \in \Pi$, and $T[q] \in \{\$\}$. We define the following α encoding to quickly assist in these types of questions.

Definition 13. alphabet (α) encoding: *Let the set of integer constants be* $\{SIGMA = 0, PI = 1, TERM = 2\}$. *The encoding α is constructed of individual i with $1 \leq i \leq n$ where:*

$$\alpha(T)[i] = \begin{cases} SIGMA, \text{ if } T[i] \in \Sigma \\ PI, \text{ if } T[i] \in \Pi \\ TERM, \text{ if } T[i] \in \{\$\} \end{cases}$$

The α encoding is designed to store the alphabet type of each symbol in T for future access. Consider the variable $\alpha T = \alpha(T)$. Then, we can simply answer the question $T[q] \in \Pi$ by comparing the integers $\alpha T[q]$ and PI. Since α can be constructed simultaneously with **sencode**, then the time bound of Lemma 3 also holds for α.

Lemma 7. *With an indexed alphabet, the construction of $\alpha(T)$ requires $O(n)$ time where $n = |T|$.*

The previous formulation permits us to introduce the s-match related functions Ψ and ψ in Algorithm 2. Note that the Ψ and ψ functions correctly implement s-matching by incorporating elements of Definition 6, Proposition 2, Definition 9, and Definition 13. Specifically, function $\psi(a, b, j)$ compares the symbols $T[a]$ and $T[b]$ as they occur in s-suffixes at symbol j, where **true** is returned when the symbols s-match. This comparison is accomplished in constant time.

Lemma 8. *Each call to ψ executes in $O(1)$ time.*

The $\Psi(i, j, k)$ function, which uses a sequence of constant time calls to ψ, returns the number of symbols m such that $\mathtt{sencode}(T[i + k - 1...i + k + m - 2]) = \mathtt{sencode}(T[j + k - 1...j + k + m - 2])$.

Lemma 9. *Each call to Ψ executes in $O(m)$ time, where m is the length of the current s-match.*

We emphasize the inability to obtain a correct *s-border* solution by trivially plugging an s-string into a *border* or *p-border* construction algorithm. To put

Algorithm 2a. s-matching function Ψ **Algorithm 2b.** s-matching function ψ

```
1   char T[n]  /* given */              boolean ψ(int a,int b,int j){
2   char prevT[n]=prev(T)                  boolean match=false
3   char complT[n]=compl(T)                if(1≤a≤n ∧ 1≤b≤n ∧ αT[a]=αT[b]) {
4   char αT[n]=α(T)                           if(αT[a]=SIGMA∧T[a]=T[b]) match=true
5                                             else if(αT[a]=PI){ /* distances */
6   int Ψ(int i,int j,int k){                   if(prevT[a]=prevT[b]=0) match=true
7     int a,b,m=-1,q=k-1                        else if(prevT[a]<j ∧ prevT[b]<j ∧
8     do{                                          prevT[a]=prevT[b]) match=true
9         q++                                   else if(prevT[a]≥j∧prevT[b]≥j) match=true
10        a=i+q-1,b=j+q-1                        else if(complT[a]<j ∧ complT[b]<j ∧
11        m++                                       complT[a]=complT[b]) match=true
12    }while(ψ(a,b,q));                          else if(complT[a]≥j∧complT[b]≥j) match=true
13    return m }                             }
14                                        }return match }
```

this problem into perspective for an s-string T, let $U = \mathsf{sencode}(T[1...n]) \circ \$_1 \circ \mathsf{sencode}(T[2...n]) \circ \$_2 \circ \mathsf{sencode}(T[3...n]) \circ \$_3 \circ ... \circ \$_{n-1} \circ \mathsf{sencode}(T[n])$ where $\{\$_1, \$_2, ..., \$_{n-1}\}$ is the set of unique terminal symbols where $\$_i \notin \{\Sigma \cup \Pi \cup \{\$\}\}$. Notice that U contains each s-suffix and thus, Lemma 6 does not apply. Let $\mathcal{B} = \mathsf{border}(U)$ compute the traditional *border* array $\mathcal{B}$ for text U. Since U clearly represents each s-suffix, the resulting array $\mathcal{B}$ contains the correct results for $\mathcal{B}_s$ within the multitude of elements in $\mathcal{B}$. However, the problems with this approach are (1) computing the *s-border* elements $\mathcal{B}_s[i]$ will require us to postprocess the resulting $\mathcal{B}$ to find the maximum border ending at symbol i in the original T and (2) the construction of $\mathcal{B}_s$ can do no better than the length of U, which is of length $O(n^2)$. Note that the *s-border* array only has n elements. From the previous example with running time $O(n^2)$ and the naïve $O(n^3)$ approach in Algorithm 1, we are motivated to further investigate properties of *s-border*. This leads to an improved algorithm for constructing the *s-border*.

4.2   Improved Algorithm

A key property used to construct the traditional *border* array $\mathcal{B}$ is the property that $\mathcal{B}[i+1] \le \mathcal{B}[i] + 1$. This property helps progress matches by oracling the previous array element and comparing the subsequent symbols. We prove that even though the s-suffixes change from Lemma 6, this property still holds when considering the *s-border* array, which is defined on prefixes of suffixes under $\mathsf{sencode}$. That is, the *way* in which the $\mathsf{sencode}$ is defined, which is a combination of prev and compl, does not invalidate this traditional *border* property. In Fig. 1, we illustrate that a prefix named *prefix* of an s-suffix is such that $prefix = \mathsf{sencode}(T)[1...|prefix|]$. More specifically, the way in which distances refer previously in the text allows us to treat *prefix* as a *valid* encoding itself. This is exactly what is needed for the *s-border* construction. Such is not true for encodings like forw from Definition 4. For construction algorithms, we emphasize that one must always consider the impact of the encoding scheme (see [18,22]).

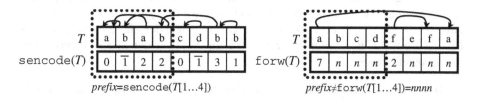

Fig. 1. Displaying the intricacies of prefixes for s-string encoded suffixes using $\Sigma = \emptyset$, $\Pi = \{a, b, c, d, e, f\}$, and $\Gamma = \{(a, b), (c, d)\}$, where n is fixed: $n = 8$

Lemma 10. *Given an s-string T of length n, the individual s-border elements $\mathcal{B}_s[i]$ are such that $\mathcal{B}_s[i+1] \le \mathcal{B}_s[i] + 1 \; \forall \; i, 1 \le i < n$.*

Proof. Initially, $\mathcal{B}_s[1] = 0$ by Definition 12. Consider that $\mathcal{B}_s[i] = k$ for some $i, 1 < k < n$. Without loss of generality, assume that $k > 2$. Then, by Definition 12, $\mathtt{sencode}(T[1...k]) = \mathtt{sencode}(T[i - k + 1...i])$ is the maximum s-border of $T[1...i]$. Consider $j = i + 1$. What we know is that (1) already $\mathtt{sencode}(T[1...k]) = \mathtt{sencode}(T[i - k + 1...i])$ and (2) $\mathtt{prev}$ (from Definition 3) and $\mathtt{compl}$ (from Definition 7) are defined such that the dynamically changing elements $T[i] \in \Pi$ of Lemma 6 are encoded to point to *previous* elements in the s-suffix, which means that appending elements to the s-suffix does not modify the encodings of the already existing s-suffix (see Fig. 1). From (2) and Definition 12, there cannot exist any s-border of $T[1...j]$ longer than one symbol beyond the maximum s-border at i, that is, the s-match of length $\mathcal{B}_s[i] + 1$. Let $a = \mathtt{sencode}(T[1...1 + k])[k + 1]$ and $b = \mathtt{sencode}(T[j - k...j])[k + 1]$. It follows that if $a = b$, then also $\mathtt{sencode}(T[1...1 + k]) = \mathtt{sencode}(T[j - k...j])$. Thus, $\mathcal{B}_s[j] = k + 1$. If $a \neq b$, then it follows that $0 \leq \mathcal{B}_s[j] \leq \mathcal{B}_s[i] = k$. Therefore, $\mathcal{B}_s[j] \leq k + 1$. ☐

The previous lemma gives us the ability to offer the improvement $\mathtt{construct\_}\mathcal{B}_s\mathtt{\_improved}$ in Algorithm 3. This algorithm makes use of the s-match related functions in Algorithm 2. Using the Ψ and ψ functions, the technique behind Algorithm 3 is to start from the left of the s-string, find the longest s-matches, and populate the elements of $\mathcal{B}_s$. A heuristic is used to determine whether or not s-matching may yield new elements of $\mathcal{B}_s$. The following theorem formalizes the algorithm and its running time.

Theorem 1. *Given an n-length s-string T, $\mathtt{construct\_}\mathcal{B}_s\mathtt{\_improved}$ constructs the s-border array $\mathcal{B}_s$ in time $O(\max\{n, b\phi\})$, where b is the length of the longest s-border and ϕ is based on the s-string.*

Proof. We first prove that the technique behind the algorithm is correct. Let $\mathcal{B}_s = \{0, 0, ..., 0\}$. Maintain a pointer to index h such that the *s-border* subarray $\mathcal{B}_s[1...h - 1]$ is always complete. Initially, $h = 2$ since $\mathcal{B}_s[1] = 0$ by Definition 12. Next, we find the longest s-match of say m symbols between $T[1...n]$ and $T[j...n]$ where $j = 2$. Let $q = m$. By Lemma 10, we know that while $m > 0$, then we can assign $\mathcal{B}_s[j + m - 1] = m$ and let $m = m - 1$. The previously populated elements are the longest s-borders for indices $j + q - 1$, $j + q - 2$, etc. because we have started at $j = 2$ and no s-border can be longer for these elements. Thus, $\mathcal{B}_s[1...j + q - 1]$ is now complete. Set $h = j + q$. We continue at $j = 3$ considering the following cases.

- If $j = h$, we continue the same process to find $\mathcal{B}_s[j]$ and populate other $\mathcal{B}_s$ elements exactly like the initial part of the proof for $j = h$ because only $\mathcal{B}_s[1...h - 1 = j - 1]$ is complete.
- When $j < h$ and an s-match exists between the symbols at $h - j + 1$ in the s-suffixes at 1 and j, i.e. $\mathtt{sencode}(T[1...n])[h - j + 1] = \mathtt{sencode}(T[j...n])[h - j + 1]$, then the following s-match of m symbols can be conducted: $\mathtt{sencode}(T[1...m]) = \mathtt{sencode}(T[j...j + m - 1])$. (Otherwise, no longer s-border is possible.) Since $\mathcal{B}_s[1...j...h - 1]$ is already complete with the longest

s-borders, only when the m exceeds the complete section of $\mathcal{B}_s$ is there a newly introduced s-border, i.e. when $j + m - 1 \geq h$. So, any newly introduced s-border to the incomplete part of $\mathcal{B}_s$ must be maximum because in previous steps $j - 1, j - 2$, etc. we have considered, but not found, s-borders that could be only longer. Considering s-borders for future $j + 1, j + 2$, etc. will only introduce shorter s-borders than the current s-border of length m. Now that we have new maximum s-borders, populate only the new findings. It follows from Lemma 10 that we can let $q = m$ and assign the other known maximum s-borders, that is, while $m > 0 \ \wedge j + m - 1 \geq h$, assign $\mathcal{B}_s[j + m - 1] = m$ and let $m = m - 1$. Finally, set $h = j + q$ to signify that now $\mathcal{B}_s[1...h - 1 = j + q - 1]$ is complete.
- Otherwise, no s-matching is necessary because it is not possible to introduce a longer s-border with the current s-suffixes considered.

The previous cases are considered for subsequent $j = 4, 5, ..., n$. In each case, we are finding the longest s-matches between the s-suffixes at 1 and j with each $\mathcal{B}_s$ element populated at most once using the first appropriate value. Since j increases and we populate $\mathcal{B}_s$ using the earliest relevant s-match found, subsequent j will only produce smaller s-borders in sencode due to the fact that appending symbols in prev (from Definition 3) and compl (from Definition 7) does not modify the encodings of the already existing s-suffix (see Fig. 1). Thus, the algorithm correctly computes $\mathcal{B}_s$.

We now analyze the running time via the displayed Algorithm 3. Assume an indexed alphabet. Then, $\text{prev}(T)$, $\text{compl}(T)$, $\text{sencode}(T)$, and $\alpha(T)$ are constructed in $O(n)$ time by Lemma 3 and Lemma 7. Since ψ executes in constant time via Lemma 8, the running time of the entire algorithm is clearly dependent on the s-matching of Ψ in line 9. This line is responsible for a sequence of symbol comparisons to conduct s-matches of which will require $O(b)$ comparisons in the worst case with b as the length of the maximum s-match by Lemma 9 and in this case, b is the also the longest s-border. This line is called when either (1) $j = h$ or (2) $j < h$ and there exists an s-match between two symbols. In case (1), at most b symbols may be matched, but h will be incremented so $h = j + b$ will force case (2). Case (2) is where any additional *rematching* is performed, that is, matching symbols in T that may have already been visited in case (1). So, the *total* comparisons by line 9 during an execution of the algorithm in case (1) is in $O(n)$. Further, the time required for the comparisons in case (1) is absorbed by the time bound required by the construction of the initial encodings. Let ϕ be the number of times that case (2) is executed. Then, the algorithm executes in $O(\max\{n, b\phi\})$ time. □

Depending on the s-string, there are cases in which Algorithm 3 will execute in linear time. That is, when either b or ϕ is small, the construct $\mathcal{B}_s$_improved algorithm executes in $O(n)$ time for an n-length s-string T. This is a significant improvement from the previously discussed solutions requiring time $O(n^2)$ and $O(n^3)$, but there are still cases when the algorithm will require more than linear time. Below, we discuss yet a further improvement to the $\mathcal{B}_s$ construction algorithm.

4.3 Further Improvement

We now investigate another fundamental property used in the traditional *border* construction. That is, the way in which the next longest border is found. Consider the longest border b_1 of T. If b_2 is the longest border of b_1, then b_2 is the next longest border of T. With this property, we can oracle previous *border* elements, $\mathcal{B}[e]$ with $e = \mathcal{B}[i] > 0$, to find the next longest border of $T[1...i]$. This property is used when the current longest border cannot be extended further and so, we can try the next longest border of which the final symbols may possibly match. Should the last symbol of the second longest border not match, we can oracle the third longest border, etc. So, the oracle may be recursive to the vth level: $\mathcal{B}^1[i] = \mathcal{B}[i]$, $\mathcal{B}^2[i] = \mathcal{B}[\mathcal{B}[i]]$,..., $\mathcal{B}^v[i] = \mathcal{B}[\mathcal{B}^{v-1}[i]]$. Even with the changing s-suffixes by Lemma 6, we prove that this property also holds for s-strings and the s-border.

Lemma 11. *Given an s-string T of length n, we find the length q_v of the vth longest s-border by $q_v = \mathcal{B}_s^v[i]$ while $\mathcal{B}_s[i] > 0$ and $q_v = 0$ otherwise.*

Proof. Consider the *s-border* array $\mathcal{B}_s$ for some element $\mathcal{B}_s[i] = q_1 > 0$. Then, $\mathtt{sencode}(T[1...q_1]) = \mathtt{sencode}(T[i - q_1 + 1...i])$ is the maximum s-border, i.e. the first longest s-border, of $T[1...i]$ by Definition 12. Now, consider the second longest s-border of $T[1...i]$ of length q_2, $0 < q_2 < q_1$. Let $\mathcal{B}_{s_1} = \mathtt{sborder}(T[1...q_1])$ compute the *s-border* array for the input s-string. From the previous discussion of Fig. 1, it follows now that the first longest s-border of $T[1...q_1]$ is also the second longest s-border of $T[1...i]$, i.e. $\mathcal{B}_{s_1}[q_1]$. Since $\mathtt{sencode}(T[1...\mathcal{B}_{s_1}[q_1]]) = \mathtt{sencode}(T[1...\mathcal{B}_s[q_1]])$ by Definition 8 and Proposition 2, the element is *already* known from the original $\mathcal{B}_s$ array element $\mathcal{B}_s[q_1]$, i.e. $\mathcal{B}_s[\mathcal{B}_s[i]] = \mathcal{B}_s^2[i]$. Thus, additional constructions of $\mathcal{B}_{s_j}$ via $\mathtt{sborder}$ are excessive and unnecessary. For the vth longest border of $T[1...i]$, we must take the first longest s-border of $T[1...i]$, i.e. $q_1 = \mathcal{B}_s[i] > 0$, then the first longest s-border of $T[1...q_1]$, i.e. $q_2 = \mathcal{B}_s[q_1] > 0$, then the first longest s-border of $T[1...q_2]$, i.e. $q_3 = \mathcal{B}_s[q_2] > 0$, ..., then the first longest s-border of $T[1...q_{v-1}]$, i.e. $q_v = \mathcal{B}_s[q_{v-1}] > 0$. Overall, $q_v = \mathcal{B}_s^v[i]$. In any case, when $\mathcal{B}_s[j] = 0$ for some j, then no such longest s-border and subsequent s-borders can exist. So, $q_v = \mathcal{B}_s^v[i] = 0$. □

With the previously proven properties in Lemma 10 and Lemma 11, we are now able to propose a further improved solution in Algorithm 4 to compute the *s-border*. This is analogous to traditional *border* construction with the core difference being *how* the individual s-suffix symbols are observed and compared. Essentially, the proofs of Lemma 10 and Lemma 11 in addition to the s-match related functions in Algorithm 2 "evolve" the traditional *border* properties and construction algorithm to now construct the *s-border* array in $O(n)$ time.

Theorem 2. *Given an n-length s-string T, there is an algorithm that constructs the s-border array $\mathcal{B}_s$ in $O(n)$ time.*

Proof. Algorithm $\mathtt{construct\_}\mathcal{B}_s$ builds the required *s-border* array. The correctness of the algorithm follows from the proofs relating *s-border* properties to traditional *border* construction properties in Lemma 10 and Lemma 11 and also, the correctness of the s-matching functions in Algorithm 2, which are developed using the theoretical foundations in Definition 6, Proposition 2, Definition 9, and Definition 13. We now analyze the running time of construct_$\mathcal{B}_s$ from Algorithm 4. The key to the analysis is observing how many times Ψ in line 7 executes in relation to how quickly the array $\mathcal{B}_s$ is filled. Respectively, the variables that correspond to these events are m and h. Initially, $h = 2$, $j = 2$, and $k = 1$. Say that originally Ψ executes m_1 comparisons. Then, by Lemma 9, there exists a current longest s-match of length m_1, i.e. $\mathtt{sencode}(T[1...m_1]) = \mathtt{sencode}(T[j...j + m_1 - 1])$. Now, m_1 elements of $\mathcal{B}_s$ are populated and then h is advanced beyond the populated elements: $h = j + m_1$. Since $\mathtt{sencode}(T[1...m_1 + 1]) \neq \mathtt{sencode}(T[j...j + m_1])$, then either (1) line 12 is executed as an attempt to extend the next longest s-border starting at element k or (2) line 13 resets the algorithm to consider the s-suffix starting at h because no longer s-border exists. When case (1) executes, there are at most m_1 next longest s-borders to try. From Lemma 11, the next longest s-border is known to match and so Ψ continues the s-match at k so that no *rematching* is done. That is, now $j = h - k + 1$ and subsequent s-matches of length m_2, m_3, etc., generally m_g, via Ψ are performed by $\mathtt{sencode}(T[i + k - 1...i + k + m_g - 2]) = \mathtt{sencode}(T[j + k - 1...j + k + m_g - 2])$ rather than rematching the already known s-border by $\mathtt{sencode}(T[i...i + k + m_g - 2]) = \mathtt{sencode}(T[j...j + k + m_g - 2])$. In other words, each m_g is the number of symbols that the s-match is extended, rather the the complete length of the s-match. When case (2) executes, even less work is done. Thus, Ψ performs a *total* of $O(n)$ comparisons during the execution of the algorithm. Since advances in h directly correspond to the s-match comparisons by Ψ and since there are at most a *total* of $O(n)$ next longest s-borders amortized across the $O(n)$ *total* work by Ψ, then the theorem holds.   □

Algorithm 3. Improved $\mathcal{B}_s$ construction

```
1   char  prevT[n],complT[n],αT[n]
2   int[] construct_Bₛ_improved(char T[n]){
3     int h=2,i=1,j,k=1,m,q,x
4     int Bₛ[n]={0,0,...,0}
5     prevT=prev(T),complT=compl(T),αT=α(T)
6     for j=2 to n {
7       x=h-j+1,m=q=0
8       if(j=h ∨ (j<h ∧ ψ(i+x-1,j+x-1,x))){
9         q=m=Ψ(i,j,k)
10        while(m>0 ∧ j+m-1≥h){
11          Bₛ[j+m-1]=m, m--
12        }h=j+q
13      }
14    }return Bₛ }
```

Algorithm 4. Further improved $\mathcal{B}_s$ construction

```
    char  prevT[n],complT[n],αT[n]
    int[] construct_Bₛ(char T[n]){
      int h=2,i=1,j=2,k=1,m,q,w=0
      int Bₛ[n]={0,0,...,0}
      prevT=prev(T),complT=compl(T),αT=α(T)
      while(h≤n){
        q=m=Ψ(i,j,k)
        while(m>0 ∧ j+m-1≥h){
          Bₛ[j+m-1]=m+w, m--
        }h=j+q
        if(w+q>0 ∧ Bₛ[w+q]>0){
          k=Bₛ[w+q]+1,j=h-k+1,w=k-1
        }else{ k=1,j=h,w=0 }
      }return Bₛ }
```

5   Generalization

From the previous section, the facts that s-suffix symbols are oracled efficiently and *s-border* shares fundamental construction properties used in traditional *border* construction leads to a $O(n)$ construction of *s-border* for an n-length s-string T. This result is significant not only for the *s-border*, but also for *p-border* and the traditional *border*. Such is the case because by modifying the alphabet for s-strings, the s-match problem becomes tailored for p-matching and even traditional matching. The following lemmas formalize the generalization possibilities for the $\texttt{construct}\mathcal{B}_s$ algorithm.

Lemma 12. *Given an n-length s-string T, the algorithm $\texttt{construct}\mathcal{B}_s$ constructs the p-border array $\mathcal{B}_p$ in $O(n)$ time.*

Proof. Set the alphabet of complement symbols to null: $\Gamma = \emptyset$. Now, if $T[i] = \pi_1 \in \Pi$, then no such complement $\pi_2 \in \Pi$ exists and so $\texttt{compl}(T)[i] = 0$ by Definition 7. At the same time, the $T[i] \in \Pi$ will be $0 \leq \texttt{prev}(T)[i] < n$ by Definition 3. So, either $\texttt{sencode}(T)[i] = T[i]$ for $T[i] \in (\Sigma \cup \{\$\})$ or $\texttt{sencode}(T)[i] = \texttt{prev}(T)[i]$ for $T[i] \in \Pi$ by Definition 8. Already $\texttt{prev}(T)[i] = T[i]$ for $T[i] \in (\Sigma \cup \{\$\})$ by Definition 3. So, under these conditions $\texttt{sencode}(T) = \texttt{prev}(T)$ and s-matching by Proposition 2 is equivalent to p-matching by Proposition 1. Then, the *s-border* problem of Definition 12 is reduced to the *p-border* problem of Definition 11. Therefore, $\texttt{construct}\mathcal{B}_s$ constructs the *p-border* array $\mathcal{B}_p$ in $O(n)$ time by Theorem 2.   □

Lemma 13. *Given an n-length s-string T, the algorithm $\texttt{construct}\mathcal{B}_s$ constructs the traditional border array $\mathcal{B}$ in $O(n)$ time.*

Proof. Collect the parameter symbols into the set of constant symbols: $\Sigma = \Sigma \cup \Pi$. Now, set the alphabets of parameter and complement symbols to null: $\Pi = \Gamma = \emptyset$. In these conditions, $\texttt{sencode}(T)[i] = T[i]$ for $T[i] \in (\Sigma \cup \Pi \cup \{\$\})$ by Definition 8. Then, s-matching by Proposition 2 is reduced to traditional matching and the *s-border* problem of Definition 12 is reduced to the traditional *border* of Definition 10. Therefore, $\texttt{construct}\mathcal{B}_s$ constructs the *border* array $\mathcal{B}$ in $O(n)$ time by Theorem 2.   □

We summarize the foregoing results in the following theorem:

Theorem 3. *Given an n-length s-string T, the algorithm $\texttt{construct}\mathcal{B}_s$ constructs the p-border array $\mathcal{B}_p$ and the traditional border array $\mathcal{B}$ each in $O(n)$ time.*

6   Conclusions

In this paper, we introduce the structural border array ($\mathcal{B}_s$) for structural strings (s-strings). We provide numerous algorithms that continually improve our $\mathcal{B}_s$ construction for the n-length text T by exploiting the properties of the *s-border* array, ultimately arriving to an $O(n)$ solution. Finally, we provide a connection between $\mathcal{B}_s$, the traditional *border* array ($\mathcal{B}$), and the parameterized border array ($\mathcal{B}_p$) by showing that each array can be constructed with the same $\mathcal{B}_s$ construction algorithm.

References

1. Smyth, W.: Computing Patterns in Strings. Pearson, New York (2003)
2. Baker, B.: A theory of parameterized pattern matching: Algorithms and applications. In: STOC 1993, pp. 71–80 (1993)
3. Idury, R., Schäffer, A.: Multiple matching of parameterized patterns. Theor. Comput. Sci. 154, 203–224 (1996)
4. I, T., Inenaga, S., Bannai, H., Takeda, M.: Counting Parameterized Border Arrays for a Binary Alphabet. In: Dediu, A.H., Ionescu, A.M., Martín-Vide, C. (eds.) LATA 2009. LNCS, vol. 5457, pp. 422–433. Springer, Heidelberg (2009)
5. I, T., Inenaga, S., Bannai, H., Takeda, M.: Verifying a Parameterized Border Array in $O(n^{1.5})$ Time. In: Amir, A., Parida, L. (eds.) CPM 2010. LNCS, vol. 6129, pp. 238–250. Springer, Heidelberg (2010)
6. Baker, B.: Finding clones with dup: Analysis of an experiment. IEEE Trans. Software Eng. 33(9), 608–621 (2007)
7. Zeidman, B.: Software v. software. IEEE Spectr. 47, 32–53 (2010)
8. Shibuya, T.: Generalization of a suffix tree for RNA structural pattern matching. Algorithmica 39(1), 1–19 (2004)
9. Beal, R.: Parameterized Strings: Algorithms and Data Structures. MS Thesis. West Virginia University (2011)
10. Kosaraju, S.: Faster algorithms for the construction of parameterized suffix trees. In: FOCS 1995, pp. 631-637 (1995)
11. Cole, R., Hariharan, R.: Faster suffix tree construction with missing suffix links. SIAM J. Comput. 33(1), 26–42 (2003)
12. Lee, T., Na, J., Park, K.: On-line construction of parameterized suffix trees for large alphabets. Inf. Process. Lett. 111(5), 201–207 (2011)
13. Gusfield, D.: Algorithms on Strings, Trees and Sequences: Computer Science and Computational Biology. Cambridge University Press, Cambridge (1997)
14. Adjeroh, D., Bell, T., Mukherjee, A.: The Burrows-Wheeler Transform: Data Compression, Suffix Arrays and Pattern Matching. Springer, New York (2008)
15. I, T., Deguchi, S., Bannai, H., Inenaga, S., Takeda, M.: Lightweight Parameterized Suffix Array Construction. In: Fiala, J., Kratochvíl, J., Miller, M. (eds.) IWOCA 2009. LNCS, vol. 5874, pp. 312–323. Springer, Heidelberg (2009)
16. Deguchi, S., Higashijima, F., Bannai, H., Inenaga, S., Takeda, M.: Parameterized suffix arrays for binary strings. In: PSC 2008, pp. 84-94 (2008)
17. Beal, R., Adjeroh, D.: p-Suffix Sorting as Arithmetic Coding. In: Iliopoulos, C.S., Smyth, W.F. (eds.) IWOCA 2011. LNCS, vol. 7056, pp. 44–56. Springer, Heidelberg (2011)
18. Beal, R., Adjeroh, D.: p-Suffix Sorting as Arithmetic Coding. JDA 16, 151–169 (2012)
19. Amir, A., Farach, M., Muthukrishnan, S.: Alphabet dependence in parameterized matching. Inf. Process. Lett. 49, 111–115 (1994)
20. Baker, B.: Parameterized pattern matching by Boyer-Moore-type algorithms. In: SODA 1995, pp. 541–550 (1995)
21. Fredriksson, K., Mozgovoy, M.: Efficient parameterized string matching. Inf. Process. Lett. 100(3), 91–96 (2006)
22. Beal, R., Adjeroh, D.: Parameterized longest previous factor. Theor. Comput. Sci. 437, 21–34 (2012)

Computing the Partial Word Avoidability Indices of Ternary Patterns[*]

Francine Blanchet-Sadri[1], Andrew Lohr[2], and Shane Scott[3]

[1] Department of Computer Science, University of North Carolina,
P.O. Box 26170, Greensboro, NC 27402–6170, USA
blanchet@uncg.edu
[2] Department of Mathematics, Mathematics Building,
University of Maryland, College Park, MD 20742, USA
alohr1@umd.edu
[3] School of Mathematics, Georgia Institute of Technology,
686 Cherry Street, Atlanta, GA 30332–0160, USA
scottsha@ksu.edu

Abstract. We study pattern avoidance in the context of partial words. The problem of classifying the avoidable unary patterns has been solved, so we move on to binary, ternary, and more general patterns. Our results, which are based on morphisms (iterated or not), determine all the ternary patterns' avoidability indices or at least give bounds for them.

1 Introduction

Pattern avoidance is a topic of interest in Combinatorics on Words. A *pattern* is a sequence over an alphabet of variables, which are denoted by A, B, C, etc. An occurrence of a pattern is obtained by replacing the variables with arbitrary non-empty words, such that two occurrences of the same variable are replaced by the same word. A pattern p is *unavoidable* if every infinite word has an occurrence of p; otherwise, p is *avoidable*. More precisely, p is *k-unavoidable* if every infinite word over a k-letter alphabet has an occurrence of p; otherwise, p is *k-avoidable*. The *avoidability index* of p is the smallest integer k such that p is k-avoidable (if no such integer exists, the avoidability index is ∞).

Deciding the avoidability of a pattern can be done easily [1,11], but deciding whether a given pattern is k-avoidable has remained an open problem. An alternative is the problem of classifying all the patterns over a fixed number of variables, i.e., to find the avoidability indices of all the patterns over a fixed number of variables. For the lower bounds, we use the so-called backtracking algorithm from [7], while for the upper bounds, we provide *HD0L systems*. For a finite alphabet Σ, a morphism $f : \Sigma^* \to \Sigma^*$, and $a_0 \in \Sigma$, the tuple (Σ, f, a_0) is called a *D0L system (Deterministic 0-sided Lindenmeyer system)* and the D0L

[*] This material is based upon work supported by the National Science Foundation under Grant No. DMS–1060775. We thank Sean Simmons from the Massachusetts Institute of Technology for his very valuable comments and suggestions.

S. Arumugam and B. Smyth (Eds.): IWOCA 2012, LNCS 7643, pp. 206–218, 2012.
© Springer-Verlag Berlin Heidelberg 2012

language generated by the system is the set $\{f^n(a_0) \mid n \in \mathbb{N}\}$. For example, the Thue–Morse morphism $t(a) = ab$ and $t(b) = ba$ gives the D0L system $(\{a, b\}, t, a)$ generating the language $\{\varepsilon, a, ab, abba, abbabaab, abbabaabbaababba, \ldots\}$. For a D0L system (Σ, f, a_0), the *fixed point* is $f^\omega(a_0) = \lim_{n \to \infty} f^n(a_0)$, provided the limit exists. The Thue–Morse word is $t^\omega(a)$. Now, for a morphism $g : \Sigma_1^* \to \Sigma_2^*$ with alphabets Σ_1, Σ_2 and a D0L system (Σ_1, f, a_0), the tuple $(\Sigma_1, f, a_0, \Sigma_2, g)$ is called an *HD0L system* whose generated language is the set $\{g \circ f^n(a_0) \mid n \in \mathbb{N}\}$.

The problem of determining the avoidability indices of all the binary patterns has been completely solved (see Chapter 3 of [9]). Binary patterns fall into three categories: the patterns ε, A, AB, ABA, and their complements, are unavoidable (or have avoidability index ∞); the patterns AA, AAB, $AABA$, $AABB$, $ABAB$, $ABBA$, $AABAA$, $AABAB$, their reverses, and complements, have avoidability index 3; all other patterns, and in particular all binary patterns of length six or more, have avoidability index 2. Ternary patterns, as well as more general patterns, have also been the subject of investigation [6,9,10].

Recently, Blanchet-Sadri et al. [4] determined all the "non-trivial" avoidability indices of the binary patterns in *partial words*, or sequences that may have some undefined positions, called holes and denoted by $\diamond$'s, that match every letter of the alphabet over which they are defined (we also say that $\diamond$ is *compatible* with each letter of the alphabet). For example, $a\diamond bca\diamond b$ is a partial word with two holes over the alphabet $\{a, b, c\}$, and $aabcabb$ is a *full word* created by filling in the first hole with a and the second one with b. They showed that, if no variable of the pattern is substituted by a partial word consisting of only one hole, the avoidability index of the pattern remains the same as in the full word case, and they started the classification in the non-restricted to non-trivial case.

In this paper, we investigate the problem of classifying all the avoidable ternary patterns, those over three variables A, B, C, with respect to partial word avoidability. First, we complete the classification of all the binary patterns that was started by Blanchet-Sadri et al., i.e., we prove that the avoidability index of the pattern $ABABA$ is two and the one of the pattern $ABBA$ is three. Next, we classify the avoidability indices of almost all of the ternary patterns and show that only four are left in order to complete the classification (for those we give lower and upper bounds).

The contents of our paper is as follows: In Section 2, we give some background on partial words and patterns (for more information, see [2,9]). In Section 3, we complete the classification of the avoidability indices of binary patterns. In Section 4, we make some observations for general pattern avoidance. In Section 5, we describe an algorithm to search for an HD0L system avoiding a given pattern. In Section 6, we discuss the classification of the ternary patterns. Finally in Section 7, we conclude with some remarks. Note that, due to the 14-page restriction, we cannot put in an appendix our ternary lexicon which lists the partial word avoidability indices for the ternary patterns, or at least lists bounds for them (the lexicon will appear in a future expanded version of our paper).

2 Preliminaries

Let Σ be an *alphabet*, a non-empty finite set of symbols. Each element $a \in \Sigma$ is a *letter*. A *(full) word* over Σ is a concatenation of letters from Σ while a *partial word* over Σ is a concatenation of symbols from $\Sigma_\diamond = \Sigma \cup \{\diamond\}$, the alphabet Σ being augmented with the "hole" symbol $\diamond$ (a full word is a partial word without holes). We denote by $u[i]$ the symbol at position i of a partial word u. The *length* of u, $|u|$, is the number of symbols in u. The *empty word* ε is the unique word of length zero. The set of all full words (resp., non-empty full words) over Σ is denoted by Σ^* (resp., Σ^+), while the set of all partial words (resp., non-empty partial words) over Σ is denoted by $\Sigma_\diamond^*$ (resp., $\Sigma_\diamond^+$). The set of all full (resp., partial) words over Σ of length n is denoted by Σ^n (resp., $\Sigma_\diamond^n$).

A partial word u is a *factor* of a partial word v if there exist x, y such that $v = xuy$ (the factor u is *proper* if $u \neq \varepsilon$ and $u \neq v$). We say that u is a *prefix* of v if $x = \varepsilon$ and a *suffix* of v if $y = \varepsilon$. We denote by $\mathrm{Pref}(v)$ the set of all prefixes of v and by $\mathrm{Suf}(v)$ the set of all suffixes of v. If u and v are two partial words of equal length, then u is *compatible* with v, denoted $u \uparrow v$, if $u[i] = v[i]$ whenever $u[i], v[i] \in \Sigma$. If u, v are non-empty compatible partial words, then uv is called a *square*. Moreover, a full word compatible with a factor of a partial word v is called a *subword* of v. For example, $\diamond b \diamond$ is a factor of $abb\diamond b\diamond\diamond ba$ and bbb is a subword compatible with that factor.

Let Δ be an alphabet with $\Sigma \cap \Delta = \emptyset$. We call the letters of Δ *pattern variables* and denote them by A, B, C, etc. A *pattern* is a word over the alphabet $\Sigma \cup \Delta$. A factor $u \in \Sigma^+$ of a pattern is called a *pattern constant*. For example, AA is the square pattern, $aAaAa$ is the overlap pattern. Other patterns include the binary pattern $ABBA$ and the ternary pattern $AABAACACCBAACA$. We denote by $\mathrm{alph}(p)$ the set of distinct variables in pattern p. For a partial word $w \in \Sigma_\diamond^*$ and pattern $p \in (\Sigma \cup \Delta)^*$, we say that w *meets* p or p *occurs in* w if there exists some non-erasing morphism $\varphi : (\Sigma \cup \Delta)^* \to \Sigma^*$, which acts as the identity over Σ, such that $\varphi(p)$ is compatible with a factor of w. We say w *avoids* p when it does not meet p. For example, $abab$ meets AA, $acbcaba$ avoids $aAaAa$, and $ababaabc\diamond a\diamond cd\diamond\diamond aba$ meets $ABBA$. These definitions also apply to infinite partial words w over Σ which are functions from $\mathbb{N}$ to $\Sigma_\diamond$.

A pattern p is called k-*avoidable* if for every $h \in \mathbb{N}$ there is a partial word with h holes over a k-letter alphabet avoiding p, or, equivalently, if there is an infinite partial word over a k-letter alphabet with infinitely many holes which avoids p. We say that p is *avoidable* if it is k-avoidable for some k. For example, AB is unavoidable, AA is unavoidable in partial words, AA is 3-avoidable in full words, and AAA is 2-avoidable [4]. For a given pattern p, we define the *avoidability index* $\mu(p)$ as the minimal k such that p is k-avoidable. If p is unavoidable, we write $\mu(p) = \infty$. For example, $\mu(AB) = \infty$, $\mu(AABB) = 3$, and every binary pattern p of length six or greater satisfies $\mu(p) = 2$ [4].

For a given pattern p, can we determine $\mu(p)$? A concept useful to answer this question is *division of patterns*. If p occurs in a pattern q, then p *divides* q.

For instance, $p = ABA\underline{C}BABC$ divides $q = ABA\underline{BC}BAB\underline{BC}$ (replacing C by BC gives q from p). If p divides q and an infinite partial word avoids p then it also avoids q, and so $\mu(q) \leq \mu(p)$.

3   Completion of the Classification of Binary Patterns

The algorithms described later have provided us with the morphisms necessary to complete the classification of the avoidability indices for binary patterns.

Let $\Sigma = \{a, b\}$, let $t : \Sigma^* \to \Sigma^*$ be the Thue–Morse morphism, and let $\chi : \Sigma^* \to \Sigma_\diamond^*$ be the morphism defined by $\chi(a) = a$ and $\chi(b) = baaa\diamond babbb$.

Theorem 1. *The pattern $ABABA$ is 2-avoidable by $\chi \circ t^\omega(a)$.*

Next, let $\Sigma = \{a, b, c\}$ and $\theta : \Sigma^* \to \Sigma^*$ be the generalized Thue–Morse morphism given by $\theta(a) = abc$, $\theta(b) = ac$, and $\theta(c) = b$.

Lemma 1. *The word $\theta^\omega(a)$ avoids both AA and $bAbcAb$.*

Now, $ABBA$ is 2-unavoidable for full words, which must also be true for partial words. We can prove that $ABBA$ is 3-avoidable by considering the morphism $\varphi : \Sigma^* \to \Sigma_\diamond^*$ given by $\varphi(a) = cccbc$, $\varphi(b) = ca\diamond bcbba$, and $\varphi(c) = baa$. The proof, based on an analysis of cases, depend on Lemmas 1, 2, and 3.

Lemma 2. *Let u and v be length five or greater factors of $\varphi(x)$, with x a full word over Σ. If u and v are compatible, then they are also equal.*

Lemma 3. *The square subwords of $\varphi \circ \theta^\omega(a)$ are $aa, bb, cc, acac, baba, cbcb$.*

Theorem 2. *The pattern $ABBA$ is 3-avoidable by $\varphi \circ \theta^\omega(a)$.*

Taken together with the results of [4,5], the complete classification of the binary patterns is summarized in the following theorem.

Theorem 3. *For partial words, binary patterns fall into three categories:*

1. *The binary patterns ε, A, AA, AAB, $AABA$, $AABAA$, AB, ABA, and their complements, are unavoidable (or have avoidability index ∞).*
2. *The binary patterns $AABAB$, $AABB$, $ABAB$, $ABBA$, their reverses, and complements, have avoidability index 3.*
3. *All other binary patterns, and in particular all binary patterns of length six or more, have avoidability index 2.*

4   Observations for General Pattern Avoidance

The following definitions are useful for our purposes. Let Σ be an alphabet.

For a letter $a \in \Sigma$ and a subset $I \subseteq \mathbb{N}$, we define the function $\mathrm{fill}_I^a : \Sigma_\diamond^* \to \Sigma^*$, where for $w \in \Sigma_\diamond^*$: $\mathrm{fill}_I^a(w)[i] = a$ if $w[i] = \diamond$ and $i \in I$, $\mathrm{fill}_I^a(w)[i] = w[i]$ otherwise. We write $\mathrm{fill}_\mathbb{N}^a$ as simply fill_a. For a word $w \in \Sigma^*$ and a subset $I \subseteq \mathbb{N}$, we define the function $\text{dig}_I : \Sigma^* \to \Sigma^*_\diamond$, where $\text{dig}_I(w)[i] = \diamond$ if $i \in I$; $\text{dig}_I(w)[i] = w[i]$ otherwise. By dig_j for $j \in \mathbb{N}$, we mean $\text{dig}_{\{j\}}$.

A k-unavoidable pattern p is (h,k)-*deep* if there exists $m \in \mathbb{N}$ such that every partial word w over a k-sized alphabet meets p whenever w has at least h holes separated pairwise from each other and from the first and final position of the word by factors of length m or greater. We call $h : \mathbb{N} \setminus \{0,1\} \to \mathbb{N}$ the *depth function* of an unavoidable pattern p if for all k, p is $(h(k),k)$-deep and is not (j,k)-deep for any $j < h(k)$. When the depth function of p is bounded, we call its supremum d, the *depth* of p, and say that p is d-*deep*. A k-unavoidable pattern p is k-*shallow* if p is $(0,k)$-deep or $(1,k)$-deep. If p is k-shallow for all k, we call p *shallow*. We say that p is k-*non-shallow* if it is not k-shallow.

Note that if p is k-shallow it is unavoidable; however, the converse is, in general, false. Every shallow pattern has depth 1 or 0. Naturally, any pattern which is k-unavoidable in the full word case is $(0,k)$-deep and therefore k-shallow. Further, if p is a (h_1,k)-deep pattern and p meets pattern q then q is (h_2,k)-deep for some $h_2 \le h_1$. In particular, if $q \mid p$ and p is k-shallow then q is k-shallow. If a pattern p is (h_1,k_1)-deep, then it is also (h_2,k_1)-deep for all $h_2 \ge h_1$ and (h_1,k_2)-deep for all $k_2 \le k_1$. Hence the depth function is always non-decreasing, and if the depth exists, the depth function is ultimately constant.

The following lemma classifies the depth of binary patterns.

Lemma 4. *The k-unavoidable binary patterns fall into five categories with respect to depth:*

1. *The patterns ε, A, AB, and ABA are shallow with depth 0.*
2. *The patterns AA and AAB are shallow with depth 1.*
3. *The pattern $AABA$ is 3-shallow, 4-non-shallow, and has depth 2.*
4. *The pattern $AABAA$ is 2-shallow and 3-non-shallow, and has depth function h satisfying $h(2) = 0$ and, for all $k \ge 3$, $h(k) = k+1$.*
5. *The patterns $AABAB$, $AABB$, $ABAB$, $ABBA$ are 2-shallow.*

Proof. For Statement 2, it is known that in the full word case AA and AAB are 2-unavoidable but 3-avoidable, hence they are $(0,2)$-deep, but not $(0,k)$-deep for any $k \ge 3$. Any word of length at least three with a hole contains an occurrence of AAB and any word of length two or greater with a hole contains an occurrence of AA. Hence they are also $(1,k)$-deep for all k, and therefore shallow. □

We can prove the following theorem.

Theorem 4. *Let $p_0, \ldots, p_n$ be k-unavoidable patterns over Δ. Let $A_1, \ldots, A_n$ be variables which are not in Δ. Then $p_0 A_1 p_1 \cdots A_n p_n$ is k-unavoidable if any of the following conditions hold:*

1. *$\text{alph}(p_i)$ and $\text{alph}(p_j)$ are pairwise disjoint for all $i \ne j$;*
2. *there exists some k-shallow pattern p such that $p_0, \ldots, p_n$ are factors of p; further, if p is $(0,k)$-deep, so is $p_0 A_1 p_1 \cdots A_n p_n$.*

Proof. For Condition 2, let $p_0, p_1, \ldots, p_n$ be k-unavoidable patterns over Δ, let p be a k-shallow pattern such that $p_0, \ldots, p_n$ are factors of p, and let $A_1, A_2, \ldots, A_n$ be variables not in Δ. Let Σ be a k-letter alphabet, and let w be an infinite partial word over Σ with infinitely many holes. Let $m \in \mathbb{N}$ be the integer implied by the k-shallowness of p.

Write $w = w_0' w_0 w_1' w_1 \cdots$, where the w_i's are length m factors with at least one hole and the w_i''s are factors of w. There are at most $(k+1)^m$ possible w_i, so at least one must occur infinitely often; call it x. Then $w = y_0 x y_1 x y_2 \cdots x y_{n+1}$, where y_i are factors of w. Because p is k-shallow, we have that x meets pattern p, so there is some non-erasing morphism $h : (\Delta \cup \{A_1, \ldots, A_n\})^* \to \Sigma^*$ such that $h(p)$ is compatible with a factor of x. Thus, for some x_i, x_i', x_i'', we may write $x = x_i x_i' x_i''$ where $x_i' \uparrow h(p_i)$, and $w = y_0 x_0 x_0' x_0'' y_1 x_1 x_1' x_1'' y_2 \cdots x_n x_n' x_n'' y_{n+1}$.

This clearly has an occurrence of $q = p_0 A_1 p_1 \cdots A_n p_n$, for let $f : (\Delta \cup \{A_1, \ldots, A_n\})^* \to \Sigma^*$ be the morphism defined by $f(B) = \text{fill}_a(x_{i-1}'' y_i x_i)$ if $B = A_i$, and $f(B) = h(B)$ otherwise, where $a \in \Sigma$. Then w has factors compatible with $f(q)$, so w meets q.

If p is $(0, k)$-deep, then the same argument holds with any filling of the holes in w and with w_i any length m factor, and it follows that q is $(0, k)$-deep. $\square$

We can deduce the following corollaries.

Corollary 1. *The sequence of patterns defined recursively by $p_0 = A_0 A_0$ and $p_{n+1} = p_n A_{n+1} p_n$ is 2-unavoidable.*

Corollary 2. *Let p be a pattern of only distinct variables over Δ and $i < |p|$ such that $p_0, p_1, \ldots, p_n \in \Delta^*$ are compatible with factors of $\text{dig}_i(p)$. Let $A_1, \ldots, A_n$ be distinct variables not in Δ. Then $p_0 A_1 p_1 \cdots A_n p_n$ is unavoidable.*

Applying Theorem 4 as well as Corollary 2 to the patterns in Lemma 4 imply, for instance, that the ternary pattern $AABAAC$, its reversal, its permutations, and its factors are unavoidable; the pattern $AABACAAB$ (resp., $AABAACAAB$), its reversal, its permutations, and its factors are 3-unavoidable (resp., 2-unavoidable). There are many patterns that can be classified this way!

In the rest of this section, we construct partial words avoiding patterns avoidable for full words. Let p be a pattern over $\Delta = \{A_1, \ldots, A_n\}$. When we discuss ternary patterns, we write $A = A_1$, $B = A_2$, and $C = A_3$. Suppose that p is avoided by w, an infinite full word over a k-letter alphabet $\Sigma = \{a_1, a_2, \ldots, a_k\}$. There are a finite number of length three factors of w, so at least one has infinitely many non-overlapping occurrences. Then there exists an infinite integer sequence $\langle i_m \rangle$ where $|i_m - i_{m'}| \geq 3$ and $w[i_m - 1..i_m + 1] = w[i_{m'} - 1..i_{m'} + 1]$ for all distinct m, m'. Let $\langle j_m \rangle$ be an infinite subsequence of $\langle i_m \rangle$ such that $j_m > 2j_{m-1} + 5$, and form the partial word w' from w by replacing $w[j_m - 1..j_m + 1]$ with $a_{k+1} \diamond a_{k+2}$. Then w' is an infinite partial word with infinitely many holes over the alphabet $\Sigma \cup \{a_{k+1}, a_{k+2}\}$. It turns out that w' and its reverse, $\text{rev}(w')$, have many useful properties and avoid many patterns between them.

We refer to $A_{i,j}$ as the jth occurrence of A_i in p, though we drop these subscripts when they are clear from the context. For a factor q of p, we write q_i

for an occurrence of q beginning at index i of p. We define a relation on the set of factors of p, $\text{Fact}(p)$, by $q_i \lessdot q_j$ if q_i is an abelian factor of q_j and there are non-overlapping occurrences of q_i and q_j. For example, if $p = ABCDCB$ then $B \lessdot B$, $B \lessdot AB$, $BC \lessdot DCB$, and $CB \lessdot BC$.

Assume that for some non-erasing morphisms $h, g : (\Delta \times \{1, \ldots, |p|\})^* \to (\Sigma \cup \{a_{k+1}, a_{k+2}\})^*_\diamond$ we have $w' = u_1 h(p) v_1$, where $h(A_{i,j}) \uparrow h(A_{i,\ell})$ for all $1 \leq j, \ell \leq |p|$, and for some factor w'' of $\text{rev}(w')$ we have $w'' = u_2 g(p) v_2$. This is equivalent to w' and $\text{rev}(w')$ meeting p. When w' or $\text{rev}(w')$ avoid p, we arrive at a contradiction and have that p is $(k+2)$-avoidable. Write

- $\overset{\diamond}{q_i}$ when $h(q_i)$ is a hole;
- $\overset{\lrcorner}{q_i}$ when a_{k+1} is a suffix of $h(q_i)$;
- $\overset{\llcorner}{q_i}$ when a_{k+2} is a prefix of $h(q_i)$;
- $\overset{\lrcorner\diamond}{q_i}$ when $a_{k+1}\diamond$ is a suffix of $h(q_i)$;
- $\overset{\diamond\llcorner}{q_i}$ when $\diamond a_{k+2}$ is a prefix of $h(q_i)$;
- $\overset{\sqcup}{q_i}$ when for some proper factor u of w', u is a factor of $h(q_i)$ and $h(q_i)$ is a factor of $\diamond a_{k+2} u a_{k+1} \diamond$;
- $\overset{1}{q_i}$ when $h(q_i)$ has length one.

We can deduce the following formal system.

Theorem 5. *The following rules of inference hold:*

(a) $\overset{\diamond}{A}_{i,j} \implies \forall \ell : \overset{1}{A}_{i,\ell}$

(b) $\overset{\diamond}{A}_{i,j} A_{\ell,m} \implies \overset{\diamond}{A}_{i,j} \overset{\llcorner}{A}_{\ell,m}$

(c) $A_{i,j} \overset{\diamond}{A}_{\ell,m} \implies \overset{\lrcorner}{A}_{i,j} \overset{\diamond}{A}_{\ell,m}$

(d) $\neg \overset{\lrcorner}{q_i} \overset{\llcorner}{q_j}$

(e) $\overset{\lrcorner}{A}_{i,j} \implies \forall \ell : \overset{\lrcorner}{A}_{i,\ell} \vee \overset{\diamond}{A}_{i,\ell}$

(f) $\overset{\llcorner}{A}_{i,j} \implies \forall \ell : \overset{\llcorner}{A}_{i,\ell} \vee \overset{\diamond}{A}_{i,\ell}$

(g) $\overset{\llcorner}{q_i} \wedge (\overset{\lrcorner\diamond}{q_j} \vee \overset{\lrcorner}{q_j}) \implies \overset{\sqcup}{q_i}$

(h) $(\overset{\lrcorner}{q_i} \vee \overset{\diamond\llcorner}{q_i}) \wedge \overset{\lrcorner}{q_j} \implies \overset{\sqcup}{q_i}$

(i) $\overset{\lrcorner\diamond}{q_i} \vee \overset{\diamond\llcorner}{q_i} \implies \forall \ell : \overset{\lrcorner\diamond}{q_\ell} \vee \overset{\diamond\llcorner}{q_\ell}$

(j) $\overset{\lrcorner\diamond}{q_i} \wedge \overset{\diamond\llcorner}{q_j} \implies \forall \ell : \overset{\sqcup}{q_\ell} \vee h(q_\ell) \in \{\diamond a_{k+2}, a_{k+1}\diamond\}$

(k) $\exists A_{i,j} : \overset{\diamond}{A}_{i,j}, \overset{\lrcorner\diamond}{A}_{i,j}, \text{ or } \overset{\diamond\llcorner}{A}_{i,j}$

(l) $\overset{\sqcup}{q_i} \implies \neg q_i \lessdot q_j$

(m) $A_{i,j} A_{\ell,m} A_{i,j+2} \implies \neg \overset{\diamond}{A}_{\ell,m}$

We obtain the following corollary.

Corollary 3. *The pattern $ABACBC$ is 4-avoidable.*

Proof. Set $p = ABACBC$. There are 2-letter full words avoiding $ABACBC$ [10], i.e., p is $(0,2)$-deep. Theorem 5 *(k)* shows one of the variables has an image length of one or two with a hole. We examine cases according to which variable satisfies this property. Note that $\operatorname{rev}(p)$ and p have identical lexical form, so it suffices to consider only cases where the holed variable is one of the first three variables. Note that by Theorem 5 *(m)*, we never have $\overset{\diamond}{B}$.

Consider for instance $\overset{\diamond}{A}BACBC$. By *(a)* and *(b)*, we have $\overset{\diamond}{A}\overset{\llcorner\,1}{B}ACBC$, and it follows by *(f)* that $\overset{\diamond}{A}\overset{\llcorner\,1}{B}A\overset{\llcorner}{C}BC$. Then we have $\overset{\diamond}{A}\overset{\llcorner\,1}{B}A\overset{\lrcorner\diamond}{C}\overset{\llcorner\lrcorner\diamond}{B}C$ or $\overset{\diamond}{A}\overset{\llcorner\,1}{B}A\overset{\diamond}{C}\overset{\llcorner\,1}{B}C$. But, by *(g)*, $\overset{\diamond}{A}\overset{\llcorner\,1}{B}A\overset{\lrcorner\diamond}{C}\overset{\llcorner\lrcorner\diamond}{B}C$ has $\overset{\sqcup}{B}C$ and $BC < BAC$, and by *(h)*, $\overset{\diamond}{A}\overset{\llcorner\,1}{B}A\overset{\diamond}{C}\overset{\llcorner\,1}{B}C$ has $\overset{\sqcup}{B}$ and $B < BC$, so both stand in contradiction to Theorem 5 *(l)*.

Several near identical constructions using fewer variables can avoid many other patterns with specific structures occurring in their factors.

Theorem 6. *Let p be a pattern over alphabet Δ with a squared variable factor AA for some $A \in \Delta$. There is a word w over a four-letter alphabet such that:*

1. *If there are factors Aq_1A and q_2 of p such that $q_1 < q_2$, then either the image of q_1 in w consists of a single letter or p is 4-avoidable.*
2. *If there are factors q_2 and AAq_1A or Aq_1AA such that $q_1 < q_2$, then p is 4-avoidable.*
3. *If there are factors q_2 and AAq_1BB such that $q_1 < q_2$ for some $B \in \Delta$, then p is 3-avoidable.*

5 An Algorithm to Search for an HD0L System Avoiding a Given Pattern

We first describe an algorithm used to attempt to find an HD0L system avoiding a given pattern over a given alphabet size k. The algorithm begins by generating a list of D0L systems. Our D0L generation algorithm on full words first generates a list of all words of a given fixed length that avoid the pattern using the backtracking algorithm. Then, for each of these words, say w, it calculates all possible morphisms, say f, such that $w \in \operatorname{Pref}(f^{\omega}(a))$. We determine f by iterating over all legal lengths of images of letters under f, for which w uniquely defines the morphism. As w is only a finite prefix of $f^{\omega}(a)$, the algorithm does not consider many D0Ls which do avoid p, but have letter images on the order of or larger than w. This restriction also means that, so long as the first letter appears somewhere in the image of a letter other than as the first letter of its image, then every letter on which f is defined appears infinitely often in $f^{\omega}(a)$.

At this point, the algorithm has found many thousands of D0Ls which avoid p for a finite prefix, but may not avoid p in general. Though these could be verified by the HD0L system checking algorithm of [7], it would be entirely unfeasible to check each of these individually. However, checking the length n prefix of $f^{\omega}(a)$ for an occurrence of p takes our algorithm $O(n^{i+2})$, where i is the number of variables. By continuing to check while letting n grow very large, we have multiple rounds of elimination, each one considering longer and longer prefixes. This means that for the longest length prefixes that we check, very few morphisms are left, offsetting the much greater computational cost for each.

Typically by length $n = 1000$, only a handful are left due to the length restriction on the word w that we used to generate the morphisms. Once only the morphisms whose fixed point avoid it for a very long length are left, we run the HD0L system checking algorithm of [7] on these remaining D0Ls to ensure that they avoid p. Note that for the computationally complex steps of this procedure, there is very little shared data, and none of it is being modified during those steps, so, concurrency is very good.

Algorithm 1. generate a D0L system to avoid a pattern

Require: length is an integer that must be tuned between potentially missing a D0L and speed, mesh is a list of integers, the lengths at which the candidate morphisms are tested

Ensure: program prints each D0L that it finds that avoids the pattern within the first $\max\{i \in mesh\}$ letters of its fixed point

1: **for all** $w \in backTrack(length)$ **do**
2:     **for all** $f \in D0L - for - word(w)$ **do**
3:         **for all** $i \in mesh$ **do**
4:             **if** $f^\omega(a)[0..i]$ *meets* p **then**
5:                 *break to a new* f
6:         **if** $H\text{-}D0Lchecker(f, id)$ **then**
7:             **print** f

To generate an HD0L system avoiding a pattern p, we first run the D0L generation algorithm on an alphabet of a greater size, since we know that the inner morphism must avoid the pattern on its own if we have any hope of the HD0L system avoiding the pattern p. We then separately generate outer morphisms by generating a set of long "seed" words with holes avoiding p using a modification of the backtracking algorithm in which we start with a hole in the middle and try to add letters alternating sides. If in this generation phase, we are unable to add any letter to one side, then we know that the pattern is not avoidable with infinitely many holes. Each seed word w is paired with each D0L, say f.

By iterating image sizes for the letters of w, an outer morphism g is determined such that w is a finite prefix of $g \circ f^\omega(a)$. Then, we apply a refining procedure similar to the D0L case, in which a longer and longer prefix of $g \circ f^\omega(a)$ is checked for an occurrence of p. After greatly reducing the number of HD0L systems we have, we verify those remaining with the partial word HD0L system checking algoritm described in [3].

Note, in order to assure that the generated HD0L system contains infinitely many holes, it suffices to know that the seed word contains at least (in practice, exactly) one, meaning that the image on one of the letters in the inner alphabet contains at least one hole, and that every letter of the underlying D0L system occurs infinitely often.

Algorithm 2. generate a morphism whose fixed point is given

Require: w is a prefix of the morphisms we are trying to find, b is a partial function from Σ to Σ^+ (initially defined for no $a \in \Sigma$), i and j are integers (initially 0)

Ensure: program prints each morphism f it finds that can be uniquely defined by w, up to the lengths of the images of letters; note also that $f^\omega(a)$ contains infinitely many of each letter in Σ

1: **while** $j < len(w)$ **do**
2:   **if** $w[i] \in Domain(b)$ **then**
3:     **if** $b(w[i]) = w[j..j + Len(b(i)))$ **then**
4:       $i \leftarrow i + 1$
5:       $j \leftarrow j + Len(b(w[i]))$
6:     **else**
7:       **return**
8:   **else**
9:     **for** $k = 1..len(w) - j$ **do**
10:       $D0L - for - word(w, i \leftarrow i + 1, j \leftarrow j + k, b \leftarrow f$ where $f(x) = b(x)$ for $x \in Domain(b)$ and $f(w[i]) = w[j..j + k))$
11:     **return**
12: **if** $Domain(b) = \Sigma$ and $\exists a \in \Sigma \setminus w[0]$ such that $b(a) = u_1 w[0] u_2$ or $b(w[0])[1..len(b(w[0]))) = u_1 w[0] u_2$ **then**
13:   print b
14: **return**

Algorithm 3. generate a HD0L system to avoid a pattern

Require: length is an integer that must be tuned between potentially missing a HD0L and speed, mesh is a list of integers, the lengths at which the candidate HD0LS are tested

Ensure: program prints each HD0L that it finds that avoids the pattern within the first $\max\{i \in mesh\}$ letters of its fixed point

1: **for all** $w \in randomizedBackTrack(length)$ **do**
2:   **for all** $f \in D0L - for - pattern(p)$ **do**
3:     **for all** $h \in HD0L - for - word(w, f)$ **do**
4:       **for all** $i \in mesh$ **do**
5:         **if** $h(f^\omega(a))[0..i]$ meets p **then**
6:           break to a new f
7:       **if** $H\text{-}D0Lchecker(f, h)$ **then**
8:         print f, h

Algorithm 4. generate HD0L systems to avoid a pattern, given a prefix and an inner morphism

---

Require: f is an inner morphism, w is a prefix of the final word, b is a partial function $\Sigma \to \Sigma^+$, i and j are integers (initially 0)

Ensure: program prints each fully defined HD0L that it finds that uses the given D0L to have a fixed point with given prefix

1: **while** $j < len(w)$ **do**
2:    **if** $f^\omega(a)[i] \in Domain(b)$ **then**
3:       **if** $b(f^\omega(a)[i]) = w[j..j + Len(b(i)))$ **then**
4:          $i \leftarrow i + 1$
5:          $j \leftarrow j + Len(b(f^\omega(a)[i]))$
6:       **else**
7:          **return**
8:    **else**
9:       **for** $k = 1..len(w) - j$ **do**
10:          $D0L - for - word(w, i \leftarrow i + 1, j \leftarrow j + k, b \leftarrow g$ where $g(x) = b(x)$ for $x \in Domain(b)$ and $g(f^\omega(a)[i]) = w[j..j + k))$
11:       **return**
12: **if** $Domain(b) = \Sigma$ **then**
13:    **print** b
14: **return**

---

6   Classification of the Ternary Patterns

In classifying the avoidability indices of the ternary patterns, it is useful to consider the directed tree of patterns T, where the root of T is labelled by ε and each node has children labelled by every canonical pattern formed by appending A, B, C to the parent node's pattern, with all edges directed from parent to child. We have a partial order relation defined on the set of canonical ternary patterns by $q > p$ if there is a path in T from the node labelled by pattern q to the node labelled   by   pattern p. Because $q > p$ means $q \mid p$, we have that $\mu(q) \geq \mu(p)$. The classification is complete when every node of T is appended with the avoidability index of the pattern labelling it.

First, we use unavoidability rules to rule out known 2-unavoidable patterns, and proceed via a depth-first search to find 2-avoidable patterns which are identified as such using division arguments from the binary patterns and the HD0L finding algorithm described in Section 5. Once a pattern p is known to have avoidability index two, we know its children, grandchildren, etc., also have avoidability index two. We find by exhaustion that every ternary pattern with length twelve or greater is 2-avoidable. This leaves us with finitely many ternary patterns to classify.

Next, for any remaining pattern p, we use division arguments and our results to establish bounds on the avoidability index of p.

Finally, we try running the algorithms of Section 5 on successively larger outer alphabet sizes, starting at the known lower bound, and going up to one less than the known upper bound in search of an HD0L system which avoids p. Because the algorithm for finding HD0Ls has so many tuning parameters, the implementation used attempted to tweak these parameters, if no HD0L was found.

Here, as an example, is one branch of the tree T, starting with $ABCABA$:

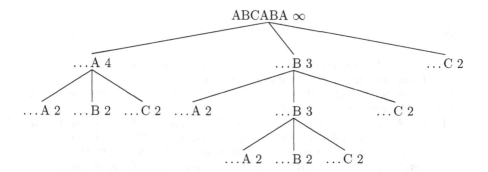

7   Concluding Remarks, Conjectures, and Open Problems

Note that there are ternary patterns with avoidability index 5 for partial words (for instance $AABCABA$), while no such ternary pattern exists for full words [7]. Indeed, to our knowledge the only known patterns with an index of 5 for full words require at least nine variables (for instance

$$ABVACWBAXBCYCDAZDCD$$

is such a pattern [8]).

Frequently, the lower bound is provided by Theorem 4 from patterns of known depth. The conditions on Theorem 4 can most likely be significantly weakened. We conjecture in particular that if p is k-shallow and p_1 and p_2 are (h_1, k)-deep and (h_2, k)-deep respectively, then $p_1 A p_2$ is $(h_1 + h_2, k)$-deep. In general, what relation does the depth of $p_1 A p_2$ have with the depth of p_1 and p_2? Classification of the depths of patterns may give insight.

Every 0-deep pattern that is may be seen to be written in the form of Corollary 2. We conjecture that every unavoidable pattern may be written in this form and that Corollary 2 may be implemented into an algorithm which decides the partial word avoidability of a pattern.

We believe the sequence of Corollary 1 has maximal length 2-unavoidable pattern p_n with $|p_n| = 3 \times 2^{n-1} - 1$. This would mean that any classification of the patterns using k variables by our method would need never explicitly calculate morphisms for any pattern $3 \times 2^{k-1}$ or longer.

In addition, a World Wide Web server interface at

$$\texttt{www.uncg.edu/cmp/research/patterns2}$$

has been established for automated use of our Pattern Avoidance Automated Archive. Given as input a pattern over any alphabet of variables, the Archive attempts to determine the avoidability index or bounds of it, using the algorithms described in our paper. The Archive first checks for unavoidability. If no reason to suspect unavoidability is found, it attempts to generate HD0Ls which avoid it. Note that the HD0L finder is not implemented for patterns with more than three distinct variables. Suggested HD0Ls are also output, and can be verified using our HD0L verification algorithm found there.

References

1. Bean, D.R., Ehrenfeucht, A., McNulty, G.: Avoidable patterns in strings of symbols. Pacific Journal of Mathematics 85, 261–294 (1979)
2. Blanchet-Sadri, F.: Algorithmic Combinatorics on Partial Words. Chapman & Hall/CRC Press, Boca Raton, FL (2008)
3. Blanchet-Sadri, F., Black, K., Zemke, A.: Unary Pattern Avoidance in Partial Words Dense with Holes. In: Dediu, A.-H., Inenaga, S., Martín-Vide, C. (eds.) LATA 2011. LNCS, vol. 6638, pp. 155–166. Springer, Heidelberg (2011)
4. Blanchet-Sadri, F., Mercaş, R., Simmons, S., Weissenstein, E.: Avoidable binary patterns in partial words. Acta Informatica 48(1), 25–41 (2011)
5. Blanchet-Sadri, F., Mercaş, R., Simmons, S., Weissenstein, E.: Erratum to: Avoidable binary patterns in partial words. Acta Informatica 49, 53–54 (2012)
6. Cassaigne, J.: Unavoidable binary patterns. Acta Informatica 30, 385–395 (1993)
7. Cassaigne, J.: Motifs évitables et régularités dans les mots. PhD thesis, Paris VI (1994)
8. Clark, R.J.: The existence of a pattern which is 5-avoidable but 4-unavoidable. International Journal of Algebra and Computation 16, 351–367 (2006)
9. Lothaire, M.: Algebraic Combinatorics on Words. Cambridge University Press, Cambridge (2002)
10. Ochem, P.: A generator of morphisms for infinite words. RAIRO-Theoretical Informatics and Applications 40, 427–441 (2006)
11. Zimin, A.I.: Blocking sets of terms. Mathematics of the USSR-Sbornik 47, 353–364 (1984)

Computing a Longest Common Palindromic Subsequence

Shihabur Rahman Chowdhury, Md. Mahbubul Hasan,
Sumaiya Iqbal, and M. Sohel Rahman

AℓEDA Group
Department of CSE, BUET, Dhaka - 1000, Bangladesh
{shihab,mahbub86,sumaiya,msrahman}@cse.buet.ac.bd

Abstract. The *longest common subsequence (LCS)* problem is a classic and well-studied problem in computer science. Palindrome is a string, which reads the same forward as it does backward. The *longest common palindromic subsequence (LCPS)* problem is an interesting variant of the classic LCS problem which finds the longest common subsequence between two given strings such that the computed subsequence is also a palindrome. In this paper, we study the LCPS problem and give efficient algorithms to solve this problem. To the best of our knowledge, this is the first attempt to study and solve this interesting problem.

Keywords: Longest common subsequence, Palindromes, Dynamic programming, Range query.

1 Introduction

The *longest common subsequence (LCS)* problem is a classic and well-studied problem in computer science with a lot of variants arising out of different practical scenarios. In this paper, we introduce and study the *longest common palindromic subsequence (LCPS)* problem: given a pair of strings X and Y over the alphabet Σ, the goal of the LCPS problem is to compute a LCS Z of X and Y such that, Z is a palindrome. In what follows, for the sake of convenience we will assume, that X and Y have equal length, n. But our result can be easily extended to handle two strings of different length.

String and sequence algorithms related to palindromes have attracted stringology researchers since long [2, 4, 6–8]. The LCPS problem also seems to be a new interesting addition to the already rich list of problems related to palindromes. To the best of our knowledge, there exists no research work in the literature on computing longest common palindromic subsequences. However, the problem of computing palindromes and variants in a single sequence has received much attention in the literature. Manacher discovered an on-line sequential algorithm that finds all *'initial'*[1] palindromes in

---

[1] A string $X[1 \ldots n]$ is said to have an initial palindrome of length k if the prefix $S[1 \ldots k]$ is a palindrome.

S. Arumugam and B. Smyth (Eds.): IWOCA 2012, LNCS 7643, pp. 219–223, 2012.

a string [7]. Gusfield gave a linear-time algorithm to find all *'maximal' palindromes* in a string [3]. Authors in [8] solved the problem of finding all palindromes in SLP (Straight Line Programs)-compressed strings. Very recently, Tomohiro *et. al.* worked on pattern matching problems involving palindromes [9].

In this paper, we propose two methods for finding an LCPS, given two strings. Firstly we present a dynamic programming algorithm to solve the problem with time complexity $O(n^4)$, where n is the length of the strings (Section 3). Then, we present another algorithm that runs in $\mathcal{O}(\mathcal{R}^2 \log^2 n \log \log n)$ time (Section 4). Here, the set of all ordered pair of matches between two strings is denoted by $\mathcal{M}$ and $|\mathcal{M}| = \mathcal{R}$. Due to space constraints all the proofs are omitted.

2   Preliminaries

We assume a finite alphabet, Σ. For a string X, we denote its substring $x_i \ldots x_j$ ($1 \le i \le j \le n$) by $X_{i,j}$. For two strings X and Y, if a common subsequence Z of X and Y is a palindrome, then Z is said to be a *common palindromic subsequence (CPS)*. A *CPS* of two strings having the maximum length is called the *Longest Common Palindromic Subsequence (LCPS)* and we denote it by $LCPS(X, Y)$. The set of all matches between two strings X and Y is denoted by $\mathcal{M}$ and it is defined as, $\mathcal{M} = \{(i, j) : 1 \le i \le n, 1 \le j \le n$ and $x_i = y_j\}$. And we have, $|\mathcal{M}| = \mathcal{R}$. We define, $\mathcal{M}_\sigma$ as a subset of $\mathcal{M}$ such that all matches within this set match to a single character $\sigma \in \Sigma$. We have $|\mathcal{M}_\sigma| = \mathcal{R}_\sigma$. Each member of $\mathcal{M}_\sigma$ is called a *σ-match*.

3   A Dynamic Programming Algorithm

We observe that the natural classes of sub-problems for LCPS correspond to pairs of *substrings* of the two input sequences. Based on this observation we present the following theorem which proves the optimal substructure property of the LCPS problem.

Theorem 1. *Let X and Y are two sequences of length n, and $X_{i,j}$ and $Y_{k,\ell}$ are two substrings of X and Y respectively. Let $Z = z_1 z_2 \ldots z_u$ be the LCPS of the two substrings, $X_{i,j}$ and $Y_{k,\ell}$. Then, the following statements hold,*

1. *If $x_i = x_j = y_k = y_\ell = a$ ($a \in \Sigma$), then $z_1 = z_u = a$ and $z_2 \ldots z_{u-1}$ is an LCPS of $X_{i+1,j-1}$ and $Y_{k+1,\ell-1}$.*
2. *If $x_i = x_j = y_k = y_l$ condition does not hold then, Z is an LCPS of $(X_{i+1,j}$ and $Y_{k,\ell})$ or $(X_{i,j-1}$ and $Y_{k,\ell})$ or $(X_{i,j}$ and $Y_{k,\ell-1})$ or $(X_{i,j}$ and $Y_{k+1,\ell})$.*

Based on Theorem 1 we give the following recursive formulation for the length of $LCPS(X, Y)$:

$$lcps[i,j,k,\ell] = \begin{cases} 0 & i > j \text{ or } k > \ell \\ 1 & (i = j \text{ and } k = \ell) \\ & \text{and} \\ & (x_i = x_j = y_k = y_\ell \\ 2 + lcps[i+1, j-1, k+1, \ell-1] & (i < j \text{ and } k < \ell) \\ & \text{and} \\ & x_i = x_j = y_k = y_\ell \\ \max(lcps[i+1, j, k, \ell], lcps[i, j-1, k, \ell], \\ lcps[i, j, k+1, \ell], lcps[i, j, k, \ell-1]) & (i \leq j \text{ and } k \leq \ell) \\ & \text{and} \\ & (x_i = x_j = y_k = y_\ell) \\ & \text{does not hold} \end{cases} \quad (1)$$

$lcps[i, j, k, \ell]$ is the length of the *LCPS* of $X_{i,j}$ and $Y_{k,\ell}$. The length of $LCPS(X, Y)$ will be stored at $lcps[1, n, 1, n]$. We can compute this length in $\mathcal{O}(n^4)$ time using a bottom up dynamic programming.

4   A Second Approach

We shall first reduce our problem to a geometry problem and then solve it with the help of modified version of range tree data structure. First, we make the following claim.

Claim 1. *Any common palindromic subsequence $Z = z_1 z_2 \ldots z_u$ of two strings X and Y can be decomposed into a set of σ-match pairs ($\sigma \in \Sigma$).*

It follows from Claim 1 that constructing a CPS of X and Y can be seen as constructing an appropriate set of σ-match pairs between them. An arbitrary σ-match pair, $\langle (i, k), (j, \ell) \rangle$ (say m_1), from among all σ-match pairs between X and Y, can be seen as inducing a substring pair in them. Now we want to construct a CPS Z with length u, placing m_1 at the two ends of Z. Clearly we have $z_1 = z_u = x_i = x_j = y_k = y_\ell$. To compute Z, we will have to recursively select σ-match pairs between $X_{i,j}$ and $Y_{k,\ell}$. This will yield a set of σ-match pairs corresponding to Z. If we consider all possible σ-match pairs as the two end points of Z, then the longest one obtained will be an *LCPS* of X and Y. Each match between X and Y can be visualized as a point on a $n \times n$ rectangular grid with integer coordinates. Any σ-match pair defines two corner points of a rectangle and thus induces a rectangle in the grid. Now, our goal is to take a σ-match pair as the two ends of a CPS and recursively construct the set of σ-match pairs from within the induced rectangle. In particular we take the following steps to compute $LCPS(X, Y)$:

1. Identify an induced rectangle (say Ψ_1) by a pair of σ-matches. Then, pair up σ-matches within Ψ_1 to obtain another rectangle (say Ψ_2) and so on until we encounter either of the following two terminating conditions:
   T1. If there is no point within any rectangle. This corresponds to the case when there is no match between the substrings.
   T2. If it is not possible to take any pair of σ-matches within any rectangle. In this case we pair a match with itself, it corresponds to the single character case in our Dynamic Programming solution.
2. We repeat the above step for all possible σ-match pairs ($\forall \sigma \in \Sigma$). At this point, we have a set of nested rectangle structures. An increase in the nesting depth of the rectangle structures as it is being constructed, corresponds to adding a pair of symbols[2] to the resultant palindromic subsequence. Hence, the set of rectangles with maximum nesting depth gives us an *LCPS*.

Now the problem reduces to the following interesting geometric problem: *Given a set of nested rectangles defined by the σ-match pairs $\forall \sigma \in \Sigma$, we need to find the set of rectangles having the maximum nesting depth.* We refer to this problem as the Maximum Depth Nesting Rectangle Structures (MDNRS) problem.

We assume, without the loss of generality that (i, k) and (j, ℓ) correspond to the lower left corner and upper right corner of the rectangle $\Psi\langle(i, k), (j, \ell)\rangle$. Now, a rectangle $\Psi'(\langle(i', k'), (j', \ell')\rangle)$ will be nested within rectangle $\Psi(\langle(i, k), (j, l)\rangle)$ *iff* the following condition holds:

   $i' > i$ and $k' > k$ and $j' < j$ and $\ell' < \ell \Leftrightarrow (i', k', -j', -\ell') > (i, k, j, \ell)$.

Now we convert a rectangle $\Psi(\langle(i, k), (j, \ell)\rangle)$ to a 4-D point $P_\Psi(i, k, -j, -\ell)$ and say that, a point (x, y, z, w) is chained to another point (x', y', z', w') *iff* $(x, y, z, w) > (x', y', z', w')$. Then, a rectangle $\Psi'(\langle(i', k'), (j', \ell')\rangle)$, is nested within a rectangle $\Psi(\langle(i, k), (j, \ell)\rangle)$ *iff* the point $P_{\Psi'}(i', k', -j', -\ell')$ is chained to the point $P_\Psi(i, k, -j, -\ell)$. Hence, the MDNRS problem in 2-D reduces to finding the set of corresponding points in 4-D having the maximum chain length. We refer to this problem as Maximum Chain Length (MCL) Problem. We solve the MCL problem in $O(\mathcal{R}^2 \log^2 n \log \log n)$ time using a modified version of 3-D range tree data structure [1]. A d-dimension range tree, $\mathcal{T}$ is in the form of multi-level trees using an inductive definition on d. Any update and query operation in $\mathcal{T}$ can be done in $\mathcal{O}(\log^d n)$ time. So in 3-D, our query and update performs in $O(\log^3 n)$ time where the array is of $n \times n \times n$ size. We process the 4-D points (x, y, z, w) in non-increasing order of the highest dimension w. For each point (x, y, z, w) we query in $\mathcal{T}$ for maximum value at (x', y', z') where $x' > x$, $y' > y$ and $z' > z$. The obtained value is incremented and stored at the point (x, y, z). We can update the value at (x, y, z) in 3-D Range tree and query for the maximum in $O(\log^3 n)$ time. For the $\mathcal{O}(\mathcal{R}^2)$ points it will take $\mathcal{O}(\mathcal{R}^2 \log^3 n)$ time to solve the MCL problem. In the deepest level of our range tree we are doing a 1-D range maximum query, with the query range always having the form $[x, n]$. According to Rahman *et. al.* such queries can be answered using a *Van Emde Boas* data structure in $\mathcal{O}(\log \log n)$ time [5]. Using this technique, the running time to solve the MCL (which in turn solves the LCPS problem) problem reduces to $O(\mathcal{R}^2 \log^2 n \log \log n)$.

---

[2] If condition T2 is reached, only a symbol shall be added.

5 Conclusion and Future Works

We have presented a $\mathcal{O}(n^4)$ time dynamic programming algorithm for solving the LCPS problem. Then, we have identified and studied some interesting relation of the problem with computational geometry. Then we solved the problem using a modified range tree data structure in $\mathcal{O}(\mathcal{R}^2 \log^2 n \log \log n)$ time. However, our results can be easily extended for the case where the two input strings are of different lengths. Further research can also be carried out towards studying different other variants of the LCPS problem.

References

1. Bentley, J.L., Friedman, J.H.: Data structures for range searching. ACM Comput. Surv. 11, 397–409 (1979)
2. Breslauer, D., Galil, Z.: Finding all periods and initial palindromes of a string in parallel. Algorithmica 14, 355–366 (1995)
3. Gusfield, D.: Algorithms on Strings, Trees, and Sequences: Computer Science and Computational Biology. Cambridge University Press, New York
4. Hsu, P.-H., Chen, K.-Y., Chao, K.-M.: Finding All Approximate Gapped Palindromes. In: Dong, Y., Du, D.-Z., Ibarra, O. (eds.) ISAAC 2009. LNCS, vol. 5878, pp. 1084–1093. Springer, Heidelberg (2009)
5. Iliopoulos, C., Rahman, M.: A new efficient algorithm for computing the longest common subsequence. Theory of Computing Systems 45, 355–371 (2009)
6. Kolpakov, R., Kucherov, G.: Searching for gapped palindromes. Theoretical Computer Science, 5365–5373 (November 2009)
7. Manacher, G.: A new linear-time on-line algorithm for finding the smallest initial palindrome of a string. Journal of the ACM 22, 346–351 (1975)
8. Matsubara, W., Inenaga, S., Ishino, A., Shinohara, A., Nakamura, T., Hashimoto, K.: Efficient algorithms to compute compressed longest common substrings and compressed palindromes. Theoretical Computer Science 410, 900–913 (2009)
9. Tomohiro, I., Inenaga, S., Takeda, M.: Palindrome Pattern Matching. In: Giancarlo, R., Manzini, G. (eds.) CPM 2011. LNCS, vol. 6661, pp. 232–245. Springer, Heidelberg (2011)

Multiset, Set and Numerically Decipherable Codes over Directed Figures

Michał Kolarz and Włodzimierz Moczurad

Institute of Computer Science, Faculty of Mathematics and Computer Science,
Jagiellonian University, Łojasiewicza 6, 30-348 Kraków, Poland
michael.kolarz@gmail.com, wkm@ii.uj.edu.pl

Abstract. Codes with various kinds of decipherability, weaker than the usual unique decipherability, have been studied since multiset decipherability was introduced in mid-1980s. We consider decipherability of directed figure codes, where directed figures are defined as labelled polyominoes with designated start and end points, equipped with catenation operation that may use a merging function to resolve possible conflicts. This is one of possible extensions generalizing words and variable-length codes to planar structures.

Here, verification whether a given set is a code is no longer decidable in general. We study the decidability status of figure codes depending on catenation type (with or without a merging function), decipherability kind (unique, multiset, set or numeric) and code geometry (several classes determined by relative positions of start and end points of figures). We give decidability or undecidability proofs in all but two cases that remain open.

1 Introduction

The classical notion of a code requires that an encoded message should be decoded uniquely, i.e. the exact sequence of codewords must be recovered. In some situations, however, it might be sufficient to recover only the multiset, the set or just the number of codewords. This leads to three kinds of decipherability, known as *multiset* (MSD), *set* (SD) and *numeric decipherability* (ND), respectively. The original exact decipherability is called *unique decipherability* (UD).

Multiset decipherability was introduced by Lempel in [15], whilst numeric decipherability originates in [10] by Head and Weber. The same authors in [11] develop what they call "domino graphs" providing a useful technique for decipherability verification. A paper by Guzman [9] defined set decipherability and presented a unifying approach to different decipherability notions using varieties of monoids. Contributions by Restivo [19] and Blanchet-Sadri and Morgan [4] settle Lempel's conjectures for some MSD and SD codes. Blanchet-Sadri in [3] characterizes decipherability of three-word codes, whilst Burderi and Restivo relate decipherability to the Kraft inequality [6] and to coding partitions [5]. A paper by Salomaa et al. [20], although not directly concerned with decipherability, uses ND codes (dubbed *length codes*) to study prime decompositions of languages.

S. Arumugam and B. Smyth (Eds.): IWOCA 2012, LNCS 7643, pp. 224–235, 2012.
© Springer-Verlag Berlin Heidelberg 2012

Extensions of classical words and variable-length word codes have also been widely studied. For instance, Aigrain and Beauquier introduced polyomino codes in [1]; two-dimensional rectangular pictures were studied by Giammarresi and Restivo in [8], whilst in [16] Mantaci and Restivo described an algorithm to verify tree codes. The interest in picture-like structures is not surprising, given the huge amounts of pictorial data in use. Unfortunately, properties related to decipherability are often lost when moving to a two-dimensional plane. In particular, decipherability testing (i.e. testing whether a given set is a code) is undecidable for polyominoes and similar structures, cf. [2,17].

In [14] we introduced directed figures defined as labelled polyominoes with designated start and end points, equipped with catenation operation that uses a merging function to resolve possible conflicts. This setting is similar to symbolic pixel pictures, described by Costagliola et al. in [7], and admits a natural definition of catenation. The attribute "directed" is used to emphasize the way figures are catenated; this should not be confused with the meaning of "directed" in e.g. directed polyominoes. We proved that verification whether a given finite set of directed figures is a UD code is decidable. This still holds true in a slightly more general setting of codes with weak equality (see [18]) and is a significant change in comparison to previously mentioned picture models, facilitating the use of directed figures in, for instance, encoding and indexing of pictures in databases. On the other hand, a directed figure model with no merging function, where catenation of figures is only possible when they do not overlap, has again undecidable UD testing [12,13].

In the present paper we extend the previous results by considering not just UD codes, but also MSD, SD and ND codes over directed figures. We prove decidability or undecidability for each combination of the following orthogonal criteria: catenation type (with or without a merging function), decipherability kind (UD, MSD, SD, ND) and code geometry (several classes determined by relative positions of start and end points of figures). Two combinations remain open, however.

We begin, in Section 2, with definitions of directed figures and their catenations. Section 3 defines decipherability kinds and shows the relationship between codes of that kinds. In Section 4 main decidability results for decipherability verification are given.

2 Preliminaries

Let Σ be a finite, non-empty alphabet. A *translation* by vector $u = (u_x, u_y) \in \mathbb{Z}^2$ is denoted by tr_u, $\mathrm{tr}_u : \mathbb{Z}^2 \ni (x,y) \mapsto (x + u_x, y + u_y) \in \mathbb{Z}^2$. By extension, for a set $V \subseteq \mathbb{Z}^2$ and an arbitrary function $f : V \to \Sigma$ define $\mathrm{tr}_u : \mathcal{P}(\mathbb{Z}^2) \ni V \mapsto \{\mathrm{tr}_u(v) \mid v \in V\} \in \mathcal{P}(\mathbb{Z}^2)$ and $\mathrm{tr}_u : \Sigma^V \ni f \mapsto f \circ \mathrm{tr}_{-u} \in \Sigma^{\mathrm{tr}_u(V)}$.

Definition 1 (Directed figure, cf. [14]). *Let $D \subseteq \mathbb{Z}^2$ be finite and non-empty, $b, e \in \mathbb{Z}^2$ and $\ell : D \to \Sigma$. A quadruple $f = (D, b, e, \ell)$ is a directed figure (over Σ) with*

$$\begin{aligned}
\text{domain} && \text{dom}(f) &= D, \\
\text{start point} && \text{begin}(f) &= b, \\
\text{end point} && \text{end}(f) &= e, \\
\text{labelling function} & \ \text{label}(f) &= \ell.
\end{aligned}$$

Translation vector *of f is defined as* $\text{tran}(f) = \text{end}(f) - \text{begin}(f)$. *Additionally, the empty directed figure ε is defined as* $(\emptyset, (0,0), (0,0), \{\})$, *where* $\{\}$ *denotes a function with an empty domain. Note that the start and end points need not be in the domain.*

The set of all directed figures over Σ is denoted by Σ°. Two directed figures x, y are *equal* (denoted by $x = y$) if there exists $u \in \mathbb{Z}^2$ such that

$$y = (\text{tr}_u(\text{dom}(x)), \text{tr}_u(\text{begin}(x)), \text{tr}_u(\text{end}(x)), \text{tr}_u(\text{label}(x))).$$

Thus, we actually consider figures up to translation.

Example 1. A directed figure and its graphical representation. Each point of the domain, (x, y), is represented by a unit square in $\mathbb{R}^2$ with bottom left corner in (x, y). A circle marks the start point and a diamond marks the end point of the figure. Figures are considered up to translation, hence we do not mark the coordinates.

$$(\{(0,0), (1,0), (1,1)\}, (0,0), (2,1), \{(0,0) \mapsto a, (1,0) \mapsto b, (1,1) \mapsto c\})$$

Definition 2 (Catenation, cf. [14]). *Let* $x = (D_x, b_x, e_x, \ell_x)$ *and* $y = (D_y, b_y, e_y, \ell_y)$ *be directed figures. If* $D_x \cap \text{tr}_{e_x - b_y}(D_y) = \emptyset$, *a catenation of x and y is defined as*

$$x \circ y = (D_x \cup \text{tr}_{e_x - b_y}(D_y), b_x, \text{tr}_{e_x - b_y}(e_y), \ell),$$

where

$$\ell(z) = \begin{cases} \ell_x(z) & \text{for } z \in D_x, \\ \text{tr}_{e_x - b_y}(\ell_y)(z) & \text{for } z \in \text{tr}_{e_x - b_y}(D_y). \end{cases}$$

If $D_x \cap \text{tr}_{e_x - b_y}(D_y) \neq \emptyset$, *catenation of x and y is not defined.*

Definition 3 (m-catenation, cf. [14]). *Let* $x = (D_x, b_x, e_x, \ell_x)$ *and* $y = (D_y, b_y, e_y, \ell_y)$ *be directed figures. An m-catenation of x and y with respect to a merging function* $m : \Sigma \times \Sigma \to \Sigma$ *is defined as*

$$x \circ_m y = (D_x \cup \text{tr}_{e_x - b_y}(D_y), b_x, \text{tr}_{e_x - b_y}(e_y), \ell),$$

where

$$\ell(z) = \begin{cases} \ell_x(z) & \text{for } z \in D_x \setminus \text{tr}_{e_x - b_y}(D_y), \\ \text{tr}_{e_x - b_y}(\ell_y)(z) & \text{for } z \in \text{tr}_{e_x - b_y}(D_y) \setminus D_x, \\ m(\ell_x(z), \text{tr}_{e_x - b_y}(\ell_y)(z)) & \text{for } z \in D_x \cap \text{tr}_{e_x - b_y}(D_y). \end{cases}$$

Notice that when $x \circ y$ is defined, it is equal to $x \circ_m y$, regardless of the merging function m.

Example 2. Let π_1 be the projection onto the first argument.

The "non-merging" catenation is not defined for the above figures. Note that the result of $(m\text{-})$catenation does not depend on the original position of the second argument.

Observe that $\circ$ is associative, whilst $\circ_m$ is associative if and only if m is associative. Thus for associative m, $\Sigma_m^\diamond = (\Sigma^\diamond, \circ_m)$ is a monoid (which is never free).

Abusing this notation, we also write $X^\diamond$ (resp. $X_m^\diamond$) to denote the set of all figures that can be composed by $\circ$ catenation (resp. $\circ_m$ m-catenation) from figures in $X \subseteq \Sigma^\diamond$. When some statements are formulated for both $\circ$ and $\circ_m$, we use the symbol $\bullet$ and "$x \bullet y$" should then be read as "$x \circ y$ (resp. $x \circ_m y$)". Similarly, "$x \in X_\bullet^\diamond$" should be read as "$x \in X^\diamond$ (resp. $x \in X_m^\diamond$)".

From now on let m be an arbitrary associative merging function.

3 Codes

In this section we define a total of eight kinds of directed figure codes, resulting from the use of four different notions of decipherability and two types of catenation. Note that by a *code* (over Σ, with no further attributes) we mean any finite non-empty subset of $\Sigma^\diamond \setminus \{\varepsilon\}$.

Definition 4 (UD code). *Let X be a code over Σ. X is a* uniquely decipherable code, *if for any $x_1, \ldots, x_k, y_1, \ldots, y_l \in X$ the equality $x_1 \circ \cdots \circ x_k = y_1 \circ \cdots \circ y_l$ implies that $(x_1, \ldots, x_k)$ and $(y_1, \ldots, y_l)$ are equal as sequences, i.e. $k = l$ and $x_i = y_i$ for each $i \in \{1, \ldots, k\}$.*

Definition 5 (UD m-code). *Let X be a code over Σ. X is a* uniquely decipherable m-code, *if for any $x_1, \ldots, x_k, y_1, \ldots, y_l \in X$ the equality $x_1 \circ_m \cdots \circ_m x_k = y_1 \circ_m \cdots \circ_m y_l$ implies that $(x_1, \ldots, x_k)$ and $(y_1, \ldots, y_l)$ are equal as sequences.*

In the remaining definitions, we use the obvious abbreviated notation.

Definition 6 (MSD code and m-code). *Let X be a code over Σ. X is a* multiset decipherable code *(resp. m-code), if for any $x_1, \ldots, x_k, y_1, \ldots, y_l \in X$ the equality $x_1 \bullet \cdots \bullet x_k = y_1 \bullet \cdots \bullet y_l$ implies that $\{\!\{x_1, \ldots, x_k\}\!\}$ and $\{\!\{y_1, \ldots, y_l\}\!\}$ are equal as multisets.*

Definition 7 (SD code and m-code). *Let X be a code over Σ. X is a* set decipherable code *(resp. m-code), if for any $x_1, \ldots, x_k, y_1, \ldots, y_l \in X$ the equality $x_1 \bullet \cdots \bullet x_k = y_1 \bullet \cdots \bullet y_l$ implies that $\{x_1, \ldots, x_k\}$ and $\{y_1, \ldots, y_l\}$ are equal as sets.*

Definition 8 (ND code and m-code). *Let X be a code over Σ. X is a numerically decipherable code (resp. m-code), if for any $x_1, \ldots, x_k, y_1, \ldots, y_l \in X$ the equality $x_1 \bullet \cdots \bullet x_k = y_1 \bullet \cdots \bullet y_l$ implies $k = l$.*

Proposition 1. *If X is a UD (resp. MSD, SD, ND) m-code, then X is a UD (resp. MSD, SD, ND) code.*

Proof. Assume X is not a UD (resp. MSD, SD, ND) code. Then for some $x_1, \ldots, x_k, y_1, \ldots, y_l \in X$ we have $x_1 \circ \cdots \circ x_k = y_1 \circ \cdots \circ y_l$ with $(x_1, \ldots, x_k)$ and $(y_1, \ldots, y_l)$ not satisfying the final condition of the respective definition. But then, irrespective of m, $x_1 \circ_m \cdots \circ_m x_k = y_1 \circ_m \cdots \circ_m y_l$ and X is not a UD (resp. MSD, SD, ND) m-code. $\qquad\square$

Note that the converse does not hold. A code may, for instance, fail to satisfy the UD m-code definition with $x_1 \circ_m \cdots \circ_m x_k = y_1 \circ_m \cdots \circ_m y_l$ and still be a UD code simply because some catenations in $x_1 \circ \cdots \circ x_k$ and $y_1 \circ \cdots \circ y_l$ are not defined.

Example 3. Take $X = \{x = \boxed{\otimes}\}$. X is not a UD m-code, since $x \circ_m x = x$. It is a trivial UD code, though, because $x \circ x$ is not defined.

Proposition 2. *Every UD code is an MSD code; every MSD code is an SD code and an ND code. Every UD m-code is an MSD m-code; every MSD m-code is an SD m-code and an ND m-code.*

Proof. Obvious. Examples may be given to show that all those inclusions are strict.

Before proceeding with the main decidability results, note that for UD, MSD and ND m-codes there is an "easy case" that can be verified quickly just by analyzing the translation vectors of figures. This is reflected in the followng theorem.

Theorem 1 (Necessary condition). *Let $X = \{x_1, \ldots, x_n\}$ be a code over Σ. If there exist $\alpha_1, \ldots, \alpha_n \in \mathbb{N}$, not all equal to zero, such that $\sum_{i=1}^{n} \alpha_i \mathrm{tran}(x_i) = (0,0)$, then X is not an ND m-code (and consequently neither an MSD nor UD m-code).*

Proof. Let

$$x = \underbrace{x_1 \circ_m \cdots \circ_m x_1}_{\alpha_1} \circ_m \underbrace{x_2 \circ_m \cdots \circ_m x_2}_{\alpha_2} \circ_m \cdots \circ_m \underbrace{x_n \circ_m \cdots \circ_m x_n}_{\alpha_n}.$$

Now consider the powers of x (with respect to $\circ_m$), x^i for $i \geq 1$. Since $\mathrm{tran}(x) = (0,0)$, each of the powers has the same domain. There is only a finite number of possible labellings of this domain, which implies that regardless of the merging function and labelling of x, there exist $a, b \in \mathbb{N}$, $a \neq b$ such that $x^a = x^b$. Hence X is not an ND m-code. $\qquad\square$

Definition 9 (Two-sided and one-sided codes). *Codes that satisfy the condition of Theorem 1 will be called* two-sided. *Codes that do not satisfy it will be called* one-sided.

These conditions can be intrepreted geometrically as follows: Translation vectors of a two-sided code do not fit in an open half-plane. For a one-sided code, there exists a line passing through $(0,0)$ such that all translation vectors are on one side of it.

Example 4. The following set of figures is a two-sided code, with translation vectors $(1,2)$, $(1,-2)$ and $(-2,0)$:

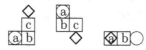

It is a one-sided code, if the rightmost figure is removed.

Corollary 1. *If X is an ND m-code, then X is one-sided.*

4   Decidability of Verification

In this section we summarize all non-trivial decidability results for the decipherability verification. We aim to prove the decidability status for for each combination of the following orthogonal criteria: catenation type (with or without a merging function), decipherability kind (UD, MSD, SD, ND) and code geometry (one-sided, two-sided, two-sided with parallel translation vectors). Two combinations remain open, however.

Because of space constraints, we only present outlines of the (un)decidability proofs. Moreover, proofs that have already appeared in our previous work and algorithms are omitted altogether; references to respective papers are given. Note, however, that in all decidable non-trivial cases there exist algorithms to test the decipherability in question; the algorithms effectively find a double factorization of a figure if the answer is negative.

4.1   Positive Decidability Results

Theorem 2 (see [14], Section 4). *Let X be a one-sided code over Σ. It is decidable whether X is a UD m-code.*

Theorem 3 (see [12], Section 3). *Let X be a one-sided code over Σ. It is decidable whether X is a UD code.*

Generalizing Theorems 2 and 3, we obtain a similar result for one-sided MSD, SD and ND codes and m-codes.

Theorem 4. *Let X be a one-sided code over Σ. It is decidable whether X is a {UD, MSD, SD or ND} {code or m-code}.*

Proof (outline). For $x_1,\ldots,x_k, y_1,\ldots,y_l \in X_{\bullet}^{\diamond}$ we definie a *configuration* as a pair of sequences $((x_1,\ldots,x_k),(y_1,\ldots,y_l))$. A *successor* of such configuration is either $((x_1,\ldots,x_k,z),(y_1,\ldots,y_l))$ or $((x_1,\ldots,x_k),(y_1,\ldots,y_l,z))$ for some $z \in$

X. If a configuration C_2 is a successor of C_1, we write $C_1 \prec C_2$. By $\prec^*$ we denote the transitive closure of $\prec$. For a configuration $C = ((x_1, \ldots, x_k), (y_1, \ldots, y_l))$ let us denote:

$$L(C) = \{x_1, \ldots, x_k\},$$
$$L_\bullet(C) = x_1 \bullet \ldots \bullet x_k,$$
$$R(C) = \{y_1, \ldots, x_l\},$$
$$R_\bullet(C) = y_1 \bullet \ldots \bullet y_l.$$

Now consider a starting configuration $((x), (y))$, for $x, y \in X$, $x \neq y$. Assume that there exists a configuration C such that $L_\bullet(C) = R_\bullet(C)$ and $((x), (y)) \prec^* C$. Now we have:

- X is not a UD code (resp. UD m-code),
- if $L(C) = R(C)$ as multisets then X is not an MSD code (resp. MSD m-code),
- if $L(C) = R(C)$ as sets then X is not an SD code (resp. SD m-code),
- if $|L(C)| = |R(C)|$ then X is not an ND code (resp. ND m-code).

A configuration C' such that $C' \prec^* C$ and $L_\bullet(C) = R_\bullet(C)$ for some C, is called a *proper configuration*.

Our goal is either to show that there exists no proper configuration, or to find such configuration(s). In the former case, X is a code (resp. m-code) of each kind. In the latter case, if we find one of such configurations, X is already not a UD code (resp. UD m-code). To verify whether X is an MSD, SD or ND code (resp. m-code), we have to check the above conditions for *all* possible proper configurations.

Notice that we do not need all of the information contained in configurations; only the labellings that can be changed by future catenations need to be stored. Hence instead of a configuration C we can consider a *reduced configuration* defined as a pair $(\pi_{RC}(L_\bullet(C), R_\bullet(C)), \pi_{RC}(R_\bullet(C), L_\bullet(C)))$ where

$$\pi_{RC}(z, z') = (\text{end}(z), \text{label}(z) \mid_{D(z)})$$

and $D(z)$ is an appropriately chosen domain.

Now, geometric considerations allow us to define "bounding areas" for figures and hence bound the span of configurations to be checked. Hence the number of reduced configurations, up to translation, is finite and there is a finite number of proper configuration to check. Consequently, we can verify whether X is a UD, MSD, SD or ND code (resp. m-code). □

Combined with Theorem 1, this proves the decidability for all UD, MSD and ND m-codes. The case of two-sided SD m-codes remains unsolved, however.

Two-sided codes with parallel translation vectors constitute an interesting special case.

Theorem 5 (see [13], Section 4). *Let $X = \{x_1, x_2, \ldots, x_n\}$ be a two-sided code over Σ with parallel translation vectors, i.e. there exists a vector $\tau \in \mathbb{Z}^2$*

such that $\mathrm{tran}(x_i) = \alpha_i \tau$ *for some* $\alpha_i \in \mathbb{Z}$ $(i = 1, 2, \ldots, n)$, *with* α_i's *not all positive and not all negative. It is decidable whether* X *is a UD code.*

This can again be generalized to two-sided MSD, SD and ND codes with parallel translation vectors:

Theorem 6. *Let* X *be a two-sided code over* Σ *with parallel translation vectors. It is decidable whether* X *is a UD, MSD, SD or ND code.*

Proof (outline). Let $X \subseteq \Sigma^{\diamond}$ be finite and non-empty and let $\mathrm{begin}(x) = (0,0)$ for each $x \in X$. Since translation vectors of elements of X are parallel, there exists a shortest vector $\tau \in \mathbb{Z}^2$ such that

$$\forall x \in X : \mathrm{tran}(x) \in \mathbb{Z}\tau = \{j\tau \mid j \in \mathbb{Z}\} \ .$$

If all translation vectors of elements of X are $(0,0)$, then the decidability problem is trivial.

We define the following *bounding areas*:

$$\begin{aligned}
\mathrm{B_L} &= \{u \in \mathbb{Z}^2 \mid 0 > u \cdot \tau\}, \\
\mathrm{B_0} &= \{u \in \mathbb{Z}^2 \mid 0 \le u \cdot \tau < \tau \cdot \tau\}, \\
\mathrm{B_R} &= \{u \in \mathbb{Z}^2 \mid \tau \cdot \tau \le u \cdot \tau\} \ .
\end{aligned}$$

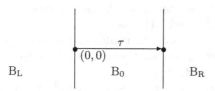

Fig. 1. Bounding areas $\mathrm{B_L}$, $\mathrm{B_0}$ and $\mathrm{B_R}$

For a non-empty figure $x \in \Sigma^{\diamond}$, *bounding hulls* of x are sets:

$$\mathrm{hull}(x) = \bigcup_{n=m\ldots M} \mathrm{tr}_{n\tau}(\mathrm{B_0}),$$

$$\mathrm{hull}^*(x) = \bigcup_{n=-M\ldots-m} \mathrm{tr}_{n\tau}(\mathrm{B_0}),$$

where

$$\begin{aligned}
m &= \min\{n \in \mathbb{Z} \mid \mathrm{tr}_{n\tau}(\mathrm{B_0}) \cap (\mathrm{dom}(x) \cup \{\mathrm{begin}(x), \mathrm{end}(x)\}) \ne \emptyset\}, \\
M &= \max\{n \in \mathbb{Z} \mid \mathrm{tr}_{n\tau}(\mathrm{B_0}) \cap (\mathrm{dom}(x) \cup \{\mathrm{begin}(x), \mathrm{end}(x)\}) \ne \emptyset\} \ .
\end{aligned}$$

In addition, for the empty figure, $\mathrm{hull}(\varepsilon) = \emptyset$ and $\mathrm{hull}^*(\varepsilon) = \emptyset$.

Our goal is either to find a figure $x \in X^\diamond$ that has two different factorizations over elements of X, or to show that such a figure does not exist. If it exists, without loss of generality we can assume it has the following two different x- and y-*factorizations*:

$$x = \dot{x}_1 \ddot{x}_1 \cdots \ddot{x}_{k-1} \dot{x}_k \ddot{x}_k = \dot{y}_1 \ddot{y}_1 \cdots \ddot{y}_{l-1} \dot{y}_l \ddot{y}_l$$

where $\dot{x}_1 \neq \dot{y}_1$, $\mathrm{begin}(\dot{x}_1) = \mathrm{begin}(\dot{y}_1) = (0,0)$ and for $i \in \{1,\ldots,k\}$ and $j \in \{1,\ldots,l\}$ we have:

$$\begin{aligned}
\dot{x}_i &\in X & &\text{and } \mathrm{hull}(\dot{x}_i) \cap B_0 \neq \emptyset, \\
\ddot{x}_i &\in X^\diamond \cup \{\varepsilon\} & &\text{and } \mathrm{hull}(\ddot{x}_i) \cap B_0 = \emptyset, \\
\dot{y}_j &\in X & &\text{and } \mathrm{hull}(\dot{y}_j) \cap B_0 \neq \emptyset, \\
\ddot{y}_j &\in X^\diamond \cup \{\varepsilon\} & &\text{and } \mathrm{hull}(\ddot{y}_j) \cap B_0 = \emptyset \;.
\end{aligned}$$

Now we consider all possible pairs of sequences $((\dot{x}_i)_i, (\dot{y}_j)_j)$ as starting configurations. Observe that there can be only a finite number of such configurations, since

$$\bigcup_{i=1\ldots k} \mathrm{dom}(\dot{x}_i) \subseteq \bigcup_{x \in X} (\mathrm{hull}(x) \cup \mathrm{hull}^*(x)),$$

$$\bigcup_{i=1\ldots l} \mathrm{dom}(\dot{y}_i) \subseteq \bigcup_{x \in X} (\mathrm{hull}(x) \cup \mathrm{hull}^*(x))$$

and the set on the right hand side is bounded in the direction of τ. Also note that if there is no starting configuration for X, then obviously X is a UD code and consequently an MSD, SD and ND code.

We consider independently all starting configurations constructed for X and we proceed with L- and R-configurations, which encode possible factorizations, restricted to B_L and B_R, respectively.

Consider the R-configuration (L-configuration is handled in a similar way). We say that an R-configuration is *terminating* if it satisfies the following conditions:

- the domain and labelling of the x-part of the R-configuration match the domain and labelling of its y-part,
- if a location of the end point of the whole figure is encoded in the R-configuration, then its location is the same in both x- and y-parts,
- all points that should be linked together are trivially linked, since they are the same points.

Now observe that all parts of an R-configuration are bounded: domains are contained in the area restricted by the widest hull of elements of X; multisets describing start/end points to be matched cannot be infinite, since eventually all points must be linked. There are only finitely many such configurations. Either we find a terminating R-configuration, or we consider all configurations that can be obtained from a given starting configuration.

Note that if for some starting configuration we obtain a pair of terminating L- and R-configurations, then X is not a UD code (it can still be an MSD, SD or ND code, though). On the other hand, if we show that for all starting configurations such pair of terminating L- and R-configurations cannot be reached, then X is a UD code (and hence an MSD, SD and ND code).

Similarly as in Theorem 4, to verify whether X is an MSD, SD or ND code, we have to check the following conditions for all possible pairs C of terminating L- and R-configurations:

- if $\pi_x(C) = \pi_y(C)$ as multisets then X is not an MSD code,
- if $\pi_x(C) = \pi_y(C)$ as sets then X is not an SD code,
- if $|\pi_x(C)| = |\pi_y(C)|$ then X is not an ND code,

where $\pi_x(C)$ and $\pi_y(C)$ denote respective multisets of elements used in the construction of C. Note that computation of $\pi_x(C)$ and $\pi_y(C)$ requires the history of C to be kept; this does not spoil the finiteness of the part of C that has to be kept.   □

4.2   Negative Decidability Results

Theorem 7 (see [12], Section 2). *Let X be a two-sided code over Σ. It is undecidable whether X is a UD code.*

This result can again be extended to other decipherability kinds:

Theorem 8. *Let X be a two-sided code over Σ. It is undecidable whether X is a UD, MSD, SD or ND code.*

Proof (outline). We prove Theorem 8 for UD codes. It is clear that exactly the same reasoning can be applied to MSD and SD codes. For ND codes we use some additional techniques, described at the end of this proof.

Let $\Sigma = \{a\}$. For $h, h_N, h_E, h_S, h_W \in \mathbb{Z}_+$ such that $h_N, h_E, h_S, h_W \leq h$ and $b, e \in \{N, E, S, W\}$ we define a *directed hooked square* $\mathrm{DHS}_h(h_N, h_E, h_S, h_W)_e^b$ to be a directed figure $f \in \Sigma^\circ$ with "hooks" of appropriate length. See Figure 2 for a sample directed figure.

Now we encode the Post Correspondence Problem (PCP) in a set of directed figures over $\Sigma = \{a\}$. PCP can be stated as follows: Let $A = \{a_1, \ldots, a_p\}$ be a finite alphabet, $x_1, \ldots, x_k, y_1, \ldots, y_k \in A^+$ such that $x_i \neq y_i$ for $i \in \{1, \ldots, k\}$. Find a sequence $i_1, \ldots, i_n \in \{1, \ldots, k\}$, $n \geq 2$, such that $x_{i_1} \cdots x_{i_n} = y_{i_1} \cdots y_{i_n}$.

We describe a set of directed figures X such that a given PCP instance has a solution if and only if X is a UD code. The set X is built with directed hooked squares.

Note that the above reasoning does not work for ND codes, since both factorizations have exactly the same number of figures. However, it can be adapted for ND codes: we replace basic directed hooked squares for both "x-part" and "y-part" with 25 squares. In the "x-part" the 25 squares will be connected (into one figure), while in the "y-part" they will be disconnected.   □

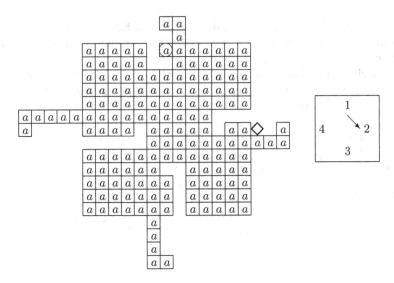

Fig. 2. DHS$_4(1, 2, 3, 4)_E^N$; full and reduced graphical representation

4.3   Summary of Decidability Results

The following table summarizes the status of decipherability decidability. Decidable cases are marked with a +, undecidable ones with a −. Combinations that are still open are denoted with a question mark.

		UD	MSD	ND	SD
1	One-sided codes	+	+	+	+
2	One-sided m-codes	+	+	+	+
3	Two-sided codes	−	−	−	−
4	Two-sided m-codes	+	+	+	?
5	Two-sided codes with parallel vectors	+	+	+	+
6	Two-sided m-codes with parallel vectors	+	+	+	?

5   Final Remarks

Note that the positive decidability cases depicted in lines 4 and 6 are trivial. By Theorem 1, two-sided UD, MSD or ND m-codes do not exist. For other decidable combinations, respective proofs lead to effective verification algorithms.

On the other hand, the case of two-sided SD m-codes is non-trivial; both SD and not-SD codes of this kind exist. However, none of the proof techniques we have used so far can be adapted to this case.

References

1. Aigrain, P., Beauquier, D.: Polyomino tilings, cellular automata and codicity. Theoretical Computer Science 147(1-2), 165–180 (1995)

2. Beauquier, D., Nivat, M.: A codicity undecidable problem in the plane. Theoretical Computer Science 303(2-3), 417–430 (2003)
3. Blanchet-Sadri, F.: On unique, multiset, set decipherability of three-word codes. IEEE Transactions on Information Theory 47(5), 1745–1757 (2001)
4. Blanchet-Sadri, F., Morgan, C.: Multiset and set decipherable codes. Computers and Mathematics with Applications 41(10-11), 1257–1262 (2001)
5. Burderi, F., Restivo, A.: Coding partitions. Discrete Mathematics and Theoretical Computer Science 9(2), 227–240 (2007)
6. Burderi, F., Restivo, A.: Varieties of codes and kraft inequality. Theory of Computing Systems 40(4), 507–520 (2007)
7. Costagliola, G., Ferrucci, F., Gravino, C.: Adding symbolic information to picture models: definitions and properties. Theoretical Computer Science 337, 51–104 (2005)
8. Giammarresi, D., Restivo, A.: Two-dimensional finite state recognizability. Fundamenta Informaticae 25(3), 399–422 (1996)
9. Guzmán, F.: Decipherability of codes. Journal of Pure and Applied Algebra 141(1), 13–35 (1999)
10. Head, T., Weber, A.: The Finest Homophonic Partition and Related Code Concepts. In: Privara, I., Ružička, P., Rovan, B. (eds.) MFCS 1994. LNCS, vol. 841, pp. 618–628. Springer, Heidelberg (1994)
11. Head, T., Weber, A.: Deciding multiset decipherability. IEEE Transactions on Information Theory 41(1), 291–297 (1995)
12. Kolarz, M.: The code problem for directed figures. Theoretical Informatics and Applications RAIRO 44(4), 489–506 (2010)
13. Kolarz, M.: Directed Figure Codes: Decidability Frontier. In: Thai, M.T., Sahni, S. (eds.) COCOON 2010. LNCS, vol. 6196, pp. 530–539. Springer, Heidelberg (2010)
14. Kolarz, M., Moczurad, W.: Directed figure codes are decidable. Discrete Mathematics and Theoretical Computer Science 11(2), 1–14 (2009)
15. Lempel, A.: On multiset decipherable codes. IEEE Transactions on Information Theory 32(5), 714–716 (1986)
16. Mantaci, S., Restivo, A.: Codes and equations on trees. Theoretical Computer Science 255, 483–509 (2001)
17. Moczurad, W.: Brick codes: families, properties, relations. International Journal of Computer Mathematics 74, 133–150 (2000)
18. Moczurad, W.: Directed Figure Codes with Weak Equality. In: Fyfe, C., Tino, P., Charles, D., Garcia-Osorio, C., Yin, H. (eds.) IDEAL 2010. LNCS, vol. 6283, pp. 242–250. Springer, Heidelberg (2010)
19. Restivo, A.: A note on multiset decipherable code. IEEE Transactions on Information Theory 35(3), 662–663 (1989)
20. Salomaa, A., Salomaa, K., Yu, S.: Variants of codes and indecomposable languages. Information and Computation 207(11), 1340–1349 (2009)

A Sequential Recursive Implementation of Dead-Zone Single Keyword Pattern Matching

Bruce W. Watson[1], Derrick G. Kourie[2], and Tinus Strauss[2]

[1] FASTAR, Dept. of Informatics, Stellenbosch University, South Africa
bruce@fastar.org
[2] FASTAR, Dept. of Computer Science, University of Pretoria, South Africa
{derrick,tinus}@fastar.org

Abstract. Earlier publications provided an abstract specification of a *family* of single keyword pattern matching algorithms [18] which search unexamined portions of the text in a divide-and-conquer fashion, generating *dead-zones* in the text as they progress. These dead zones are area of text that require no further examination. Here the results are described of implementing in C++ a sequential recursive version of the algorithm family, where all instances of a single keyword p in a text S are sought—the *online* keyword matching problem where S may not be precomputed.

We show that each step may involve a *window shift* of up to $2 \times |p| - 1$ characters—almost twice as much (and therefore potentially almost twice as fast) as the maximum of $|p|$ characters possible with the Boyer-Moore family of algorithms. Our counterintuitive improvement over Boyer-Moore algorithms is achieved by simultaneously shifting left *and* right. Ongoing benchmarking shows [12] that such bidirectional shifts are highly efficient—and we make specific comparisons here to Horspool's algorithm [9], regarded as one of the most efficient algorithms of the Boyer-Moore family.

1 Introduction and Background

At the Second Prague Stringology Club Workshop (1997), a paper was presented describing a new family of string pattern matching algorithms [17]. The proceedings of the first years of the Prague Workshop were not widely distributed, and a related paper describing this family of algorithms was published in 2003 in the South African Computer Journal [18].

The family of algorithms was articulated as a single abstract algorithm, itself derived from a naïvely formulated primitive statement of requirements. These requirements are given as first order predicate calculus formulae, which enunciate the algorithm's pre- and post-conditions. The primitive algorithmic form is then refined in a correctness-by-construction fashion [11] to produce the resulting abstract algorithm. Each refinement step involves the strengthening of the post-condition and/or the weakening of the precondition. This approach is consonant with, for example, Meyer's notion of refinement in the object-oriented context

S. Arumugam and B. Smyth (Eds.): IWOCA 2012, LNCS 7643, pp. 236–248, 2012.

[13], where a subclass inherits an abstract routine from a superclass and refines it by strengthening the post-condition and/or weakening the precondition.

Section 2 is dedicated to a high-level description of the abstract algorithm. The abstraction itself can be concretized in many ways[1] and that is why it may be regarded as representing a *family* of algorithms. For reasons that will become apparent in Section 2, we shall henceforth characterise this family as *dead-zone* (DZ) pattern matching algorithms, and the abstract algorithm to be described may be thought of as standing at the root of a taxonomy of DZ algorithms.

Our contribution here is an optimised implementation of the family of algorithms (along with some specific family members), as well as performance characterisation (in terms of match attempts), and a revised algorithm skeleton compared to previous publications [18]. The implementation is described in Section 3. It will be seen that our implementation investigates the consequences of several possible search orders through the pattern. It also investigates several ways of computing shift values at the end of a matching cycle (after a match attempt). Because C++ is well-suited to handling such variations via its templates and object-oriented features, it was the language of choice for the current sequence of experiments. Moreover, because the base abstraction of DZ algorithms was articulated in terms of recursive calls, the current C++ implementation retained this recursive approach.

Our implementation choices reflect the fact that, in this paper, our interest is not in raw speed so much as in the number of match probes. (Execution performance is revisited in [12].) We will show in Section 4 that the dead-zone algorithm can adjudicate that there is no match of pattern p in string S in less than $\left\lceil \frac{|S|}{|p|} \right\rceil$ match probes. This is less than the Horspool algorithm [9], broadly considered to be one of the most efficient algorithms.

Section 5 closes with a reflection on the consequences of our results, and outlines our intended future research agenda on dead-zone pattern matching.

Throughout this paper, we contrast the DZ algorithms with Horspool's Boyer-Moore-style algorithm. We are aware of the multitude of more recent algorithms, many of which are variants on Horspool's or use alternative match orders, newer hardware instructions, etc. Horspool's algorithm remains competitive against those newer algorithms, and most of the later techniques can also be applied to the DZ family of algorithms. There are several outstanding overviews of keyword pattern matching algorithms, for example [6,7,14,8,3,4].

2   The Abstract Algorithm

The recursive version of the abstract DZ algorithm derived in [18] has been reproduced in Algorithm 1 in a slightly modified form. The intention in this section is to give a high-level intuitive account of this algorithm. See [18] for a detailed account of its correctness

---

[1] Just as the Boyer-Moore algorithm skeleton can be concretized with a variety of shift functions, etc.

The recursive procedure is called *dzmat*. It searches in text S for all occurrences of pattern p. However, it does not search all of S. Instead, its search is limited to a range of integer indices[2] into S designated as $[live\_low, live\_high)$. The nomenclature *live* was chosen to indicate that searching in this interval is still a live concern because the algorithm has yet to explore whether some indices in the *live* zone correspond to matches in S. Note that the algorithm assumes that *live* is maximal; any extension of its boundaries, either to the left or to the right, would be into *dead* territory.

In earlier versions of the abstract algorithm, a variable *dead* was used to represent a set of "dead" indices—"dead" in the sense that it has already been established whether or not S indexed by an integer in *dead* will lead to a match. Note that some indices in *dead* may be match positions in S that have already been reported. In [18], variable *dead* was used to rigorously express the invariants; as we omit those proofs here, we omit *dead* from the current version of *dzmat*. Nevertheless, we continue to characterise the family of algorithms that are based on growing a dead-zone of indices as *DZ algorithms*.

Algorithm 1 (Abstract DZ Matcher)
proc $dzmat(live\_low, live\_high) \rightarrow$
    if $(live\_low \geq live\_high) \rightarrow$ **skip**
    $[\!]$ $(live\_low < live\_high) \rightarrow$
        $j := \lfloor (live\_low + live\_high)/2 \rfloor;$
        $i := 0;$
        $\{$ **invariant:** $(\forall k : k \in [0,i) : p_{mo(k)} = S_{j+mo(k)}) \}$
        do $((i < |p|)$ **cand** $(p_{mo(i)} = S_{j+mo(i)})) \rightarrow$
            $i := i + 1$
        od;
        $\{$ **postcondition:** $(\forall k : k \in [0,i) : p_{mo(k)} = S_{j+mo(k)})$
                     $\land$ *if* $i < |p|$ *then* $p_{mo(i)} \neq S_{j+mo(i)} \}$
        if $i = |p| \rightarrow print('Match \; at \; ', j)$
        $[\!]$ $i < |p| \rightarrow$ **skip**
        fi;
        $new\_dead\_left := j - shift\_left(i,j) + 1;$
        $new\_dead\_right := j + shift\_right(i,j);$
        $dzmat(live\_low, new\_dead\_left);$
        $dzmat(new\_dead\_right + 1, live\_high)$
    fi
corp

The first invocation of *dzmat* is parameterised by the initial boundaries of the *live* zone as follows: $dzmat(0, |S| - |p| + 1)$; i.e. the *live* zone encompasses all except the last $|p| - 1$ indices of S. The last $|p| - 1$ indices are already in the

---

[2] Note that we follow the convention that indexing in S and p starts at 0 and ends at position $S.length - 1$, $p.length - 1$ respectively. Also by convention, integer intervals are generally indicated as closed from below and open from above. Thus, the indices into S are in the interval $[0, |S|)$.

dead-zone as there is no possibility of a match of length $|p|$ occurring there. All those indices are therefore implicitly relegated to the dead-zone. Note that *dzmat* assumes that S and p are globally available to all its invocations, including recursive invocations.

Turning now to the algorithmic steps within the algorithm, it is immediately clear that $(live\_low \geq live\_high)$ (i.e. $live = \emptyset$) serves as the recursion base case for terminating the recursion. If, alternatively, $(live\_low < live\_high)$ (i.e. $live \neq \emptyset$) then j is computed as an index into S from which the next match attempt will take place. Note that although j is computed as the midpoint of *live*, this is not entirely necessary; as will be seen in Subsection 3.3, other starting positions within the *live* range are legitimate.

Using i to reference into p and i and j to reference into S, a loop matches symbols of p against those of S. The order in which this matching takes place is not necessarily left-to-right, i.e. p_0 against S_j, p_1 against S_{j+1}, etc. Instead, the match order is determined by a bijective function $mo : [0, |p|) \rightarrow [0, |p|)$. The abstract DZ algorithm allows for *any* permuted order to be predetermined by the implementor of the algorithm—i.e. the implementor may define mo according to personal or application domain determined preferences, and the choice will have consequences in subsequent parts of the algorithm. Subsection 3.1 specifies the match orders available in our C++ implementation.

The loop terminates upon the first mismatch, or if a complete match is found. In the latter case, j, the starting position in S of the match, is printed out.

The next step in the abstract DZ algorithm is the computation of the new portion of *dead* territory that can be inferred as a result of the matchings that have taken place. Two functions, *shift_right* and *shift_left*, are used for this purpose. A high-level explanation of the role of these functions will be given below, and more concrete detail is provided in Section 3. For the moment, note that the returned value of *shift_right* is added to j, and is considered to be the upper bound of an interval that can be added into the (implicit) dead-zone region. Similarly, the returned value of *shift_left* is subtracted from $j + 1$ and is regarded as the lower bound of this interval—i.e. the closed interval $[new\_dead\_left, new\_dead\_right]$ is seen as augmenting the dead-zone of S as determined to date. However, no explicit bookkeeping of this dead-zone region is needed. Instead, two recursive calls are made to *dzmat*. The first probes the remaining live zone in the (contiguous) interval $[live\_low, new\_dead\_left)$, and the second the rest of the live zone in the contiguous interval $[new\_dead\_right + 1, live\_high)$.

To gain some insight into the role of the two shift functions, consider the visual representation of the search just after calling these functions as given in Figure 1. The figure shows how matching has progressed until a mismatch after i iterations has been found. To make things interesting, it is assumed that the match ordering as determined by mo resulted in the probing of the two shaded areas in the live zone of S. Based on this information, the right shift function has determined that there are no match opportunities in $[j + 1, new\_dead\_right)$ to be forfeited. Subsequent matching may therefore resume at the right-shifted position $new\_dead\_right + 1$ and rightward of this. Likewise, the left shift function has determined that there are no match opportunities in $[new\_dead\_left, j)$ to be forfeited. Thus another matching activity may also resume at the left-shifted position $new\_dead\_left - 1$ and leftward of this position. Since index j has just been checked, the entire interval $[new\_dead\_left, new\_dead\_right + 1)$ can be considered to be part of the dead-zone.

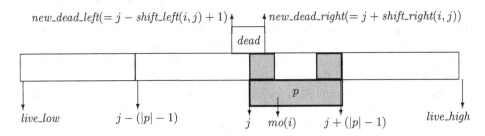

Fig. 1. *dzmat* state after calling *shift_right* and *shift_left*

Note that, as with *dzmat*, the shift functions are assumed to have global access to S and p. The abstract DZ algorithm leaves it up to the implementor to determine how these shift functions determine their respective shift distances. Clearly, in any given implementation, each shift function will depend on the order determined by *mo* for that implementation. These will typically be variants and combinations of Knuth-Morris-Pratt (KMP), Boyer-Moore (BM), and Horspool shift functions [10,2,9]. Furthermore, virtually all of the shift functions in the BM-type of algorithms would be adaptable for use in DZ[3]. Subsection 3.2 outlines the various shift functions developed in our current implementation. The key thing to note here, however, is that every mismatch leads to the possibility of claiming dead-zone territory *both to the left and the right* of j. This possibility of doubly claiming dead-zone "real estate" is unique amongst the various classes of pattern matching algorithms.

The next section discusses how the foregoing abstract algorithm was implemented in C++.

3   A C++ Implementation

The DZ algorithm family has been implemented in an extremely compact C++ toolkit comprising less than five hundred lines of code, most of which are comments. The toolkit (available from the authors by email) is oriented towards readability and performance while making use of the safety-aspects of C++, such as the string class, etc. (Using the C++ string class stands in contrast to C, in which strings are arrays of characters.) The high abstraction level of C++ compared to C is often viewed as a source of inefficiency, though this is

---

[3] See any of [6,7,14,3] for overviews and also [4] and [15, §4.4.1] for a systematic way of deriving such functions

almost a misplaced concern given the high quality of C++ optimising compilers. Throughout the code, we have made use of C++ idioms such as inline functions, function objects, and template parameterisation. The code is generally structured as a simplified version of the toolkit in [5,16].

Given that DZ is a *family* of algorithms, there are several axes of parameterisation, all of which are discussed in the following sections: match orders, shift functions, and the choice of match attempt point (variable j in the abstract algorithm).

3.1 Match Orders: Permutations Over $[0, |p|)$

Match orders are typically extremely simple functions. The toolkit includes:

Forward Left-to-right, best known from the KMP algorithm [10].
Reverse Right-to-left, typically from classical BM algorithms [2].
Outside-in p_0 then $p_{|p|-1}$ then p_1 then $p_{|p|-2}$, etc. An additional variant is provided which starts with the last character of p.
Inside-out The middle character of p and working outwards both left and right.
Lockstep p_0 then $p_{|p|/2}$ then p_1 then $p_{|p|/2+1}$, etc.

The C++ for the Reverse match order, implemented as a function object for performance, is shown in Listing 1.1.

Listing 1.1. Reverse match order function object

```
1  struct MO_rev {
2    inline int operator()(int i, const std::string &p) const {
3      assert(i >= 0 && i < p.length());
4      return p.length() - 1 - i;
5    }
6  };
```

3.2 Shift Functions

Shift functions have also been implemented as classes, given that they typically contain some shift tables or other precomputed information used to make shifts. The following shift functions are provided:

Naïve Always make a shift of one character left and one right. This gives a brute-force algorithm.
KMP/BM This shifter which is parameterised by the match order (class) in use. It contains two shift functions, precomputed (in the constructor of the class) to the minimum shift distances that realign the keyword, given the (partial) match information. When using the forward (resp. reverse) match order, the left- and right-shifters degenerate to the classical reverse-BM (resp. KMP) and KMP (resp. BM).
First & last A shifter resembling Horspool's, which uses the symbol in S aligned with $p_{|p|-1}$ to determine the right-shift distance. Furthermore, it uses the S-symbol aligned with p_0 to determine the left-shift distance. Just as in Horspool's original algorithm, this shifter gives particularly good results.

An example of the left-shift member function is (for the KMP/BM shifter) is shown in Listing 1.2.

Listing 1.2. Left-shift member function

```
1    int shift_left(int i, const std::string &in, int j) const {
2      assert(i >= 0 && i <= plen && j >= 0 && j < in.length());
3      return shl[i];
4    }
```

3.3 Match Attempt Point

Throughout this paper, we have discussed match attempt points which are midway between the low- and high-bounds. Just as other choices of pivot in Quicksort can lead to interesting algorithm variants, the toolkit also supports other match attempt points:

Mid-point As used in *dzmat.*
First-third Choose j at the $\frac{1}{3}$ point between *live_low* and *live_high.*
Left-most Choose *live_low.* This degenerates the algorithm to the traditional left-to-right algorithms. When combined with the classical match orders (forward and reverse), this arrives at KMP and classical BM algorithms.
Right-most Choose *live_high.* This degenerates to the classical right-to-left algorithm, in analogue to the left-most chooser.

3.4 The Pattern Matcher Class

Procedure *dzmat* is embedded as a member function within a pattern matcher class. The class is template parameterised by the match order, the shift function and the match attempt point chooser. Template parameterisation fixes these parameters at compile-time, thereby making all inlineable functions and code available to the C++ optimiser—a significant performance gain even thought the code is split out in the smaller classes described above. The pattern matcher class constructor builds a local shifter object for use in the *dzmat* member function. The only significant deviation from the abstract algorithm in Section 2 is that the C++ recursive member function also returns a $[lo, hi)$ pair designating the dead-zone as discovered in that recursive invocation. This is then used to maximally merge dead-zones over adjacent calls—something which was not done in the original algorithm in [18]. The resulting performance gains are the subject of ongoing work.

3.5 Performance Tuning Potential

The body of the matcher procedure is no more complex than implementations of BM-type algorithms (such as Horspool's), and consists of:

- Integer comparison for detecting empty ranges to terminate recursion. This is usually a two-cycle pipelined operation, and resembles the $|S|$-overrun check in BM-type algorithms, though those algorithms can make use of a 'sentinel' on the right of S to avoid this; we continue tuning DZ for a similar trick.
- Typically division by two[4] for computing the match attempt point. Division by two is also a single-cycle operation on modern processors, usually as arithmetic right-shift.
- Character comparisons for the match attempt. This can be tuned as in any similar algorithm.
- A table lookup for shift distances. These tables typically fit within cache and are of similar structure to BM-type algorithms.
- Integer arithmetic for computing the shift distances—an operation also appearing in BM-type algorithms. In our case, we require two such operations (for right- *and* left-shift), though these will occur in parallel on most modern processors with multiple arithmetic units.
- Two recursive invocations.[5]. This brings forth two tuning concerns:

  1. A reasonable question is whether there is a risk of stack-overflow in such recursive algorithms (or their 'tail-recursion optimised' versions)? When making match attempts at the mid-point (between low and high), the *depth* of the recursion is bounded by $\log_2 |S|$ because each invocation cuts the live range at least in half. This remains small even for extremely large strings[6], and the risk of stack-overflow is negligible.

  2. Recursion is often wrongly perceived as being inefficient. Modern C++ optimising compilers can eliminate tail recursion, and thereby the overhead of the procedure call, instead using a loop and customised stack for the arguments which are 'passed.' In *dzmat*, the arguments consist only of the low- and high-bounds, though the stack depth is doubled because there are *two dzmat* tail-calls. In cases where a C++ compiler is unable to make this optimisation automatically, manual tail-recursion elimination is a trivial code-transformation. The maximum number of stack entries, $\log_2 |S|$, is easy to compute at run-time by counting arithmetic right shifts of $|S|$. The stack (entries being two integers, for low and high) can be pre-allocated before matching begins. Alternatively a statically allocated array of 2048 low/high pairs will allow for all conceivable input strings[7]. Each 'recursion step' (now iteration) then involves only updating a top-of-stack pointer. The performance of such an implementation is reported in [12].

---

[4] *Typically* because the match attempt chooser may choose for first-third, or left-most, etc., though at compile-time thanks to the use of templates.

[5] These invocations are a form of *tail recursion*, meaning they occur a the end of *dzmat*. They are therefore subject to special optimisations.

[6] For example, $\log_2$(one megabyte) is only 20, and $\log_2$(one petabyte) $= 50$.

[7] where 'conceivable' means up to length 2^{1024}, which seems reasonable, given the current estimates of particles in the observable universe are $\sim 2^{266}$.

4   Results

As discussed in Section 3, the C++ implementation can be viewed as a proof-of-concept, oriented towards exploring general algorithm behaviour, such as the number of match attempts, shift distances, etc. Most of the sublinear algorithms appearing in the literature have a worst-case performance quadratic in the size of the input—$\mathcal{O}(|S|^2)$. This typically occurs in pathological cases such as searching for pattern a^m in input a^n. Their best case performance is $\mathcal{O}(\frac{|S|}{|p|})$, often occurring when S and p have disjoint alphabets. In such cases, Horspool's algorithm (and related variants) make shifts of $|p|$, giving a total number of match attempts of

$$\left\lceil \frac{|S| - |p| + 1}{|p|} \right\rceil$$

Indeed, every such left-to-right algorithm, however clever the shift function, is limited to shifts of $|p| + c$ where the constant c is the size of some lookahead—for example two or three symbols as used in the Berry-Ravindran algorithm [1].

DZ displays the same worst-case performance, on the same types of pathological cases. The best-case performance is, however, significantly better: at each step, DZ can eliminate up to $2|p| - 1$ symbols of S, yielding a match-attempt count of

$$\left\lceil \frac{|S| - |p| + 1}{2|p| - 1} \right\rceil$$

This is, of course, also $\mathcal{O}(\frac{|S|}{|p|})$, though the practical consequences can be significant, performing *half* the number of match attempts. Horspool's algorithm can shift up to $|p|$ for a *single* character/shift-table lookup; Section 3.5 shows that DZ requires two character (and one shift-table) lookups to shift up to $2|p| - 1$. This seemingly similar performance differs significantly: DZ's character lookups and shift arithmetic is easily parallelisable (even without programmer intervention, whereby modern processors use multiple logic units).

Two simple examples illustrate this, both of which are, admittedly, explicitly chosen to highlight when such a factor of two improvement (of DZ over classical algorithms such as BM) in match attempts occurs.

4.1   Matching abracadabra in The quick brown fox...

A trace of Horspool's algorithm shows it making four match attempts[8]

```
Attempting a match at 0
    The quick brown fox jumped over th̶e̶/̶l̶̶a̶z̶y̶/̶d̶o̶g̶
    abracadabra
Match got as far as i = 0. Will now shift right by 2
```

---

[8] This output is taken directly from the our implementation, with the addition of the strike-through text. Note that the last $|p| - 1$ symbols of S are already dead.

```
Attempting a match at 2
    T̸h̸e quick brown fox jumped over th̸e̸/̸l̸a̸z̸y̸/̸d̸o̸g̸
      abracadabra
Match got as far as i = 0. Will now shift right by 11

Attempting a match at 13
    T̸h̸e̸/̸q̸u̸i̸c̸k̸/̸b̸r̸o̸wn fox jumped over th̸e̸/̸l̸a̸z̸y̸/̸d̸o̸g̸
                abracadabra
Match got as far as i = 0. Will now shift right by 11

Attempting a match at 24
    T̸h̸e̸/̸q̸u̸i̸c̸k̸/̸b̸r̸o̸w̸n̸/̸f̸o̸x̸/̸j̸u̸m̸p̸ed over th̸e̸/̸l̸a̸z̸y̸/̸d̸o̸g̸
                      abracadabra
Match got as far as i = 0. Will now shift right by 11
```

DZ traces as follows, making only three match attempts (note in some cases negative range bounds due to empty ranges and over-shifting off the left end of S):

```
Invoked with a live-zone of [0,34). Attempting a match at 17
    The quick brown fox jumped over th̸e̸/̸l̸a̸z̸y̸/̸d̸o̸g̸
                      abracadabra
Match got as far as i = 0. Will now shift left/right by 11/11
New dead-zone is [7,28).
Left will be [0,7) and right will be [28,34)

Invoked with a live-zone of [0,7). Attempting a match at 3
    The qui̸c̸k̸/̸b̸r̸o̸w̸n̸/̸f̸o̸x̸/̸j̸u̸m̸p̸e̸d̸/̸o̸ver th̸e̸/̸l̸a̸z̸y̸/̸d̸o̸g̸
      abracadabra
Match got as far as i = 0. Will now shift left/right by 11/11
New dead-zone is [-7,14).
Left will be [0,-7) and right will be [14,7)

Invoked with a live-zone of [28,34). Attempting a match at 31
    T̸h̸e̸/̸q̸u̸i̸c̸k̸/̸b̸r̸o̸w̸n̸/̸f̸o̸x̸/̸j̸u̸m̸p̸e̸d̸/̸o̸ver th̸e̸/̸l̸a̸z̸y̸/̸d̸o̸g̸
                      abracadabra
Match got as far as i = 0. Will now shift left/right by 11/4
New dead-zone is [21,35).
Left will be [28,21) and right will be [35,34)
```

4.2 Matching 01234 in a³¹

This keyword/input combination gives the longest possible shift distances. Horspool's algorithm traces as follows, making six match attempts:

```
Attempting a match at 0
    aaaaaaaaaaaaaaaaaaaaaaaaaaaaa̸a̸a̸a̸a̸
    01234
```

```
Match got as far as i = 0. Will now shift right by 5

Attempting a match at 5
     b́b́b́b́b́aaaaaaaaaaaaaaaaaaaaaaab́b́b́b́
         01234
Match got as far as i = 0. Will now shift right by 5

Attempting a match at 10
    b́b́b́b́b́b́b́b́b́b́aaaaaaaaaaaaaaaaaab́b́b́b́
              01234
Match got as far as i = 0. Will now shift right by 5

Attempting a match at 15
    b́b́b́b́b́b́b́b́b́b́b́b́b́b́b́aaaaaaaaaaaaab́b́b́b́
                   01234
Match got as far as i = 0. Will now shift right by 5

Attempting a match at 20
    b́b́b́b́b́b́b́b́b́b́b́b́b́b́b́b́b́b́b́b́aaaaaaaab́b́b́b́
                        01234
Match got as far as i = 0. Will now shift right by 5

Attempting a match at 25
    b́b́b́b́b́b́b́b́b́b́b́b́b́b́b́b́b́b́b́b́b́b́b́b́b́aab́b́b́b́
                             01234
Match got as far as i = 0. Will now shift right by 5
```

The DZ trace shows it makes half as many match attempts

```
Invoked with a live-zone of [0,27). Attempting a match at 13
    aaaaaaaaaaaaaaaaaaaaaaaaaaab́b́b́b́
                   01234
Match got as far as i = 0. Will now shift left/right by 5/5
New dead-zone is [9,18).
Left will be [0,9) and right will be [18,27)

Invoked with a live-zone of [0,9). Attempting a match at 4
    aaaaaaaaab́b́b́b́b́b́b́b́b́b́aaaaaaaaaab́b́b́b́
         01234
Match got as far as i = 0. Will now shift left/right by 5/5
New dead-zone is [0,9).
Left will be [0,0) and right will be [9,9)

Invoked with a live-zone of [18,27). Attempting a match at 22
    b́b́b́b́b́b́b́b́b́b́b́b́b́b́b́b́b́b́b́b́aaaaaaaaab́b́b́b́
                        01234
Match got as far as i = 0. Will now shift left/right by 5/5
New dead-zone is [18,27).
Left will be [18,18) and right will be [27,27)
```

5 Conclusion

The foregoing clearly demonstrated that the class of DZ algorithms has the potential to significantly reduce the number of match attempts—to as little as half that required in Horspool-like algorithms. This is because it offers the possibility of claiming dead-zone territory by shifting to both the left and the right of a match attempt. In Figure 1, the maximum territory to be claimed is between $j + (|p| - 1)$ and $j - (|p| - 1)$, yielding a total of $2|p| - 1$ spaces in comparison to $|p|$ in other algorithms.

First results of an efficient implementation are reported in [12]. The performance costs are buried in matters such as recursion overhead, checking for empty ranges and midpoint determination, some of which can be optimised as shown in Section 4. The DZ algorithm clearly allows of parallel computation of the recursive parts. Experiments in this regard, among other extensions, form part of our ongoing research.

Acknowledgements. We would like to thank the anonymous referees for their constructive comments. We also received useful technical feedback from several other people, including Gonzalo Navarro, Bill Smyth and Rajeev Raman (during IWOCA), and Maxime Crochemore and Thierry Lecroq. Moreover, some of the approaches in the dead-zone algorithm derivation were inspired by discussions with Edsger Dijkstra and Tony Hoare.

References

1. Berry, T., Ravindran, S.: A fast string matching algorithm and experimental results. In: Holub, J., Simánek, M. (eds.) Proceedings of the Prague Stringology Club Workshop 1999, pp. 16–26. No. Collaborative Report DC-99-05, Czech Technical University, Prague, Czech Republic (1999)
2. Boyer, R.S., Moore, J.S.: A fast string searching algorithm. Communications of the ACM 20(10), 62–72 (1977)
3. Charras, C., Lecroq, T.: Handbook of exact string matching algorithms. Kings College Publications (2004)
4. Cleophas, L., Watson, B.W., Zwaan, G.: A new taxonomy of sublinear right-to-left scanning keyword pattern matching algorithms. Science of Computer Programming 75, 1095–1112 (2010)
5. Cleophas, L.G., Watson, B.W.: Taxonomy-Based Software Construction of SPARE Time: A case study. IEE Proceedings — Software 152(1), 29–37 (2005)
6. Crochemore, M.A., Rytter, W.: Text Algorithms. Oxford University Press (1994)
7. Crochemore, M.A., Rytter, W.: Jewels of Stringology. World Scientific Publishing Company (2003)
8. Faro, S., Lecroq, T.: 2001–2010: Ten years of exact string matching algorithms. In: Holub, J., Žďárek, J. (eds.) Proceedings of the Prague Stringology Conference 2011, pp. 1–2. Czech Technical University in Prague, Czech Republic (2011)
9. Horspool, R.N.: Practical fast searching in strings. Software — Practice & Experience 10(6), 501–506 (1980)
10. Knuth, D.E., Morris, J., Pratt, V.R.: Fast pattern matching in strings. SIAM Journal of Computing 6(2), 323–350 (1977)

11. Kourie, D.G., Watson, B.W.: The Correctness-by-Construction Approach to Programming. Springer (2012)
12. Mauch, M., Watson, B.W., Kourie, D.G., Strauss, T.: Performance assessment of dead-zone single keyword pattern matching. In: Kroeze, J. (ed.) Proceedings of the South African Institute of Computer Scientists and Information Technologists Conference, Pretoria, South Africa (October 2012)
13. Meyer, B.: Object-Oriented Software Construction, 2nd edn. Addison-Wesley (1998)
14. Smyth, W.F.: Computing Patterns in Strings. Addison-Wesley (2003)
15. Watson, B.W.: Taxonomies and Toolkits of Regular Language Algorithms. Ph.D dissertation. Eindhoven University of Technology, Eindhoven, Netherlands (1995)
16. Watson, B.W., Cleophas, L.: SPARE Parts: A C++ toolkit for String Pattern Recognition. Software — Practice & Experience 34(7), 697–710 (2004)
17. Watson, B.W., Watson, R.E.: A new family of string pattern matching algorithms. In: Holub, J. (ed.) Proceedings of the Second Prague Stringologic Workshop, pp. 12–23. Czech Technical University, Prague, Czech Republic (July 1997)
18. Watson, B.W., Watson, R.E.: A new family of string pattern matching algorithms. South African Computer Journal 30, 34–41 (2003); for rapid access, A reprint of this article appears on www.fastar.org. This journal remains the appropriate citation reference

A Catalogue of Algorithms for Building Weak Heaps

Stefan Edelkamp[1], Amr Elmasry[2,3], and Jyrki Katajainen[2]

[1] Faculty 3—Mathematics and Computer Science, University of Bremen,
PO Box 330 440, 28334 Bremen, Germany
edelkamp@tzi.de
[2] Department of Computer Science, University of Copenhagen,
Universitetsparken 1, 2100 Copenhagen East, Denmark
{elmasry,jyrki}@diku.dk
[3] Computer and Systems Engineering Department, Alexandria University,
Alexandria 21544, Egypt

Abstract. An array-based weak heap is an efficient data structure for realizing an elementary priority queue. In this paper we focus on the construction of a weak heap. Starting from a straightforward algorithm, we end up with a catalogue of algorithms that optimize the standard algorithm in different ways. As the optimization criteria, we consider how to reduce the number of instructions, branch mispredictions, cache misses, and element moves. We also consider other approaches for building a weak heap: one based on repeated insertions and another relying on a non-standard memory layout. For most of the algorithms considered, we also study their effectiveness in practice.

1 Introduction

In its elementary form, a priority queue is a data structure that stores a collection of elements and supports the operations *construct*, *find-min*, *insert*, and *extract-min*. In applications, for which this set of operations is sufficient, the basic selection that the users have to make is to choose between binary heaps [9] and weak heaps [3]. Both of these data structures are known to perform well, and the difference in performance is quite small in typical cases.

Most library implementations are based on binary heaps. However, one reason why a user might select a weak heap over a binary heap is that weak heaps are known to perform less element comparisons in the worst case. In Table 1 we summarize the results known for these two data structures.

In [4] we showed that, for weak heaps, the cost of *insert* can be improved to an amortized constant. The idea is to use a buffer that supports constant-time insertion. A new element is inserted into the buffer as long as the number of its elements is below the threshold. Once the threshold is reached, a bulk insertion is performed by moving all elements of the buffer to the weak heap. This modification increases the number of element comparisons per *extract-min* by one.

S. Arumugam and B. Smyth (Eds.): IWOCA 2012, LNCS 7643, pp. 249–262, 2012.
© Springer-Verlag Berlin Heidelberg 2012

Table 1. The worst-case number of element comparisons performed by two elementary priority queues; n denotes the number of elements stored in the data structure prior to the operation in question

data structure	construct	find-min	insert	extract-min
binary heap [6,9]	$2n$	0	$\lceil \lg n \rceil$	$2\lceil \lg n \rceil$
weak heap [3]	$n-1$	0	$\lceil \lg n \rceil$	$\lceil \lg n \rceil$

In this paper we study the construction of a weak heap in more detail. The standard algorithm for building a weak heap is asymptotically optimal, involving small constant factors. Nevertheless, this algorithm can be improved in several ways. The reason why we consider these different optimizations is that it may be possible to apply the same type of optimizations for other similar fundamental algorithms. For some applications the proposed optimizations may be significant although we do not consider any concrete applications per se.

Our catalogue of algorithms for building a weak heap include the following:

Instruction optimization: We utilize bit-manipulation operations on the word level to find the position of the least-significant 1-bit fast. Since the used bit-manipulation operations are native on modern computers, a loop can be replaced by a couple of instructions that are executed in constant time.

Branch optimization: We describe a simple optimization that reduces the number of branch mispredictions from $O(n)$ to $O(1)$. On modern computers having long pipelines, a branch misprediction can be expensive.

Cache optimization: We improve the cache behaviour of an alternative weak-heap construction algorithm by implementing it in a depth-first manner. The resulting algorithm is cache oblivious. We give a recursive implementation and point out how to convert it to an iterative implementation.

Move optimization: We reduce the bound on the number of element moves performed from $3n$ to $2n$. In the meantime, we keep the number of element comparisons performed unchanged.

Repeated insertions: We show that by starting from an empty weak heap and inserting elements repeatedly into it one by one, the overall construction only involves at most $3.5n + O(\lg^2 n)$ element comparisons.

New memory layout: We investigate an alternative weak-heap array embedding that does not need any complex ancestor computations.

The experiments discussed in the paper were carried out on a laptop (model Intel® Core™2 CPU P8700 @ 2.53GHz) running under Ubuntu 11.10 (Linux kernel 3.0.0-16-generic) using g++ compiler (gcc version 4.6.1) with optimization level -O3. The size of L2 cache was about 3 MB and that of the main memory 3.8 GB. Input elements were 4-byte integers and reverse bits occupied one byte each. All execution times were measured using the function gettimeofday in sys/time.h. Other measurements were done using the tools available in valgrind (version 3.6.1). For a problem of size n, each experiment was repeated $2^{26}/n$ times and the average was reported.

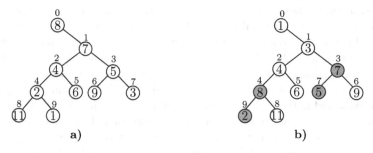

Fig. 2. a) An input of 10 integers and **b)** a weak heap constructed by the standard algorithm. The reverse bits are set for grey nodes.

Even though we only consider array-based weak heaps, it deserves to be mentioned that pointer-based weak heaps can be used to implement addressable priority queues, which also support *delete* and *decrease* operations (see [2,4]).

2   The Standard Weak-Heap Construction Procedure

A *weak heap* (see Fig. 2) is a binary tree that has the following properties:

1. The root of the entire tree has no left child.
2. Except for the root, the nodes that have at most one child are at the last two levels only. Leaves at the last level can be scattered, i.e. the last level is not necessarily filled from left to right.
3. Each node stores an element that is smaller than or equal to every element stored in the right subtree of this node.

From the first two properties we deduce that the height of a weak heap that has n elements is $\lceil \lg n \rceil + 1$. The third property is called the *weak-heap ordering* or *half-tree ordering*. In particular, this property enforces no relation between an element in a node and those stored in the left subtree of this node.

In an array-based implementation, besides the element array a, an array r of *reverse bits* is used, i.e. $r_i \in \{0,1\}$ for $i \in \{0,\ldots,n-1\}$. The array index of the left child of a_i is $2i + r_i$, the array index of the right child is $2i + 1 - r_i$, and the array index of the parent is $\lfloor i/2 \rfloor$ (assuming that $i \neq 0$). Using the fact that the indices of the left and right children of a_i are reversed when flipping r_i, subtrees can be swapped in constant time by setting $r_i \leftarrow 1 - r_i$.

The *distinguished ancestor* of a_j, $j \neq 0$, is the parent of a_j if a_j is a right child, and the distinguished ancestor of the parent of a_j if a_j is a left child. We use *d-ancestor*(j) to denote the index of such ancestor. By the weak-heap ordering, no element is smaller than the element at its distinguished ancestor.

The *join* operation combines two weak heaps into one conditioned on the following settings. Let a_i and a_j be two nodes in a weak heap such that the element at a_i is smaller than or equal to every element in the left subtree of a_j. Conceptually, a_j and its right subtree form a weak heap, while a_i and the left subtree of a_j form another weak heap. (Note that a_i cannot be a descendant of a_j.) If the element at a_j is smaller than that at a_i, the two elements are swapped and r_j is flipped. As a result, the element at a_j will be smaller than or equal to every element in its right subtree, and the element at a_i will be smaller than or equal to every element in the subtree rooted at a_j. Thus, *join* requires constant time and involves one element comparison and possibly an element swap.

In the standard bottom-up construction of a weak heap (see Fig. 3) the nodes are visited one by one in reverse order, and the two weak heaps rooted at a node and its distinguished ancestor are joined. It has been shown, for example in [5], that the amortized cost to get from a node to its distinguished ancestor is $O(1)$. Hence, the overall construction requires $O(n)$ time in the worst case. Moreover, $n - 1$ element comparisons and at most $n - 1$ element swaps are performed.

procedure: *d-ancestor*(j: index)
while (j **bitand** 1) $= r_{\lfloor j/2 \rfloor}$
$\quad | \quad j \leftarrow \lfloor j/2 \rfloor$
return $\lfloor j/2 \rfloor$

procedure: *join*(i, j: indices)
if $a_j < a_i$
$\quad | \quad swap(a_i, a_j)$
$\quad | \quad r_j \leftarrow 1 - r_j$

procedure: *construct*(a: array of n elements, r: array of n bits)
for $i \in \{0, 1, \ldots, n-1\}$
$\quad | \quad r_i \leftarrow 0$
for $j = n - 1$ **to** 1 **step** -1
$\quad | \quad i \leftarrow d\text{-}ancestor(j)$
$\quad | \quad join(i, j)$

Fig. 3. Standard construction of a weak heap

3   Instruction Optimization: Accessing Ancestors Faster

The number of instructions executed by the standard algorithm can be reduced by observing that the reverse bits are initialized to 0 and set bottom-up while scanning the nodes. Therefore, the distinguished ancestor can be computed from the array index by considering the position of the least-significant 1-bit. On most computers this position can be computed by using the native primitive operation that counts the number of trailing 0-bits in a word.

Assuming the availability of the needed hardware support, this refinement makes the analysis of the algorithm straightforward: Each distinguished ancestor is accessed in constant worst-case time, and for each node (except the root) one element comparison and at most one element swap are performed.

procedure: $d\text{-}ancestor(j:$ index)
$z \leftarrow trailing\_zero\_count(j);$
return $j \gg (z + 1)$

Fig. 4. A faster way of finding the distinguished ancestor of a node

To test the effectiveness of this idea in practice, we programmed the standard algorithm and this instruction-optimized refinement. When implementing the function $trailing\_zero\_count$, we tried both the built-in functions $\_\_builtin\_ctz$ and $\_\_builtin\_popcount$ provided by our compiler (g++); in out test environment the former led to a superior performance. As seen from the numbers in Table 5, the instruction optimization made the program faster, and the numbers in Table 6 verify that the number of instructions executed actually reduced.

Table 5. Standard vs. instruction-optimized constructions; execution time divided by n in nanoseconds; the elements were given in random order

n	standard	instruction optimized $\_\_builtin\_ctz$
2^{10}	10.49	7.86
2^{15}	10.26	7.49
2^{20}	10.61	7.83
2^{25}	10.96	8.16

Table 6. Standard vs. instruction-optimized constructions; number of instructions executed divided by n; the elements were given in random order

n	standard	instruction optimized $\_\_builtin\_ctz$	$\_\_builtin\_popcount$
2^{10} 2^{15} 2^{20} 2^{25}	22.5	12.5	16.5

4   Branch Optimization: No if Statements

Branch prediction is an important efficiency issue in pipelined processors, as upon a conditional branch being fetched the processor normally guesses the outcome of the condition and starts the execution of the instructions in one of the branches speculatively. If the prediction was wrong, the pipeline must be flushed, a new set of instructions must be fetched in, and the work done with the wrong branch of the code is simply wasted. To run programs efficiently in this kind of environment one may want to avoid conditional branches if possible.

The standard weak-heap construction algorithm has few conditional branches. By our instruction optimization we already removed the loop used for computing the distinguished ancestors. In accordance, the main body of the algorithm has two unnested loops that both end with a conditional branch; but only when stepping out of a loop a misprediction is incurred. Hence, the main issue to guarantee $O(1)$ branch mispredictions is to remove the **if** statement in the *join* procedure. To do that, we replace the conditional branch with arithmetic operations.

procedure: $join(i, j$: indices)
$smaller \leftarrow (a_j < a_i)$
$\Delta \leftarrow smaller * (j - i)$
$k \leftarrow i + \Delta$
$\ell \leftarrow j - \Delta$
$t \leftarrow a_\ell$
$a_i \leftarrow a_k$
$a_j \leftarrow t$
$r_j \leftarrow smaller$

Fig. 7. Joining two weak heaps without a conditional branch

As shown in Table 8, combining this optimization with that described in the previous section, in our test environment the running times again improved. In fact, these are the best running times among those of the algorithms presented in this paper. As confirmed in Table 9, for the branch-optimized version the number of branch mispredictions incurred is indeed negligible.

Table 8. Instruction-optimized vs. branch-optimized constructions; execution time divided by n in nanoseconds; the elements were given in random order

n	instruction optimized	branch optimized
2^{10}	7.86	6.28
2^{15}	7.49	6.39
2^{20}	7.83	6.72
2^{25}	8.16	7.08

5   An Alternative Construction: Don't Look Upwards

Another way of building a weak heap is to avoid climbing to the distinguished ancestors altogether. We still build the heap bottom-up level by level, but we fix the weak-heap ordering by considering each node and its right subtree. The idea comes from Floyd's algorithm for constructing binary heaps [6]. To mimic this algorithm we need the *sift-down* procedure, which is explained next.

Table 9. Standard vs. instruction-optimized vs. branch-optimized constructions; total number of mispredicted branches; the elements were given in random order

n	standard	instruction optimized	branch optimized
2^{10}	1 061	512	1
2^{15}	34 171	16 385	1
2^{20}	1 093 963	524 110	2
2^{25}	35 005 433	16 776 271	34

Assume that the elements at the right subtree of a_j obey the weak-heap ordering. The *sift-down* procedure is used to establish the weak-heap ordering between the element at a_j and those in the right subtree of a_j. Starting from the right child of a_j, the last node on the left spine of the right subtree of a_j is identified; this is done by repeatedly visiting left children until reaching a node that has no left child. The path from this node to the right child of a_j is traversed upwards, and *join* operations are repeatedly performed between a_j and the nodes along this path. The correctness of the *sift-down* procedure follows from the fact that, after each *join*, the element at location j is less than or equal to every element in the left subtree of the node considered in the next *join*.

With *sift-down* in hand, a weak heap can be constructed by calling the procedure on every node starting from the the lower levels upwards (see Fig. 10). Unfortunately, the running times achieved for this method are not satisfactory, as indicated in Table 11. Compared to the standard method, the slowdown is more significant for large values of n. We relate this behaviour to cache effects.

procedure: *sift-down*(j: index)
$k \leftarrow 2j + 1 - r_j$
while $2k + r_k < n$
  $k \leftarrow 2k + r_k$
while $k \neq j$
  join(j, k)
  $k \leftarrow \lfloor k/2 \rfloor$

procedure: *construct*(a: array of n elements, r: array of n bits)
for $i \in \{0, 1, \ldots, n-1\}$
  $r_i \leftarrow 0$
for $j = \lfloor n/2 \rfloor - 1$ **to** 0 **step** -1
  sift-down(j)

Fig. 10. An alternative construction of a weak heap

Table 11. Standard vs. alternative constructions; execution time divided by n in nanoseconds; the elements were given in random order

n	standard	alternative
2^{10}	10.49	11.16
2^{15}	10.26	11.66
2^{20}	10.61	18.44
2^{25}	10.96	19.96

6   Cache Optimization: Depth-First Construction

To avoid the bad cache performance of the previous method, the nodes of the heap should be visited in depth-first rather than in breadth-first order. This idea, applied for binary heaps in [1], improves the locality of accesses as the traversal tends to stay longer at nearby memory locations. The number of element comparisons obviously remains unchanged. A recursive procedure is given in Fig. 12.

procedure: $df\text{-}construct(i, j\text{: indices})$
if $j < \lfloor n/2 \rfloor$
   | $df\text{-}construct(j, 2j + 1)$
   | $df\text{-}construct(i, 2j)$
$join(i, j)$

procedure: $construct(a\text{: array of } n \text{ elements, } r\text{: array of } n \text{ bits})$
for $i \in \{0, 1, \ldots, n - 1\}$
   | $r_i \leftarrow 0$
if $n > 1$
   | $df\text{-}construct(0, 1)$

Fig. 12. Recursive depth-first weak-heap construction

Although we do not explicitly use the *sift-down* procedure here, we would like to point out that this method is a cache-optimized version of the method described in the previous section. Following the guidelines given in [1], it is not difficult to avoid recursion when implementing the procedure.

The running times given in Table 13 indicate that we are far from those of the standard method. On the other hand, the cache-miss rates given in Table 14 indicate that the depth-first construction has a better cache behaviour.

7   Move Optimization: Trading Swaps for Delayed Moves

Now we implement the aforementioned depth-first construction in a different way (see Fig. 15). The idea is to walk down the left spine of the child of the root

Table 13. Standard vs. cache-optimized constructions; execution time divided by n in nanoseconds; the elements were given in random order

n	standard	cache optimized
2^{10}	10.49	15.05
2^{15}	10.26	15.08
2^{20}	10.61	15.21
2^{25}	10.96	15.20

Table 14. The number of cache misses incurred by different algorithms as a factor of n/B, where B is the block size in words; the elements were given in random order

n	standard	instruction optimized	alternative	cache optimized
2^{10}	1.25	1.25	1.25	1.25
2^{15}	1.25	1.25	1.25	1.25
2^{20}	1.72	1.52	6.38	1.25
2^{25}	2.47	2.23	7.36	1.25

and call the procedure recursively at every node we visit. After coming back from the recursive calls, the *sift-down* operation is applied to restore the weak-heap ordering at the root. In the worst case the number of element moves performed by this algorithm is the same as that performed by the standard algorithm. For both algorithms, $n - 1$ swaps may be performed. As each swap involves three element moves (element assignments), the number of element moves is bounded by $3n$.

To reduce the bound on the number of element moves to $2n$, we postpone the swaps done during the *sift-down* operation. A similar approach was used by Wegener [8] when implementing *sift-down* in a bottom-up manner for binary heaps. The idea is to use a bit vector (of at most $\lg n$ bits) that indicates the

procedure: *df-construct*(i: index)
$j \leftarrow 2i + 1$
while $j < \lfloor n/2 \rfloor$
$\quad | \quad$ *df-construct*(j)
$\quad | \quad j \leftarrow 2j$
sift-down(i)

procedure: *construct*(a: array of n elements, r: array of n bits)
for $i \in \{0, 1, \ldots, n-1\}$
$\quad | \quad r_i \leftarrow 0$
if $n > 1$
$\quad | \quad$ *df-construct*(0)

Fig. 15. Another implementation for depth-first weak-heap construction

winners of the comparisons along the left spine; a 1-bit indicates that the corresponding element on the spine was the smaller of the two elements involved in this comparison. Of course, we still compare the next element up the spine with the current winner of the executed comparisons. After the results of the comparisons are set in the bit vector, the actual element movements are performed. More precisely, the elements corresponding to 1-bits are rotated; this accounts for at most $\mu + 2$ element moves if all the $\mu + 1$ elements on the left spine plus the root are rotated (assuming that the number of nodes on the left spine of the child of the root is μ). To calculate the number of element moves involved, we note that the sum of the lengths of all the left spines is $n - 1$. In addition, we note that the *sift-down* operation will be executed for at most $n/2$ nodes; these are the non-leaves. The $2n$ bound follows. See Tables 16 and 17.

Table 16. Standard vs. move-optimized constructions; execution time divided by n in nanoseconds; the elements were given in random/decreasing orders

n	standard		move optimized	
	random	decreasing	random	decreasing
2^{10}	10.49	7.86	22.60	19.91
2^{15}	10.26	7.60	22.02	18.43
2^{20}	10.61	7.95	22.14	18.52
2^{25}	10.96	8.30	22.11	18.53

Table 17. Standard vs. move-optimized constructions; number of moves per element; the elements were given in random/decreasing orders

n	standard		move optimized	
	random	decreasing	random	decreasing
2^{10}	1.49	2.99	1.16	1.99
2^{15}	1.49	2.99	1.16	1.99
2^{20}	1.49	3.00	1.16	2.00
2^{25}	1.50	3.00	1.16	2.00

8   Repeated Insertions: Non-linear Work, Yet a Linear Number of Element Comparisons

To insert an element e into a weak heap, we first add e to the next available array entry. To reestablish the weak-heap ordering, we use the *sift-up* procedure. As long as e is smaller than the element at its distinguished ancestor, we swap the two elements and repeat for the new location of e using the *join* procedure. Thus, *sift-up* at location j requires $O(\lg j)$ time and involves at most $\lceil \lg(1 + j) \rceil$ element comparisons.

```
procedure: sift-up(j: index)
while j ≠ 0
    i ← d-ancestor(j)
    before ← r_j
    join(i, j)
    if before = r_j
        break
    j ← i


procedure: construct(a: array of n elements, r: array of n bits)
for k = 1 to n − 1
    r_k ← 0
    if (k bitand 1)    (*)
        r_⌊k/2⌋ ← 0
    sift-up(k)
```

Fig. 18. Constructing a weak heap by repeated insertions

When constructing a weak heap using repeated insertions (see Fig. 18), we observed that, while the execution time increased with n, the number of element comparisons stayed constant per element for an increasing value of n (see Tables 19 and 20). As the worst-case input for the experiments we used the sequence of the form $\langle 0, n-1, n-2, \dots, 1 \rangle$ as adviced in [5]. Next we prove that the number of element comparisons performed is indeed linear in the worst case.

Theorem 1. *The total number of element comparisons performed while constructing a weak heap using n in-a-row insertions is at most $3.5n + O(\lg^2 n)$.*

Proof. We distinguish between two types of element comparisons done by the *sift-up* operations. An element comparison that involves the root or triggers the **break** statement to get out of the **while** loop is called a *terminal element comparison*. There is exactly one such comparison per insertion, except for the first insertion, resulting in $n-1$ terminal element comparisons during the whole construction. All other element comparisons are *non-terminal*.

Next we caclulate an upper bound for the number of non-terminal element comparisons performed. Fix a node x whose height is h, $h \in \{1, 2, \dots, \lceil \lg n \rceil + 1\}$. Consider all the non-terminal element comparisons performed between the elements at x and the distinguished ancestor of x throughout the process. For such a comparison to take place an element should have been inserted in the right subtree of x (may include x itself). Consider the first element e that is inserted in the right subtree of x at distance d from x and results in a non-terminal element comparison between the elements at x and its distinguished ancestor. That is, d can take values 0, 1, etc.

For $d = 0$, the elements at x and its distinguished ancestor are always compared unless x is the root. For $d = 1$, we have to consider the **if** statement marked with a star (*) in the program. When the inserted node is the only child of x, it is made a left child by updating the reverse bit at x. So this first insertion at distance one will never trigger a non-terminal element comparison; only the second insertion does that. Consider now the case where $d \geq 2$. Because of the non-terminal comparison between the elements at x and its distinguished ancestor the reverse bit at x is flipped, and the right subtree of x becomes its left subtree. All the upcoming insertions that will land in this subtree at distance d from x will not involve x as a distinguished ancestor. It follows that, for this given subtree, the element at x will not be compared with that at the distinguished ancestor of x until an element is inserted below x at distance $d + 1$ from x. In conclusion, the node x can be charged with at most one element comparison for each level of both its subtrees. Summing the number of non-terminating element comparisons done for all values of d, we get that the element at x is compared against the element at its distinguished ancestor at most twice its height minus two; that is at most $2h - 2$ times.

In a weak heap of size n, there are at most $\lceil n/2^h \rceil$ nodes of height h, $h \in \{1, 2, \ldots, \lceil \lg n \rceil + 1\}$. On the basis of the discussion in the preceeding paragraph it follows that the number of non-terminal element comparisons is bounded by $\lceil n/2 \rceil + \sum_{h=2}^{\lceil \lg n \rceil + 1} (2h - 2) \cdot \lceil n/2^h \rceil < 2.5n + O(\lg^2 n)$. Adding the $n - 1$ terminal element comparisons, the total number of element comparisons performed by all n insertions is at most $3.5n + O(\lg^2 n)$.   □

Observe that the number of element comparisons and that of element moves go hand in hand so that the number of element moves performed is also linear. That is, the running time is wasted in distinguished-ancestor calculations and, unfortunately, we cannot take any shortcuts since some of the reverse bits may be set above an insertion point.

Table 19. Standard vs. repeated-insertion constructions; execution time divided by n in nanoseconds; the elements were given in random/special worst-case orders

n	standard		repeated insertions	
	random	special	random	special
2^{10}	10.49	7.86	31.46	45.91
2^{15}	10.26	7.60	32.68	64.09
2^{20}	10.61	7.95	33.04	78.00
2^{25}	10.96	8.30	33.18	89.36

9   New Memory Layout: Less Work, Different Outcome

There are other possible array embeddings than the standard one (where the children of the node at location i are at locations $2i$ and $2i + 1$). Consider the following layout: For a node a_i whose depth in the tree is d, its right child is stored at location $i + 2^{d-r_i}$ and its left child at location $i + 2^{d-1+r_i}$. As with the standard layout, reverse bits are used to swap the two subtrees. To access the parent of a node a_i whose depth is d, we need an extra check. If $i \geq 2^{d-1} + 2^{d-2}$,

Table 20. Standard vs. repeated-insertion constructions; number of element comparisons divided by n; the elements were given in random/special worst-case orders

n	standard		repeated insertions	
	random	special	random	special
2^{10}	0.99	0.99	1.76	3.38
2^{15}	0.99	0.99	1.77	3.49
2^{20}	0.99	0.99	1.77	3.49
2^{25}	1	1	1.77	3.49

the parent is at location $i - 2^{d-1}$; otherwise, the parent is at location $i - 2^{d-2}$. If the depth is not available, it can be computed in constant time using the number of leading zeros. The main advantage of this dual layout is that the construction algorithm does not need distinguished-ancestor calculations.

> **procedure:** *dual-construct*(a: array of n elements, r: array of n bits)
> $\ell \leftarrow n$
> **while** $\ell > 1$
> $\quad$ **for** $i \in \{0, \ldots, \lfloor \ell/2 \rfloor\}$
> $\quad\quad$ *join*($i, \lfloor \ell/2 \rfloor + i$)
> $\quad$ $\ell \leftarrow \lfloor \ell/2 \rfloor$

Fig. 21. Constructing a weak heap for the new memory layout

10   Conclusion

The weak heap is an amazingly simple and powerful structure. If perfectly balanced, weak heaps resemble heap-ordered binomial trees [7]. A weak heap is implemented in an array with extra bits that are used for subtree swapping.

Binomial-tree parents are distinguished ancestors in a weak-heap setting. We showed that, for Dutton's construction of weak heaps [3], a distinguished ancestor can be computed in constant wost-case time. We also provided depth-first weak-heap building procedures. The standard memory layout is fastest for accessing neighbouring nodes, but other layouts may lead to a faster construction.

Contrary to binary heaps, repeated insertions lead to a constant number of element comparisons per inserted element. We proved that a sequence of n insertions in an initially empty weak heap requires at most $3.5n + O(\lg^2 n)$ element comparisons.

We would like end this paper with a warning: The experimental results reported depended on the environment where the experiments were performed and on the type of data used as input. In our experimental setup, element comparisons and element moves were cheap, and branch mispredictions and cache misses were expensive. When some of these conditions will change, the overall picture may drastically change.

Source Code

The programs used in the experiments are available via the home page of the CPH STL (http://cphstl.dk/) in the form of a PDF document and a tar file.

References

1. Bojesen, J., Katajainen, J., Spork, M.: Performance engineering case study: Heap construction. ACM J. Exp. Algorithmics 5, Article 15 (2000)
2. Bruun, A., Edelkamp, S., Katajainen, J., Rasmussen, J.: Policy-Based Benchmarking of Weak Heaps and Their Relatives. In: Festa, P. (ed.) SEA 2010. LNCS, vol. 6049, pp. 424–435. Springer, Heidelberg (2010)
3. Dutton, R.D.: Weak-heap sort. BIT 33(3), 372–381 (1993)
4. Edelkamp, S., Elmasry, A., Katajainen, J.: The weak-heap data structure: Variants and applications. J. Discrete Algorithms 16, 187–205 (2012)
5. Edelkamp, S., Wegener, I.: On the Performance of $WEAK - HEAPSORT$. In: Reichel, H., Tison, S. (eds.) STACS 2000. LNCS, vol. 1770, pp. 254–266. Springer, Heidelberg (2000)
6. Floyd, R.W.: Algorithm 245: Treesort 3. Commun. ACM 7(12), 701 (1964)
7. Vuillemin, J.: A data structure for manipulating priority queues. Commun. ACM 21(4), 309–315 (1978)
8. Wegener, I.: Bottom-up-heapsort, a new variant of heapsort beating, on an average, quicksort (if n is not very small). Theoret. Comput. Sci. 118(1), 81–98 (1993)
9. Williams, J.W.J.: Algorithm 232: Heapsort. Commun. ACM 7(6), 347–348 (1964)

On Counting Range Maxima Points in Plane

Anil Kishore Kalavagattu, Jatin Agarwal, Ananda Swarup Das,
and Kishore Kothapalli

International Institute of Information Technology, Hyderabad, India
{anilkishore,jatin.agarwal}@research.iiit.ac.in,
anandaswarup@gmail.com,
kkishore@iiit.ac.in

Abstract. We consider the problem of reporting and counting maximal points in a given orthogonal query range in two-dimensions. Our model of computation is the pointer machine model. Let P be a static set of n points in $\mathbb{R}^2$. A point is *maximal* in P if it is not dominated by any other point in P. We propose a linear space data structure that can support counting the number of maximal points inside a 3-sided orthogonal query rectangle unbounded on its right in $O(\log n)$ time. For counting the number of maximal points in a 4-sided orthogonal query rectangle, we propose an $O(n \log n)$ space data structure that can be constructed in $O(n \log n)$ time and queried upon in $O(\log n)$ time. This work proposes the first data structure for counting the number of maximal points in a query range. Das et al. proposed a data structure for the counting version in the word RAM model [WALCOM 2012].

For the corresponding reporting versions, we propose a linear size data structure for reporting maximal points inside a 3-sided query range in time $O(\log n + k)$, where k is the size of the output. We propose an $O(n \log n)$ size data structure for reporting the maximal points in a 4-sided orthogonal query range in time $O(\log n + k)$, where k is the size of the output. The methods we propose for reporting maximal points are simpler than previous methods and meet the best known bounds.

Keywords: Maxima, Plane, Orthogonal range, Reporting, Counting.

1 Introduction

Range searching is one of the widely studied topics in spatial databases and computational geometry. Informally, range searching can be stated as follows: Given a set P of objects, we wish to preprocess the data set into a data structure such that given a query object q, we can efficiently report the objects or count the number of objects in $P \cap q$, the set of objects in both P and q. The problem has wide applications in geographic information systems, CAD tools, database retrieval, and the like. The advancement of technology has led to information explosion and the number of objects inside a query range can be huge. In most cases, reporting a sample of the result set is preferred. In this regard, we borrow

S. Arumugam and B. Smyth (Eds.): IWOCA 2012, LNCS 7643, pp. 263–273, 2012.

the concept of skyline query from the database community. With the increase in the number of dimensions, the number of skyline points may also increase [3]. Finding the number of skyline points can be useful in reporting every kth skyline point, reporting the k skyline points in the middle, and many other applications. In this work, we propose some data structures to solve the problem of skyline counting and reporting. Henceforth we refer skyline points as maximal points.

1.1   Definitions

We are given a set of n points $P = \{p_1, \ldots, p_n\}$ in $\mathbb{R}^2$. We use $x(p)$ to represent the x-coordinate of a point p and similarly $y(p)$ to represent its y-coordinate. A point p_i is said to dominate another point p_j if both $x(p_i) \geq x(p_j)$ and $y(p_i) \geq y(p_j)$ hold. The set P' $(\subseteq P)$ of all the points, each of which is not dominated by any other point in P is called the dominating set of P. They are also called maximal elements, or *maxima*, of the set P [1]. The maximal points when sorted by increasing x-coordinate are also sorted by decreasing y-coordinate, and hence they are also called the *staircase* of the set. Figure 1(a) shows the maxima as a staircase.

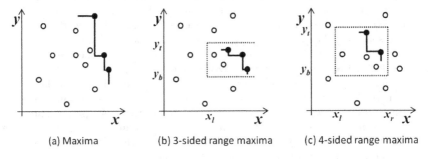

(a) Maxima   (b) 3-sided range maxima   (c) 4-sided range maxima

Fig. 1. Maxima points are shown as filled circles. Maxima of the (a.) point set, (b.) 3-sided orthogonal query range and (c.) 4-sided orthogonal query rectangle

In this work, we consider the following two problems.

Problem 1. We are given a set P of n points in $\mathbb{R}^2$. We wish to preprocess P into a data structure such that given an orthogonal query region q, we can efficiently **report** the maximal points of $P \cap q$, where $P \cap q$ is the set of points in both P and q.

Problem 2. We are given a set P of n points in $\mathbb{R}^2$. We wish to preprocess P into a data structure such that given an orthogonal query region q, we can efficiently **count** the number of maximal points of $P \cap q$, where $P \cap q$ is the set of points in both P and q.

We consider two types of query ranges for q. In the first type, q is a 3-sided orthogonal rectangle unbounded on its right, as shown in Figure 1(b). In the second type, q is a general 4-sided orthogonal rectangle, as shown in Figure 1(c).

1.2   Related Work

The maxima of a static set of two-dimensional points can be computed in optimal $O(n \log n)$ time [4]. Dynamic data structures of linear size that can support reporting all maximal points in the plane have been proposed [5,6]. Range maxima in the two-dimensional point set is studied in [7,8]. In [8] the authors proposed a static data structure of size $O(n \log n)$ that can support reporting maximal points in a given orthogonal query rectangle in $O(\log n + k)$ time, where k is the size of the output. Brodal et al. in [7] suggested a dynamic data structure of size $O(n \log n)$ and worst case query time of $O(\log^2 n + k)$. Here we consider a static set of points and computations in pointer machine model. To the best of our knowledge, we are considering the problem of counting the number of maximal points inside a query rectangle, for the first time in literature. The previous data structure of [11] has a query time of $O(\frac{\log^{\frac{3}{2}} n}{\log \log n})$ for the problem of counting the maximal points in an orthogonal query rectangle in the word RAM model. Our structures for counting the number of maximal points are built upon their corresponding reporting versions and hence they also support reporting the maximal points. Though the bounds for reporting maximal points are not improved, the methods presented here are simpler and achieve the best known bounds. Note that reporting (resp. counting) maxima of a set of points in a given query range is different from a dominance query, where the aim is to report all the points dominating a given query point q. Given a query point q, maximal points of the $north - east$ quadrant $w.r.t$ the point q are a subset of dominating points, where as a dominance query results all the points in that quadrant [2].

1.3   Our Results

We present a linear size data structure that supports reporting maximal points in a 3-sided query range unbounded on its right in $O(\log n + k)$ time, where k is the size of the output. Counting the number of maximal points in a 3-sided query range can be supported in $O(\log n)$ time using linear space. Reporting the maximal points in an orthogonal query rectangle can be done using $O(n \log n)$ space and in $O(\log n + k)$ time, where k is the size of the output. The corresponding counting can be supported in $O(\log n)$ time using $O(n \log n)$ space.

The rest of paper is organized as follows. In Section 2 we give brief details of the supporting data structures and outline of our solution. In Section 3 we give the details of the preprocessing and the query algorithms for reporting and counting maximal points in a 3-sided query. In Section 4 we extend this to a general 4-sided query. In Section 5 we propose a better solution for the counting version based on some observations. In Sections 6 and 7, we briefly discuss some direct applications of our results and discuss future work.

2   Our Solution Framework

Preliminaries: Apart from the standard range tree [12], we use a *red-black tree* [9], a balanced binary search tree that maintains a set of n elements in sorted order. It supports the following operations in $O(\log n)$ time using $O(n)$ space : INSERT(u) inserts the element u into the tree and PREDECESSOR(u) returns the largest element in the tree that is smaller than or equal to element u. We also use a persistent version of a *red-black tree*. We maintain a sorted set of elements. Initially the data structure S is empty and a new version of the data structure is created each time a new element is inserted. Thus we obtain a list of versions of S. A query on S can now be made on any of its previous versions. In [10], it is shown that after insertion of n elements, the data structure occupies $O(n)$ space and a query on any previous version takes $O(\log n)$ time. We use the terms *persistent sorted set* and *persistent red-black tree* interchangeably.

Outline of Solution: Given a set of points P, the point p_a with the maximum y-coordinate in P is a maximal point, because there is no other point dominating p_a. If we are asked to report all the maximal points in P, we can start by reporting this point p_a. The next maximal point p_b is the point with the maximum y-coordinate among the points not dominated by p_a i.e., the points in the south-east quadrant of p_a. The main idea of the solution is, instead of finding the next maximal point by computing at query time, we can preprocess and store it. For each point p_i in P, we answer the question, if p_i is a maximal point, what is the next maximal point of P immediately following p_i, in decreasing order of y-coordinate. We denote this using $nxt[p_i]$. The nxt points are shown using pointers in Figure 2 below. If there is no such point, we assign $nxt[p_i] = (+\infty, 0)$. This is shown as ∞ in the Figure 2.

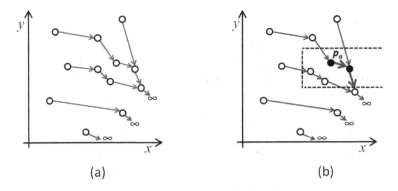

(a)   (b)

Fig. 2. (*a.*)For each of the points p, $nxt[p]$ is shown using pointers. (*b.*) Starting from the point point p_a, the maximal points (filled circles) along the nxt pointers.

3 3-Sided Range Maxima Queries

3.1 Reporting

In this subsection, we present the algorithm for reporting the maximal points in a 3-sided orthogonal query, unbounded on its right.

Preprocessing. We need to compute the array nxt such that for each point $p_i \in P$, $nxt[p_i]$ is the point with the maximum y-coordinate in the south-east(SE) quadrant of p_i. Initialize a *red-black tree* T and a persistent sorted set S. Sort the point set P in the decreasing order of x-coordinate and process the points in that order. We insert the points into T and S with their y-coordinate values as the key. For each point p_i, set $nxt[p_i] = \text{PREDECESSOR}(y(p_i))$ if there exists any, otherwise $nxt[p_i] = (+\infty, 0)$ and then insert p_i into the trees T and S. This requires $O(n \log n)$ time. Note that when we are at a point p_i, all the points with x-coordinate larger than $x(p_i)$ are already in the trees T and S, and none of the points with x-coordinate smaller than $x(p_i)$ are present in them. Hence the proof of correctness of nxt pointers follows. We now also have different versions of the data structure S. Each version S_i obtained just after inserting the point p_i has all the points p having $x(p) \geq x(p_i)$ and sorted using the y-coordinate values as the key.

Query Algorithm. Given a 3-sided orthogonal query $q = [x_l, \infty] \times [y_b, y_t]$, we find the point p_a with the maximum y-coordinate in $P \cap q$ using the persistent sorted set S as follows. Using binary search, identify the version of S that existed just after the insertion of all points having x-coordinate $\in [x_l, \infty]$. Query this version of S with PREDECESSOR(y_t) to obtain the point p_a. Initialize $p = p_a$.

While $y(p)$ is not less than y_b, report the point p and set $p \leftarrow nxt[p]$

The proof of correctness is as follows. The point $p_a \in P \cap q$ with the maximum y-coordinate is a maximal point, because there is no other point in $P \cap q$ that dominates p_a. Let $p_k = nxt[p_i]$, for any maximal point p_i reported. If $p_k \in q$, then p_k is the point with the maximum y-coordinate in the South-east quadrant of p_i and hence the next maximal point in the decreasing y-coordinate, immediately following p_i. If $p_k \notin q$, then the orthogonal rectangular region $[x(p_i), \infty] \times [y_b, y(p_i)]$ is empty and hence there exists no more maximal point following p_i. The main computation involved is in finding the first maximal point p_a and then just following the nxt pointers from there, reporting a maximal point each time. We now conclude this subsection by stating the following theorem.

Theorem 1. *Given a set P of n points in $\mathbb{R}^2$, we can preprocess P into a linear space data structure in $O(n \log n)$ time such that, given a 3-sided orthogonal* *query* $q = [x_l, \infty] \times [y_b, y_t]$, *we can report the maximal points of* $P \cap q$ *in time* $O(\log n + k)$, *where* k *is the size of the output.*

3.2   Counting

In this subsection, we present the algorithm for counting the number of maximal points in a 3-sided orthogonal query, unbounded on its right.

Preprocessing. We compute the array nxt as given in Section 3.1. Let $nxt^m[p_i] = nxt[nxt[\dots _{m\ times}[p_i]]]$, the point reached after making exactly m moves by following the nxt pointers, starting from p_i. If there is no such point, then $nxt^m[p_i] = (+\infty, -\infty)$. We compute the sparse table jmp having n rows and $(\log n + 1)$ columns. The ith row corresponds to the point p_i and for $k \in \{0, .., \log n\}$ $jmp[p_i][k] = nxt^{2^k}[p_i]$. We compute the table jmp of size $O(n \log n)$ in $O(n \log n)$ time using dynamic programming and the recurrence, $jmp[p][0] = nxt[p]$ and for $k > 0$, $jmp[p][k] = jmp[\ jmp[p][k-1]\ ][k-1]$.

Query Algorithm. Given a 3-sided orthogonal query $q = [x_l, \infty] \times [y_b, y_t]$, we first find the point p_a with the maximum y-coordinate in $P \cap q$, as described in the query algorithm of Section 3.1. The number of maximal points in $P \cap q$ is equal to the minimum m such that $y(nxt^m[p_a]) < y_b$ and can be computed in $O(\log n)$ time using MinJumps(p_a, y_b, n) as described below.

```
Program MinJumps (p, ylim, n)
    var
       k := log(n)
       jumps := 0
    while k >= 0
       if y(jmp[p][k]) >= ylim
           jumps := jumps + power(2,k)
           p := jmp[p][k]
       end-if
       k := k - 1
    end-while
    return jumps + 1
end-MinJumps
```

The above procedure essentially finds the number of maximal points encountered before we cross the lower bound y_b on the y-coordinate, by following the nxt points starting from the point p_a. We now conclude this section with the following theorem.

Theorem 2. *Given a set* P *of* n *points in* $\mathbb{R}^2$, *we can preprocess* P *into a data structure of size* $O(n \log n)$ *in time* $O(n \log n)$ *such that, given a 3-sided orthogonal query* $q = [x_l, \infty] \times [y_b, y_t]$, *we can count the number of maximal points of* $P \cap q$ *in time* $O(\log n)$.

4   4-Sided Range Maxima Queries

4.1   Reporting

In this subsection, we present the algorithm for reporting maximal points among the points in the given query orthogonal rectangle.

Preprocessing. The data structure is a standard layered range tree [12] in which the main tree stores the points sorted by increasing x-coordinate values at the leaves. Each internal node u is associated with the points in the leaves of the sub-tree rooted at u. A secondary structure A_u at each internal node u is an array storing the points associated with it in decreasing order of the y-coordinate values. The array A at each node can be constructed by carrying out merge sort using the y-coordinate values as keys. The array $nxt[u]$ is computed while merging the left child $lc(u)$ and the right child $rc(u)$ of u. Each point p from $rc(u)$ computes $nxt[u][p] = nxt[rc(u)][p]$ and for each point p from $lc(u)$, $nxt[u][p]$ is the point with maximum y-coordinate between $nxt[lc(u)][p]$ and p_{rc}, where p_{rc} is the recent most point from $rc(u)$ merged into A_u. In order to speed up query time, we use fractional cascading [12] at the cost of storing additional pointers at each node. Let w and v be the left and right children of u respectively. While merging the secondary arrays A_w and A_v to construct A_u, we create and store pointers as follows. Each index i of A_u stores two pointers, one pointing to the largest value in A_w which is smaller than or equal to $A_u[i]$ and similarly the other pointing to the largest value in A_v which is smaller than or equal to $A_u[i]$. The merging step takes linear time at each node, resulting in total $O(n \log n)$ preprocessing time.

Query Algorithm. Given an orthogonal query rectangle $q = [x_l, x_r] \times [y_b, y_t]$, the query algorithm for reporting maximal points in $p \cap q$ is as follows:

1. The range of x-coordinates in $[x_l, x_r]$ can be expressed as the disjoint union of $l = O(\log n)$ canonical subsets. Let the canonical subset of nodes be $\nu_1, \nu_2, \ldots, \nu_l$ from left to right in that order, as shown in the Figure 3(a).
2. Find the node ν_{split}, which is the least common ancestor of ν_1 and ν_l. Find the index of the point with the largest y-coordinate less than or equal to y_t in $A_{\nu_{split}}$ using binary search.
3. Process the canonical nodes in reverse order, starting from ν_l back to ν_1, as follows. Initialize $i \leftarrow l$, $y_{low} \leftarrow y_b$.
4. Consider the node ν_i. Find the point $ptop_i \in \nu_i$ with the largest y-coordinate less than or equal to y_t. This can be obtained by following the pointer from $A_{\nu_{split}}$ at the index found in step 2.
5. If there is no point $ptop_i$, skip this step. Let $p = ptop_i$.

    While $y(p) \geq y_{low}$, report the point p and set $p \leftarrow nxt_{\nu_i}[p]$

    The points reported are the maximal points of $P \cap q$, which are associated with the node ν_i.

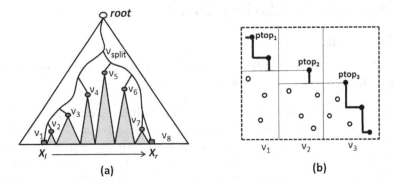

Fig. 3. (*a*.)Range tree with the canonical nodes highlighted (*b*.) Processing a 4-sided range maxima query

6. At this point, we processed the nodes $\nu_l, \nu_{l-1}, \ldots, \nu_i$. If $ptop_i$ exists, then set $y_{low} \leftarrow ptop_i$. Set $i \leftarrow i - 1$ and if $i \geq 1$, move to the node ν_i and repeat from step 4, else exit.

Theorem 3. *Given a set P of n points in $\mathbb{R}^2$, we can preprocess P into a data structure of size $O(n \log n)$ in time $O(n \log n)$ such that, given an orthogonal query rectangle $q = [x_l, x_r] \times [y_b, y_t]$, we can report the maximal points of $P \cap q$ in time $O(\log n + k)$, where k is the number of maximal points reported.*

Proof : At each of the $O(\log n)$ levels of the tree, each point is stored in array A and array nxt and has two pointers to its children. So the space required is $O(n \log n)$. At each of the $O(\log n)$ levels of the tree, at most two nodes are visited [12]. When we visit a canonical node u, we also have the pointer to the maximal point p_a with maximum y-coordinate in u and hence reporting the maximal points starting from p_a, by following the nxt pointers takes $O(m)$ time, where m is the number of maximal points reported at u. So the overall time complexity is $O(\log n + k)$ □

4.2   Counting

The basic structure is same as the one used for reporting in Section 4.1. Similar to counting described in Section 3.2, we maintain a sparse table at each internal node of the range tree. At each of the $O(\log n)$ levels of the tree, we use additional $O(nh)$ space for storing the sparse table, where h is the height of the level. The total space required is thus $O(n \log^2 n)$. Given an orthogonal query rectangle $q = [x_l, x_r] \times [y_b, y_t]$, we follow similar to the query algorithm of 4.1 and instead of the step 5, we sum up the count $\texttt{MinJumps}(ptop_i, y_{low}, n)$ at each of the $O(\log n)$ canonical nodes. The total time taken is $O(\log^2 n)$. We conclude this section with the following theorem.

Theorem 4. *Given a set P of n points in $\mathbb{R}^2$, we can preprocess P into a data structure of size $O(n \log^2 n)$ in time $O(n \log^2 n)$ such that, given an orthogonal* rectangle $q = [x_l, x_r] \times [y_b, y_t]$, we can count the number of maximal points of $P \cap q$ in $O(\log^2 n)$ time.

5   Further Improvements

As we can see in the previous section, to facilitate counting queries, space and time complexities increased by a factor of $\log n$ than the reporting version. This is due to maintenance of a sparse table to find the number of points along a maximal chain between two points. Consider a horizontal slab query $q = [-\infty, \infty] \times [y_l, y_h]$ and let the point with the maximum y-coordinate and the point with the maximum x-coordinate inside q be $p_{y_{max}}$ and $p_{x_{max}}$ respectively. Reporting maxima of q can start with the point $p_{y_{max}}$, which is a maximal point and proceed further down, till the last maximal point. The main observation is, the maximal point with the minimum y-coordinate is the point with the maximum x-coordinate in q, $p_{x_{max}}$. So, for a horizontal slab, the maximal chain starting from the point $p_{y_{max}}$ must pass through the point $p_{x_{max}}$. We state this in the following lemma.

Lemma 1. *Given any horizontal slab query $q = [-\infty, \infty] \times [y_l, y_h]$, by following the nxt pointers along the maximal chain, starting from the point with the maximum y-coordinate in q, the point with the maximum x-coordinate in q is encountered, which is also the maximal point having minimum y-coordinate.*

Using the above lemma, we can now improve the bounds for the counting version as follows.

5.1   3-Sided Range Maxima Counting Queries

We preprocess the point set P such that, given a 3-sided orthogonal query range q unbounded on its right, we can find the point $p_{y_{max}}$ in time $O(\log n)$, as explained in Section 3.1. Also, preprocess P for efficient 1D range maximum queries on x-coordinate, such that the point $p_{x_{max}}$ can be found in constant time using additional linear storage space. Along with the nxt pointer for each of the points, we also store *level* information, where for a given point p, $level[p]$ is the number of points starting from p, along the maximal chain by following the nxt pointers, till the end of the chain. This can be found while computing nxt pointers using, $level[p] = level[nxt[p]] + 1$ if $nxt[p]$ exists and $level[p] = 0$ otherwise. Using the above lemma, we know that the point $p_{x_{max}}$ is present along the maximal chain further down, starting from the point $p_{y_{max}}$. So the difference in their *levels* gives the number of points between them, along their maximal chain. The following steps summarizes the query algorithm.

1. Find the point $p_{y_{max}}$ as described in the query algorithm of Section 3.1
2. Find the point $p_{x_{max}}$ using 1D RMQ
3. Number of maximal points of $q = level[p_{y_{max}}] - level[p_{x_{max}}] + 1$

We now conclude this subsection by stating the following theorem.

Theorem 5. *Given a set P of n points in $\mathbb{R}^2$, we can preprocess P into a linear space data structure in $O(n \log n)$ time such that, given a 3-sided orthogonal query $q = [x_l, \infty] \times [y_l, y_h]$, we can count the number of maximal points of $P \cap q$ in time $O(\log n)$.*

5.2   4-Sided Range Maxima Counting Queries

We preprocess the point set P as explained in Section 4.2, except that we do not compute the sparse table jmp. The tree is a standard layered range tree, with the following additional preprocessing.

1. Fractional cascading among the secondary arrays at each node, having points sorted by y-coordinate values. Using this we can binary search for the query bounds on y-coordinate in the root and find the point with maximum y-coordinate in each of the canonical nodes in total time $O(\log n)$.
2. Each of the secondary arrays is also preprocessed for 1D range maximum queries (RMQ) such that, given two indices in a secondary array, we can find the point with the maximum x-coordinate in constant time, using linear space at each node.
3. Along with the nxt pointers for the points at each node, we also compute the $level$ values, as explained in the previous section.

For the query algorithm, given an orthogonal query rectangle $q = [x_l, x_r] \times [y_l, y_h]$, we proceed similar to that of Section 4.2, but instead of the $O(\log n)$ computations using jmp table, we find the points $p_{y_{max}}$ and $p_{x_{max}}$ inside q in each of the canonical nodes from right to left, and sum up the counts ($level[p_{y_{max}}]$ - $level[p_{x_{max}}] + 1$).

We now conclude this subsection by stating the following theorem.

Theorem 6. *Given a set P of n points in $\mathbb{R}^2$, we can preprocess P into a data structure of size $O(n \log n)$ in time $O(n \log n)$ such that, given an orthogonal query rectangle $q = [x_l, x_h] \times [y_l, y_h]$, we can count the number of maximal points of $P \cap q$ in time $O(\log n)$.*

6   Remarks

We proposed data structures for counting the number of maximal points and also reporting the maximal points in an orthogonal query range. Using the information stored in the table jmp, we can find the kth maximal point in the decreasing order of y-coordinate value, in time $O(\log n)$. Computing the nxt points with respect to the set P allow us to report the maximal points of P, starting from any other point. So the methods proposed can be easily extended to various other applications like, reporting the k maximal points in the middle, reporting every kth maximal point starting from the first maximal point etc., which can be treated as sample of the skyline points in the query range.

The rectangular visibility query problem can be solved using four 4-sided range maxima structures [7]. Our data structure described in Section 5.2 can be used to find the number of rectangularly visible points from a given query point in $O(\log n)$ time.

7 Conclusion

In this work, we studied the problem of reporting and counting maximal points in a given orthogonal query range, in the pointer machine model. We restricted the point set to static two-dimensional points. It will be interesting to see dynamic range maxima counting in plane and counting range maxima in dimensions more than two.

References

1. Bentley, J.L.: Multidimensional divide-and-conquer. Communications of the ACM 23(4), 214–229 (1980)
2. JáJá, J., Mortensen, C.W., Shi, Q.: Space-Efficient and Fast Algorithms for Multidimensional Dominance Reporting and Counting. In: Fleischer, R., Trippen, G. (eds.) ISAAC 2004. LNCS, vol. 3341, pp. 558–568. Springer, Heidelberg (2004)
3. Chan, C.-Y., Jagadish, H.V., Tan, K.-L., Tung, A.K.H., Zhang, Z.: Finding k-dominant skylines in high dimensional space. In: Proceedings of the ACM SIGMOD International Conference on Management of Data (2006)
4. Kung, H.T., Luccio, F., Preparata, F.P.: On finding the maxima of a set of vectors. Journal of the ACM 22(4), 469–476 (1975)
5. Frederickson, G.N., Rodger, S.H.: A New Approach to the Dynamic Maintenance of Maximal Points in a Plane. Discrete & Comp. Geom. 5, 365–374 (1990)
6. Janardan, R.: On the Dynamic Maintenance of Maximal Points in the Plane. Information Processing Letters 40(2), 59–64 (1991)
7. Brodal, G.S., Tsakalidis, K.: Dynamic Planar Range Maxima Queries. In: Aceto, L., Henzinger, M., Sgall, J. (eds.) ICALP 2011. LNCS, vol. 6755, pp. 256–267. Springer, Heidelberg (2011)
8. Kalavagattu, A.K., Das, A.S., Kothapalli, K., Srinathan, K.: On Finding Skyline Points for Range Queries in Plane. In: Proceedings of 23rd Canadian Conference on Computational Geometry (CCCG), pp. 343–346 (2011)
9. Bayer, R.: Symmetric Binary B-Trees: Data Structure and Maintenance Algorithms. Acta Informatica 1, 290–306 (1972)
10. Sarnak, N., Tarjan, R.E.: Planar point location using persistent search trees. Communications of the ACM 29, 669–679 (1986)
11. Das, A.S., Gupta, P., Kalavagattu, A.K., Agarwal, J., Srinathan, K., Kothapalli, K.: Range Aggregate Maximal Points in the Plane. In: Rahman, M. S., Nakano, S.-i. (eds.) WALCOM 2012. LNCS, vol. 7157, pp. 52–63. Springer, Heidelberg (2012)
12. de Berg, M., van Kreveld, M., Overmars, M., Schwarzkopf, O.: Computational Geometry: Algorithms and Applications. Springer (2000) ISBN 3-540-65620-0
13. Yu, C.C., Hon, W.K., Wang, B.F.: Improved Data Structures for Orthogonal Range Successor Queries. Computational Geometry: Theory and Applications 44, 148–159 (2011)

Indexing Highly Repetitive Collections

Gonzalo Navarro

Dept. of Computer Science, University of Chile
gnavarro@dcc.uchile.cl

Abstract. The need to index and search huge highly repetitive sequence collections is rapidly arising in various fields, including computational biology, software repositories, versioned collections, and others. In this short survey we briefly describe the progress made along three research lines to address the problem: compressed suffix arrays, grammar compressed indexes, and Lempel-Ziv compressed indexes.

1 Introduction

After a years-long race to sequence the first human genome in the early 2000's, sequencing has become a routine activity that costs a few thousand dollars[1] and large sequencing companies are producing thousands of genomes per day[2]. Maintaining databases of millions of genomes will be a real possibility very soon. With a storage requirement of about 715 MB per human genome (about 3×10^9 bases using 2 bits each), storing, say, one million genomes is perfectly realistic (around 700 TB). However, sequence analysis tools require *indexed* access to the data, where one can carry out pattern searches and mining. Such indexes require at the very least 50 bits per base (one pointer), raising the 700 TB to more than 16 PB (petabytes). Those sizes, especially if we require fast indexed access to the data, exceed today's technological possibilities within reasonable cost.

What makes this challenge affordable is that most of those genomes (if we assume they belong to the same species, say human) are very similar to each other — 99.99% similar according to typical figures (although there is some debate about the exact value). If we were able to index such a collection within space proportional to the number of differences between the genomes, and not to their total size, then a one-million genome collection could be indexed within 1.6 TB, which is perfectly feasible. Yet, we still do not know how to do this.

Other scenarios where a set of very similar sequences is indexed and pattern searches are provided on them are software repositories (where versions form a tree or graph structure) and versioned document collections (such as Wikipedia or the Internet Wayback Machine, where versions have a linear structure).

In this short survey we will cover the results achieved on the challenge of storing and indexing those "highly repetitive" sequence collections. We will focus on the following scenario, looking for a balance of generality and simplicity that leads to both useful and algorithmically interesting research:

---

[1] *The Guardian* Jan 12, 2012, "Company announces low-cost DNA decoding machine".
[2] *The New York Times* Jan 12, 2011, "DNA sequencing caught in deluge of data".

S. Arumugam and B. Smyth (Eds.): IWOCA 2012, LNCS 7643, pp. 274–279, 2012.

1. The collection has d documents of total length n, where each document is an arbitrary string of symbols over an alphabet of size σ. This is general enough to consider DNA, proteins, source code, natural language, etc. Then the uncompressed data size is $n \log \sigma$ bits (our logarithms are in base 2) and a classical index requires $O(n \log n)$ bits.

2. The collection is *repetitive*, meaning that most documents can be covered by a few chunks that appear in other documents. This captures most cases, in particular those with known linear, tree, etc. version structures as well as those with unknown version structures like DNA collections. Yet, it leaves aside special cases such as reverse complemented repetitions that are frequent in DNA, which must be dealt with separately.

3. We wish to store and index the collection in a way that provides efficient *access*, that is, extracting any substring of any document, and *searches*, that is, locating the occurrences of string patterns in the collection. We ignore more complex searches such as approximate matching, complex pattern matching, sequence mining, etc. Yet this is challenging enough to leave aside methods like encoding the differences, which may support access but not searches.

4. We wish to use space proportional to the "repetitiveness" of the collection. While we seek for techniques that work on any collection without any explicit structure, we analyze them on a simplified case where there is one uncompressible *base* sequence of length ℓ and then $d - 1$ other sequences identical to the base one, where s *edit operations* (single character insertions, deletions, and replacements) have been done on them. Ideally, we should achieve $\ell \log \sigma + O(s \log n)$ bits of space, as well as search time within $O((m + occ)\,\text{polylog}\,n)$ to find the occ occurrences of a pattern of length m.

2   Compressed Suffix Arrays

The suffix array [18] is a classical structure to support pattern searches. Assume the d documents are concatenated into a single string $T[1, n]$, using a special symbol "$\$$" to mark the end of the documents. Each string $T[i, n]$ is called a *suffix* and is identified with its starting point i. The *suffix array* $A[1, n]$ is a permutation of $[1..n]$ where all the suffixes are listed in lexicographic order. The occurrences of $P[1, m]$ in T can be seen as the suffixes of T that start with P, and these form a contiguous interval of A that can be binary searched in time $O(m \log n)$, where m owes to the time needed to compare P with a suffix of A.

The suffix array uses $n \log n$ bits, but can be compressed into $O(n \log \sigma)$ bits by means of the Ψ *function* [13,23], $\Psi(i) = A^{-1}[(A[i] \bmod n) + 1]$, which tells where the value $A[i] + 1$ appears in A. With a bitmap $D[1, n]$ that marks the points in A where the first letter of suffixes change, and a string $S[1, \sigma]$ noting the different symbols of T in order, we know that the first letter of suffix $T[A[i], n]$ is $S[rank(D, i)]$, where $rank(D, i)$ counts the number of 1s in $D[1, i]$. Moreover, the second letter is $S[rank(D, \Psi(i))]$, the third is $S[rank(D, \Psi^2(i))]$, and so on. With a preprocessing of D to solve $rank$ in constant time, we can do the binary search in $O(m \log n)$ time using Ψ, S and D, and without A or T.

In principle Ψ would also require $n \log n$ bits, but it is shown to be compressible as it is covered by σ increasing ranges. In fact, those ranges feature a much richer structure [12,21], which allows one to represent the whole index within $nH_k(T) + o(n \log \sigma)$ bits for any $k < \log_\sigma n$ (roughly), where $H_k(T) \leq \log \sigma$ is the empirical k-th order entropy of T [19,21]. This is a statistical compressibility measure sensitive to symbols distribution but blind to long-range repetitions: If we concatenate two copies of T, its k-th order entropy is $2nH_k(TT) \geq 2nH_k(T)$, so the space of such compressed indexes simply doubles [16].

What is most interesting for us is that, when the text contains long and frequent repetitions of a string, long *runs* of consecutive values appear on Ψ [10,17]. In our model of s edits, it is shown that Ψ contains ℓ runs of length d when $s = 0$, and then each edit breaks, *on average* (assuming the base text and the edits are uniformly generated), $O(\log_\sigma \ell)$ runs. As a consequence, Ψ can be represented using $\ell \log \sigma + O(s \log_\sigma \ell \log n)$ bits on average.

A further problem is that, since we do not store A, knowing the interval of A given by the binary search is not sufficient to output the occurrence positions (which would be the content of the cells of A within the range). This can be done within $O(\log n)$ time per occurrence if we use $O(n)$ further bits of space for *sampling* the suffix array [23]. Such sampling also provides extraction of any substring of T of length t in time $O(t + \log n)$. While various attempts to reduce those $O(n)$ bits have been made [17], they have not been successful in practice. Further, they [17] show that, while LZ78-based compression [25] is poor on highly repetitive sequences, LZ77-based compression [24] is extremely promising.

3   Lempel-Ziv (LZ77) Compressed Indexes

The compression that most accurately reflects the kind of repetitiveness we wish to capture is Lempel-Ziv's, particularly the LZ77 parsing [24]. Here we advance in T, and at each step generate a new *phrase* by taking the longest prefix of the remaining text that has already appeared before in T, plus one further letter. Let us call z the number of phrases generated by such a parsing. It is not hard to see that, in our model, it holds $z \leq \ell / \log_\sigma \ell + s$.

Extracting an arbitrary substring of T is not easy in an LZ77 parsing. If we call $h \leq n$ the "height" of the parse, that is, the maximum number of times a single character is copied, then a substring of length t is extracted in time $O(t h)$ (variants that speed this up [16] may produce more phrases than edits).

Searching for patterns yields further complications. An occurrence of P in T may be split across phrases, that is, a prefix $P[1, i]$ may match a suffix of a phrase, and the corresponding suffix $P[i + 1, m]$ may match a prefix of the concatenation of the following phrases. Such occurrences cut by a phrase boundary are called *primary*, whereas the others are called *secondary*. The strategy of LZ77-based indexes [15,16] is to find both types of occurrences with different means.

Let $T = Z_1 \ldots Z_z$ be the partition of T into phrases. For primary occurrences, we index all reverse phrases Z_j^{rev} and all suffixes starting at phrase boundaries, $Z_k Z_{k+1} \ldots Z_z$. Both sets are connected in a grid where the rows correspond to the lexicographically sorted Z_j^{rev} and the columns to the lexicographically sorted $Z_k Z_{k+1} \ldots Z_z$. Points in this grid connect reverse phrases Z_j^{rev} with the following suffix, $Z_{j+1} Z_{j+2} \ldots Z_z$. At query time, P is partitioned into $P[1, i]$ and $P[i+1, m]$ in the $m-1$ possible ways. We find the interval of the reverse phrases starting with $(P[1, i])^{rev}$ (i.e., phrases ending with $P[1, i]$) and the interval of the phrase-aligned suffixes starting with $P[i+1, m]$. Then each point in the grid in the intersection of the row and column ranges is precisely one primary occurrence.

For secondary occurrences, a structure to describe which portions of T are copied where, is used to track copies of areas where primary occurrences appear. Those copies are secondary occurrences, which must be recursively tracked for further secondary occurrences, until all are reported.

The most recent index of this family [16] uses $O(z \log n)$ bits of space and can search for P in time $O(m^2 h + (m + occ) \log^\epsilon z)$, for any constant $\epsilon > 0$, if we use the best grid representation that fits in this space [3]. The term m^2 owes to the search for all the partitions of P, and the h factor to the time to extract the phrase prefixes/suffixes when determining the intervals of rows and columns. The term $O(m \log^\epsilon z)$ refers to the $m - 1$ range searches on the grid, and $O(occ \log^\epsilon z)$ to the time to report each point (i.e., primary occurrence) in the grid. Secondary occurrences take just constant time each. Note that the h factor is highly undesirable, as it is limited only by n in the worst case.

4   Grammar Compressed Indexes

Grammar compression is a successful approach to factor out repetitiveness in a text. It derives a context-free grammar that generates (only) T, and such a grammar can be small if T is repetitive. While finding the smallest grammar that generates T is NP-complete and popular methods offer poor approximation ratios [22,5], an $O(\log n)$ approximation is easy to achieve by converting the LZ77 factorization, which lower-bounds the smallest grammar, into a balanced grammar [22]. A more sophisticated approximation [4] achieves ratio $O(\log(n/g^*))$, where g^* is the size of the smallest grammar. We will call g the size (total length of the rules) of our grammar that generates T, where it holds $z \leq g \leq z \log n$. In our model it is easy to obtain $g \leq \ell / \log_\sigma \ell + s \log n$. It has been shown that a grammar representation using $O(g \log n)$ bits provides access to any substring of T of length t in time $O(t + \log n)$ [2], much better than with LZ77 compression.

The concept of primary and secondary occurrences is useful here too. Given a rule $X \to ABC \ldots$, a prefix $P[1, i]$ may match a suffix of the string generated by A, and the corresponding suffix $P[i+1, m]$ may match a prefix of the string generated by $BC \ldots$, and thus P will have a primary occurrence inside the string generated by X. Occurrences of X elsewhere yield secondary occurrences of P.

In the most recent grammar-based compressed index [8], primary occurrences are handled very much as in LZ77 parsings, by indexing the reverse string generated by nonterminals and the strings generated by the following sequences of nonterminals on right hands of the grammar productions. Tracking secondary occurrences is done by using a pruned version of the parse tree of T, where all

Table 1. Simplified complexities for current approaches to index repetitive collections

Approach	Space $+\ell \log \sigma$	Time $+O(occ \log n)$	Extract time
Suffix arrays [17] (avg. space)	$O(n + s \log_\sigma \ell \log n)$	$O(m \log n)$	$O(t + \log n)$
Grammar compression [2,8]	$O(s \log^2 n)$	$O(m^2 + m \log n)$	$O(t + \log n)$
Lempel-Ziv compression [16]	$O(s \log n)$	$O(m^2 h + m \log n)$	$O(t\,h)$

the nodes labeled by a given nonterminal containing a primary occurrence are found. To find the text position of an occurrence at a node v, the grammar tree is traversed upwards from v to the root. Each label (i.e., nonterminal) found in this upward traversal is a source of further secondary occurrences, and as such is sought in the tree node labels. The grammar is transformed so that it is guaranteed that any nonterminal appears at least twice in the tree, and thus the upward traversal time is amortized with more secondary occurrences.

The resulting structure requires $O(g \log n)$ bits of space and supports searches in time $O(m^2 + (m + occ) \log^\epsilon g)$ for any constant $\epsilon > 0$ [8].

5   Conclusions

Table 1 summarizes the state of the art in very broad terms. While suffix arrays offer the ideal search time $O((m+occ) \log n)$, they are far from the ideal space of $\ell \log \sigma + O(s \log n)$, which is offered only by Lempel-Ziv indexes. These in turn are far from the ideal time complexity, both for searching and for extracting substrings. Grammar compression offers an intermediate space/time tradeoff that might be attractive. This theoretical picture coincides pretty well with practical experiences on biological and natural language sequence collections [6,7].

The most obvious challenge ahead is to combine good space and time on repetitive sequences. For example, a recent development combining grammars and LZ77 parsing [11] achieves $O(m^2 + (m + occ) \log \log z)$ search time and $O(z \log n \log \log z)$ bits of space (i.e., slightly superlinear on the LZ77 compressed size). Other close relatives of LZ77 parsings may also yield interesting indexes in particular application scenarios [14,20,9]. A further challenge is to support more complex searches, for example suffix tree functionality [1].

References

1. Abeliuk, A., Navarro, G.: Compressed Suffix Trees for Repetitive Texts. In: Calderón-Benavides, L., González-Caro, C., Chávez, E., Ziviani, N. (eds.) SPIRE 2012. LNCS, vol. 7608, pp. 30–41. Springer, Heidelberg (2012)
2. Bille, P., Landau, G., Raman, R., Sadakane, K., Rao Satti, S., Weimann, O.: Random access to grammar-compressed strings. In: Proc. 22nd SODA, pp. 373–389 (2011)
3. Chan, T., Larsen, K., Patrascu, M.: Orthogonal range searching on the RAM, revisited. In: Proc. 27th SoCG, pp. 1–10 (2011)
4. Charikar, M., Lehman, E., Liu, D., Panigrahy, R., Prabhakaran, M., Rasala, A., Sahai, A., Shelat, A.: Approximating the smallest grammar: Kolmogorov complexity in natural models. In: Proc. 34th STOC, pp. 792–801 (2002)

5. Charikar, M., Lehman, E., Liu, D., Panigrahy, R., Prabhakaran, M., Sahai, A., Shelat, A.: The smallest grammar problem. IEEE Trans. Inf. Theo. 51(7), 2554–2576 (2005)
6. Claude, F., Fariña, A., Martínez-Prieto, M., Navarro, G.: Compressed q-gram indexing for highly repetitive biological sequences. In: Proc. 10th BIBE, pp. 86–91 (2010)
7. Claude, F., Fariña, A., Martínez-Prieto, M., Navarro, G.: Indexes for highly repetitive document collections. In: Proc. 20th CIKM, pp. 463–468 (2011)
8. Claude, F., Navarro, G.: Improved Grammar-Based Compressed Indexes. In: Calderón-Benavides, L., González-Caro, C., Chávez, E., Ziviani, N. (eds.) SPIRE 2012. LNCS, vol. 7608, pp. 180–192. Springer, Heidelberg (2012)
9. Do, H.-H., Jansson, J., Sadakane, K., Sung, W.-K.: Fast relative Lempel-Ziv self-index for similar sequences. In: Proc. FAW-AAIM, pp. 291–302 (2012)
10. Fischer, J., Mäkinen, V., Navarro, G.: Faster entropy-bounded compressed suffix trees. Theor. Comp. Sci. 410(51), 5354–5364 (2009)
11. Gagie, T., Gawrychowski, P., Kärkkäinen, J., Nekrich, Y., Puglisi, S.J.: A Faster Grammar-Based Self-index. In: Dediu, A.-H., Martín-Vide, C. (eds.) LATA 2012. LNCS, vol. 7183, pp. 240–251. Springer, Heidelberg (2012)
12. Grossi, R., Gupta, A., Vitter, J.: High-order entropy-compressed text indexes. In: Proc. 14th SODA, pp. 841–850 (2003)
13. Grossi, R., Vitter, J.: Compressed suffix arrays and suffix trees with applications to text indexing and string matching. SIAM J. Comp. 35(2), 378–407 (2006)
14. Huang, S., Lam, T.W., Sung, W.K., Tam, S.L., Yiu, S.M.: Indexing Similar DNA Sequences. In: Chen, B. (ed.) AAIM 2010. LNCS, vol. 6124, pp. 180–190. Springer, Heidelberg (2010)
15. Kärkkäinen, J.: Repetition-Based Text Indexing. PhD thesis, Dept of Comp. Sci., Univ. of Helsinki, Finland (1999)
16. Kreft, S., Navarro, G.: On compressing and indexing repetitive sequences. Theor. Comp. Sci. (to appear, 2012); Earlier versions in Proc. DCC 2010 and Proc. CPM 2011
17. Mäkinen, V., Navarro, G., Sirén, J., Välimäki, N.: Storage and retrieval of highly repetitive sequence collections. J. Comp. Biol. 17(3), 281–308 (2010)
18. Manber, U., Myers, E.: Suffix arrays: a new method for on-line string searches. SIAM J. Comp., 935–948 (1993)
19. Manzini, G.: An analysis of the Burrows-Wheeler transform. J. ACM 48(3), 407–430 (2001)
20. Maruyama, S., Nakahara, M., Kishiue, N., Sakamoto, H.: ESP-Index: A Compressed Index Based on Edit-Sensitive Parsing. In: Grossi, R., Sebastiani, F., Silvestri, F. (eds.) SPIRE 2011. LNCS, vol. 7024, pp. 398–409. Springer, Heidelberg (2011)
21. Navarro, G., Mäkinen, V.: Compressed full-text indexes. ACM Comp. Surv. 39(1), article 2 (2007)
22. Rytter, W.: Application of Lempel-Ziv factorization to the approximation of grammar-based compression. Theo. Comp. Sci. 302(1-3), 211–222 (2003)
23. Sadakane, K.: New text indexing functionalities of the compressed suffix arrays. J. Alg. 48(2), 294–313 (2003)
24. Ziv, J., Lempel, A.: A universal algorithm for sequential data compression. IEEE Trans. Inf. Theo. 23(3), 337–343 (1977)
25. Ziv, J., Lempel, A.: Compression of individual sequences via variable-rate coding. IEEE Trans. Inf. Theo. 24(5), 530–536 (1978)

Range Extremum Queries

Rajeev Raman

University of Leicester, UK

Abstract. There has been a renewal of interest in data structures for *range extremum* queries. In such problems, the input comprises N *points*, which are either elements of a d-dimensional matrix, that is, their coordinates are specified by the 1D submatrices they lie in (row and column indices for $d = 2$), or they are points in $\mathbb{R}^d$. Furthermore, associated with each point is a *priority* that is independent of the point's coordinate. The objective is to pre-process the given points and priorities to answer the *range maximum query (RMQ)*: given a d-dimensional rectangle, report the points with maximum priority. The objective is to minimze the space used by the data structure and the time taken to answer the above query. This talk surveys a number of recent developments in this area, focussing on the cases $d = 1$ and $d = 2$.

1   Introduction

Range searching is one of the most fundamental problems in computer science with important applications in areas such as computational geometry, databases and string processing. The range searching problem is defined as follows. The input is a set S of N points in $\mathbb{R}^d$, where each point is associated with *satellite* data, and an associative and commutative aggregation function f defined on the satellite data. We wish to preprocess the input to create a data structure with reasonable space usage that answers queries of the following form quickly: given any axis-aligned d-dimensional rectangle R, return the value of f on the satellite data of all points in $R \cap S$. Researchers have considered range searching with respect to diverse aggregation functions such as emptiness checking, counting, reporting, minimum/maximum, etc. [13,18,24]. In general, the aggregation function is assumed to be *decomposable*: i.e. for any pair of disjoint sets $X, X' \subseteq S$, $f(X \cup X')^1$ can be computed in constant time from $f(X)$ and $f(X')$. Here, a distinction can be made between functions f that allow inverses, i.e. allow $f(X)$ to be computed in constant time from $f(X \cup X')$ and $f(X')$, and those that do not (these are said to be in the general *semigroup* model [13]). In this talk we will focus on the case where f is the maximum (or symmetrically, the minimum) function. This aggregation function is the most commonly studied among those that do not allow inverses. Although the max/min function does not have an inverse, it has special properties that allow relatively efficient solutions.

---

[1] A slight abuse of notation: f should be defined over the satellite data in the set of points.

S. Arumugam and B. Smyth (Eds.): IWOCA 2012, LNCS 7643, pp. 280–287, 2012.

1.1   Problem Variants

We consider two variants of the problem when $d = 2$: the *2D array* variant, and the *2D rank space* variant.

2D array variant. The input to this problem is a two dimensional m by n array A of $N = m \cdot n$ elements from a totally ordered set. We assume w.l.o.g. that $m \leq n$ and that all the entries of A are distinct. The query rectangle R is specified by four indices i_1, i_2, j_1, j_2 s.t. $0 \leq i_1 \leq i_2 \leq m - 1$ and $0 \leq j_1 \leq j_2 \leq n - 1$, that is, $R = [i_1 \cdots i_2] \times [j_1 \cdots j_2]$. The response to a query is the position of the maximum element in the query range, i.e., $\text{RMQ}(A, q) = \text{argmax}_{(i,j) \in R} A[i, j]$. See Fig. 1.

2D geometric rank space variant. For simplicity, we focus on the case where the points are in *rank space*. the x-coordinates of the N points are $N = \{0, \ldots, N - 1\}$, and the y-coordinates are given by a permutation $v : [N] \rightarrow [N]$, such that the points are $(i, v(i))$ for $i = 0, \ldots, N - 1$. The priorities ofthe points are given by another permutation π such that $\pi(i)$ is the priority of the point $(i, v(i))$. The query rectangle R is specifed by two points from $[N] \times [N]$ and includes the boundaries. If the original and query points are points in $\mathbb{R}^2$, then the problem can be reduced to a problem in rank space in $O(\lg N)$ time with a linear space structure [18,13]. Observe that in the 1D case, the geometric rank space and array variants coincide.

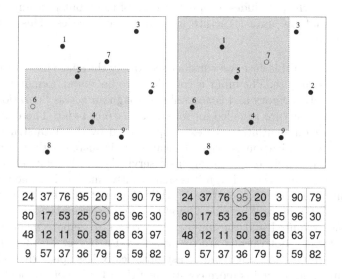

Fig. 1. Geometric 2D RMQ: a 4-sided query (top left) and a 2-sided query (top right); the hollow point is the one that is reported as the answer to the query. Array 2D RMQ: a 4-sided query (bottom left) and a 2-sided query (bottom right); the circled entry is the one reported as the answer to the query.

The algorithms we give will sometimes decompose the given problem into sub-queries which are (apparently) simpler, namely *2-sided* and *3-sided* queries. In general a 2-sided query only directly specifies two sides of the rectangle, the other two being specified by a boundary of the region spanned by the input (see Fig. 1). A 3-sided query only directly specifies three sides of the rectangle, the remaining one being specified by a boundary of the region spanned by the input. In the geometric case, a 2-sided query may be specified by a point $p = (p_x, p_y)$, and the query will return the point with maximum priority point among all points with x-coordinate $\leq p_x$ and y-coordinate $\geq p_y$ (or any of the three symmetric alternatives $(\leq p_x, \leq p_y)$, $(\geq p_x, \leq p_y)$ and $(\geq p_x, \geq p_y)$).

1.2   Models and Measures

In both the above problems, the model of computation assumed for proving upper bounds is the the word-RAM model with word size $\Theta(\lg N)$ bits[2]. In all cases, we are given the input, and we wish to pre-process the input to create a data structure, that can answer the desired query. A few measures can be used compare solutions: (a) the time (and space) used to create the data structure (the *preprocessing time/space*) (b) the time taken to answer a query, after pre-processing and (c) the space taken by the data structure. Our focus is firstly on (c) and secondarily on (b) — we are usually not concerned with (a). We consider three separate ways to measure the space usage.

General model. In this model we consider just the overall space used by the data structure, which includes a representation of the input plus any additional information that the data structure uses to answer queries. This is the least restrictive model.

Indexing model. In this case, we consider separately the space used for the input and the data structure. The input is considered to be stored "remotely" in some kind of read-only memory and is accessible through an *access* operation. In this case, the data structure is called an *index*, or *succinct index*. There are several motivations for differentiating the extra space used by the data structure from the input. One motivation is technological — it is increasingly the case that data is stored remotely (e.g. "storage as a service") and is relatively expensive to access. In this case, the index, if it is significantly smaller than the input, may be stored in a computer's local storage, while the input may be stored remotely. A well-designed index will provide a good trade-off between the amount of local storage space required and the number of accesses to remote memory. Other, more algorithmic, motivations are discussed by Barbay et al. [3]. Clearly, an index is only interesting if it requires asymptotically less space than the input. Thus, the indexing model is more constrained than the general model.

The access operation can be of several kinds. For example, in the matrix version of the problem, we access the entries of the original matrix. In the geometric version of the problem, we consider that the point coordinates are stored

---

[2] $\lg x = \log_2 x$.

remotely, and the access function is an *orthogonal range-reporting query*, which reports the coordinates of all points that lie within a query rectangle.

Encoding model. In this case, the input data is pre-processed and an encoding E is created. Once the encoding E is created, the input is no longer accessible and all queries have to be answered using E alone. Again, clearly, an encoding is interesting only if it requires aysmpotically less space than the input. Having an encoding E of the input that takes less space than the input itself may seem contradictory at first. If it is possible to deduce the input by performing queries on E, then in general interesting encodings do not exist. However, if it is *not* possible to deduce the input by asking queries from E, then we can (effectively) partition all possible inputs into equivalence classes such that each member of the equivalence class gives the same answer to all possible queries, and only indicate in which equivalence class our given input lies. If this takes significantly less memory than storing the entire input, we say that the *effective entropy* [20] of the problem is low. The encoding model is the most constrained of the three memory models we consider.

The encoding model is useful when the access operation is too expensive to use. It can also be used when say (part of) the input is generated randomly and has no particular intrinsic signficance. For example Kaplan et al. [22] show that performing RMQs using random priorities leads to interesting data structures for approximate higher-dimensional geometric queries.

2   1D Array Case

In the 1D case, the geometric and array variants coincide, and the problem seems to be well-understood. In the encoding model, it is known that any encoding must be of size at least $2N - O(\lg N)$ bits [28,16], and this bound is tight. Observe that the input cannot be reconstructed by asking RMQ queries: if the input array is $A[1..N]$ and the maximum element is in position i, then we cannot determine the relative ordering of any elements $A[j]$ and $A[k]$, where $j < i < k$, by means of RMQ queries. Furthermore, there are data structurs that encode an array in $2N + o(N)$ bits so that 1D RMQs can be answered in $O(1)$ time after $O(N)$-time preprocessing [16,14]. In the indexing model, Brodal et al. [8] showed that an index of size $O(N/c)$ bits suffices to answer queries in $O(c)$ time, and that this trade-off is optimal.

3   2D Array Case

After the initial consideration of this problem by Chazelle and Rosenberg [12], motivated by the rapid increase of applications of 1D RMQ, Amir et al. restarted the investigation of 2D RMQ [1]. In particular, they asked in what ways the 2D and 1D RMQ problems differ, and the answer turns out to be relatively complex.

Table 1. Space/time tradeoffs for 2D range maximum searching in the word RAM

Citation	Space (in words)	Query time
Gabow et al. [19]	$O(N \lg N)$	$O(\lg N)$
Chazelle'88 [13]	$O(N \lg^\epsilon N)$	$O \lg N$
Chan et al.'10 [10]	$O(N \lg^\epsilon N)$	$O \lg \lg N$
Karpinski et al.'09 [23]	$O(N(\lg \lg N)^{O(1)})$	$O(\lg \lg N)^2$
Chazelle'88 [13]	$O(N \lg \lg N)$	$O(\lg N \lg \lg N)$
Chazelle'88 [13]	$O(\frac{1}{\epsilon} N)$	$O(\lg^{1+\epsilon} N)$
Farzan et al. [8]	$O(N)$	$O(\lg N \lg \lg N)$

- In the indexing model, Brodal et al. [8], building upon Atallah and Yuan's [2] earlier result, showed that there is an index that answers 2D RMQs in $O(1)$ time, using $O(N)$ bits, and that can be constructed in $O(N)$ preprocessing time. Thus, in this model, 1D and 2D RMQs are equivalent.
- In the encoding model, Demaine et al. [15] showed that, unlike 1D RMQ, non-trivial (size $o(N \lg N)$ bits) encodings for 2D RMQs do not exist for square $n \times n$ matrices. Generalizing Demaine et al.'s result, Brodal et al. showed that for general $m \times n$ matrices with $m \leq n$, any encoding must take $\Omega(N \lg m)$ bits, which is $o(N \lg N)$ bits if $m \ll n$, and complemented this with an encoding of $O(Nm)$ bits. Golin et al. [20] showed that for random A, the expected size of an encoding is $\Theta(N)$ bits (Golin et al. also showed that the expected size of an encoding in the random 1D case is $\Theta(N)$ bits). Thus, in the encoding model, 1D and 2D RMQs are not equivalent except in special cases.
- Coming back to the indexing model, Brodal et al. [8] showed that with an index of size $O(N/c)$ bits, $\Omega(c)$ query time is needed to answer queries, and gave an index of size $O(N/c)$ bits that answered queries in $O(c \lg^2 c)$ time. Brodal et al. [7] gave an index of size $O(N/c)$ bits that takes $O(c \lg c (\lg \lg c)^2)$ query time, thus improving the time-space tradeoff, but there still remains a gap with the 1D RMQ case.

4   2D Geometric Case

Existing results on this problem are summarised in Table 1. All of this work is in the general model (note that all of the space bounds given in Table 1 are $\Omega(N \lg N)$ bits, so are uninteresting in the index or encoding models). Gabow et al. [19], basing their approach on the classic range tree [5] gave a solution that took $O(N \lg N)$ words of space. Chazelle [13] used ideas similar to the *wavelet tree* [17] in order to break through the $N \lg N$ space barrier. He gave a number of trade-offs, some of which were improved by Chan et al. [23,10]. Nekrich [25] gives a nice survey of much of this work. These approaches, however, do not seem to be able to achieve space below $O(N \lg \lg N)$ words.

However, obtaining linear space ($O(N)$ words) space is an important objective in geometric data structures. Data structures with non-linear space usage but good worst-case query times are rarely preferred in practice to space-partioning methods such as quadtrees and k-D trees, which have linear space usage (but poor worst-case query time). Chazelle's linear-space data structure used $O(\frac{1}{\epsilon}N)$ words with a query time $O(\lg^{1+\epsilon} N)$ for any fixed $\epsilon > 0$. Farzan et al. [8] recently gave a linear-space data structrue with $O(\lg N \lg \lg N)$, or "almost" logarithmic, query time. Farzan et al.'s data structure uses a number of results on 2D queries of various kinds including range counting [21] range reporting [10], range selection [9] and in fact even the 2D array RMQ results of [8,2]. However, it adds two new ingredients.

As with many other solutions to range-searching problems, the solution of Farzan et al. breaks the given 4-sided query into 2-, 3- and 4-sided sub-queries. The first new contribution of Farzan et al. is to note that to answer 2-sided range maxima queries on a problem of size m, one need not store the entire total order of priorities using $\Theta(m \lg m)$ bits: $O(m)$ bits suffice, i.e., the effective entropy of 2-sided queries is low. However, this does not help immediately, since the effective entropy is low only if the point coordinates are known, but storing the point coordinates themselves would take $\Theta(m \lg m)$ bits. To circumvent this problem, Farzan et al. represent these 2-sided problems as succinct indices: these succinct indices do not store the point coordinates themselves, but are able to access point coordinates by means of an orthogonal range reporting query. Interestingly, this component can be viewed as a stand-alone succinct index result for 2-sided 2D RMQ, and this result was used by Brodal et al. [7] to obtain trade-offs for the indexing model.

5   Conclusions and Open Problems

We have summarized some of the known results for the 2D RMQ problem, both its array and geometric variants. A number of questions are still open:

1. What is the precise complexity of encoding 2D array RMQ in the case $m \ll n$?
2. What is the precise trade-off between index size and query time, in the indexing model, for 2D array RMQ?
3. Can Farzan et al.'s solution for 2D geometric RMQ be improved to take $O(\lg N / \lg \lg N)$ time while still using $O(N)$ space (their approach intrinsically requires $\Omega(\lg N / \lg \lg N)$ time when using $O(N)$ space [26]). Solving the following combinatorial problem would help:

   Given n non-intersecting horizontal lines in 2D and an integer parameter k, $1 \leq k \leq n$. Partition the 2D plane into $O(n/k)$ rectangles such that (a) no rectangle contains more than $O(k)$ lines, or parts of lines (b) at most $O(\sqrt{k})$ lines cross the boundary of any rectangle.

   It is possible to achieve (a) and (b), if the relaxation to rectangular regions is dropped, via the planar separator theorem (see e.g. [6]). Alternatively, we can achieve (a) but not (b) using rectangular regions [4,11].
4. Can we solve geometric 2D range maxima in $O(N)$ space and $o(\lg N / \lg \lg N)$ time? A lower bound of $\Omega(\lg \lg N)$ follows from [27].

References

1. Amir, A., Fischer, J., Lewenstein, M.: Two-Dimensional Range Minimum Queries. In: Ma, B., Zhang, K. (eds.) CPM 2007. LNCS, vol. 4580, pp. 286–294. Springer, Heidelberg (2007)
2. Atallah, M.J., Yuan, H.: Data structures for range minimum queries in multi-dimensional arrays. In: Proc. 20th Annual ACM-SIAM Symposium on Discrete Algorithms, pp. 150–160. SIAM (2010)
3. Barbay, J., He, M., Munro, J.I., Rao, S.S.: Succinct indexes for strings, binary relations and multi-labeled trees. In: Bansal, N., Pruhs, K., Stein, C. (eds.) SODA, pp. 680–689. SIAM (2007)
4. Bender, M.A., Cole, R., Raman, R.: Exponential Structures for Efficient Cache-Oblivious Algorithms. In: Widmayer, P., Triguero, F., Morales, R., Hennessy, M., Eidenbenz, S., Conejo, R. (eds.) ICALP 2002. LNCS, vol. 2380, pp. 195–207. Springer, Heidelberg (2002)
5. Bentley, J.L.: Decomposable searching problems. Information Processing Letters 8(5), 244–251 (1979)
6. Bose, P., Chen, E.Y., He, M., Maheshwari, A., Morin, P.: Succinct geometric indexes supporting point location queries. In: Mathieu, C. (ed.) SODA, pp. 635–644. SIAM (2009)
7. Brodal, G.S., Davoodi, P., Lewenstein, M., Raman, R., Rao, S.S.: Two Dimensional Range Minimum Queries and Fibonacci Lattices. In: Epstein, L., Ferragina, P. (eds.) ESA 2012. LNCS, vol. 7501, pp. 217–228. Springer, Heidelberg (2012)
8. Brodal, G.S., Davoodi, P., Rao, S.S.: On space efficient two dimensional range minimum data structures. Algorithmica 63(4), 815–830 (2012)
9. Brodal, G.S., Jørgensen, A.G.: Data Structures for Range Median Queries. In: Dong, Y., Du, D.-Z., Ibarra, O. (eds.) ISAAC 2009. LNCS, vol. 5878, pp. 822–831. Springer, Heidelberg (2009)
10. Chan, T.M., Larsen, K.G., Pătraşcu, M.: Orthogonal range searching on the ram, revisited. In: Proceedings of the 27th Annual ACM Symposium on Computational Geometry, SoCG 2011, pp. 1–10. ACM, New York (2011), http://doi.acm.org/10.1145/1998196.1998198
11. Chan, T.M., Patrascu, M.: Transdichotomous results in computational geometry, I: Point location in sublogarithmic time. SIAM J. Comput. 39(2), 703–729 (2009)
12. Chazelle, B., Rosenberg, B.: Computing partial sums in multidimensional arrays. In: Proc. 5th Annual Symposium on Computational Geometry, pp. 131–139. ACM (1989)
13. Chazelle, B.: A functional approach to data structures and its use in multidimensional searching. SIAM J. Comput. 17(3), 427–462 (1988); prel. vers. FOCS 1985
14. Davoodi, P., Raman, R., Rao, S.S.: Succinct Representations of Binary Trees for Range Minimum Queries. In: Gudmundsson, J., Mestre, J., Viglas, T. (eds.) CO-COON 2012. LNCS, vol. 7434, pp. 396–407. Springer, Heidelberg (2012)
15. Demaine, E.D., Landau, G.M., Weimann, O.: On Cartesian Trees and Range Minimum Queries. In: Albers, S., Marchetti-Spaccamela, A., Matias, Y., Nikoletseas, S., Thomas, W. (eds.) ICALP 2009, Part I. LNCS, vol. 5555, pp. 341–353. Springer, Heidelberg (2009)
16. Fischer, J., Heun, V.: Space-efficient preprocessing schemes for range minimum queries on static arrays. SIAM J. Comput. 40(2), 465–492 (2011)
17. Foschini, L., Grossi, R., Gupta, A., Vitter, J.S.: When indexing equals compression: Experiments with compressing suffix arrays and applications. ACM Trans. Algorithms 2(4), 611–639 (2006)

18. Gabow, H.N., Bentley, J.L., Tarjan, R.E.: Scaling and related techniques for geometry problems. In: Proc. 16th Annual ACM Symposium on Theory of Computing, pp. 135–143. ACM (1984)
19. Gabow, H.N., Bentley, J.L., Tarjan, R.E.: Scaling and related techniques for geometry problems. In: Proc. 16th Annual ACM Symposium on Theory of Computing, pp. 135–143. ACM (1984)
20. Golin, M., Iacono, J., Krizanc, D., Raman, R., Rao, S.S.: Encoding 2D Range Maximum Queries. In: Asano, T., Nakano, S.-I., Okamoto, Y., Watanabe, O. (eds.) ISAAC 2011. LNCS, vol. 7074, pp. 180–189. Springer, Heidelberg (2011)
21. JáJá, J., Mortensen, C.W., Shi, Q.: Space-Efficient and Fast Algorithms for Multidimensional Dominance Reporting and Counting. In: Fleischer, R., Trippen, G. (eds.) ISAAC 2004. LNCS, vol. 3341, pp. 558–568. Springer, Heidelberg (2004)
22. Kaplan, H., Ramos, E., Sharir, M.: Range minima queries with respect to a random permutation, and approximate range counting. Discrete & Computational Geometry 45(1), 3–33 (2011)
23. Karpinski, M., Nekrich, Y.: Space Efficient Multi-dimensional Range Reporting. In: Ngo, H.Q. (ed.) COCOON 2009. LNCS, vol. 5609, pp. 215–224. Springer, Heidelberg (2009)
24. Mehta, D.P., Sahni, S. (eds.): Handbook of Data Structures and Applications. Chapman & Hall/CRC (2009)
25. Nekrich, Y.: Orthogonal range searching in linear and almost-linear space. Comput. Geom. 42(4), 342–351 (2009)
26. Patrascu, M. (data) structures. In: FOCS, pp. 434–443. IEEE Computer Society Press (2008)
27. Patrascu, M., Thorup, M.: Time-space trade-offs for predecessor search. In: Kleinberg, J.M. (ed.) STOC, pp. 232–240. ACM (2006)
28. Vuillemin, J.: A unifying look at data structures. Communications of the ACM 23(4), 229–239 (1980)

Design and Analysis of a Tree-Backtracking Algorithm for Multiset and Pure Permutations

Ray Jinzhu Chen[1], Kevin Scott Reschke[2], and Hailong Hu[3]

[1] Fast Switch, Dublin, OH USA
rchen@dataverify.com
[2] Department of Computer Science, Stanford University, Stanford, CA, USA
[3] Software School, Xiamen University, Xiamen, China

Abstract. A tree-backtracking-based technique, permutation tree generation with Anterior-items-in-an-array for Remaining Distinct Elements (ARDE), is introduced for multiset and pure permutations. We analyze the algorithm at the assembly level and obtain its time formula for pure permutations. We **mathematically** prove that our time formula is 11.5% faster than the corresponding formula of the previous fastest algorithm for pure permutations for any length $N > 3$. We also offer related source codes and executable files on our web site for others to use.

Keywords: tree-backtracking, tree generation, multiset permutation.

1 PureARDE Pure Permutation Tree (See Fig. 1 and 2)

Without loss of generality, here we consider N distinct elements numbered $0, 1, \cdots, N - 1$.

Definition 1. *(Permutation array a)* $a[\ ] = \{a[0], a[1], \cdots, a[N - 1]\}$ *is the integer array used for receiving and storing the output permutation.*

Definition 2. *("PureARDE Permutation Tree Structure") A PureARDE Permutation Tree is a pure permutation generation tree. A node value exists inside the node. A node value at level p indicates a[p]. A path from any leaf to the root is a permutation* $a[\ ] = \{a[0], a[1], \cdots, a[N - 1]\}$.

Definition 3. *(Position and tree level variable, p) p represents the current position in a[ ] as well as the current node level in the permutation tree.*

Definition 4. *(Next position variable np) np(np = p − 1) is the next position to be processed in a[ ].*

Definition 5. *("Remaining positions") The remaining positions np, np−1, · · · , 1, 0 are the positions that we have not processed for the current output.*

S. Arumugam and B. Smyth (Eds.): IWOCA 2012, LNCS 7643, pp. 288–292, 2012.
© Springer-Verlag Berlin Heidelberg 2012

Definition 6. *(Remaining elements r) Multiset r is comprised of the remaining elements for the remaining positions.*

Definition 7. *(RDE: Remaining Distinct Elements) The set RDE is comprised of the remaining distinct elements for the remaining positions. For pure permutation, RDE = r, since there are no duplicated elements in r.*

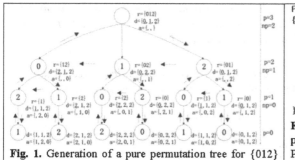

Fig. 1. Generation of a pure permutation tree for {012} with PureARDE

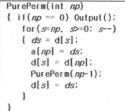

```
PurePerm(int np)
{ if(np == 0) Output();
  for(s=np, s>=0; s--)
  { ds = d[s];
    a[np] = ds;
    d[s] = d[np];
    PurePerm(np-1);
    d[s] = ds;
  }
}
```

Fig. 2. C code for pure permutation based on PureARDE (see [1] for details)

Definition 8. *(Distinct integer array d) d[ ] = {d[0], d[1], ···, d[N − 1]} is the distinct integer array used to track RDE. We use the anterior items in d[ ] to dynamically represent the current RDE.*

Definition 9. *(PureARDE: Pure-permutation Anterior-items-in-an-array for RDE) We always keep the RDE for a node at tree level p as the first p (p = np+1) items, d[0], d[1], ···, d[np] of d[ ]. We call the items PureARDE. In Fig. 1, the PureARDE are the underlined elements in d[ ] for each node. Note: PureARDE can be directly referred to in d[ ] by the first np + 1 indices in a loop to quickly generate nodes and to backtrack. Fig. 2 shows the C code.*

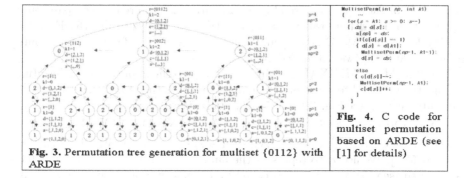

Fig. 3. Permutation tree generation for multiset {0112} with ARDE

Fig. 4. C code for multiset permutation based on ARDE (see [1] for details)

2   ARDE Multiset Permutation Tree (See Fig. 3 and 4)

Definition 10. *(Variable k) Integer variable k indicates the number of remaining distinct integers to fill the remaining positions* $a[np], a[np-1], \cdots, a[0]$.

Definition 11. *(Variable k1) We use variable k1 to store the value of* $k-1$.

Definition 12. *(**ARDE:** **A**nterior-items-in-an-array for **RDE**) In multiset permutation, we always keep the RDE for a node at tree level p as the first* k $(k = k1 + 1)$ *items,* $d[0], d[1], \cdots, d[k1]$, *of* $d[\ ]$. *We call the items ARDE. In Fig. 3, the ARDE are the underlined elements in* $d[\ ]$ *for each node.*

Definition 13. *(c[ ]) Integer array* $c[\ ]$ *is used to dynamically record the current counts of the integers in ARDE. (In Fig. 3, the current counts are the underlined items in* $c[\ ]$ *).*

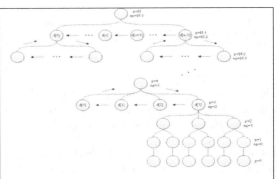

v: A unit that represents the clock cycle time in a pipelined implementation.
μ: The average cost of a memory reference.
CMP, JMP, SUB, INCL, PUSHJ, PUT, GET, SET — each costs v time units.
LDO, STO, STB, LDB — each costs μ+v time units.
BZ, BP, BNN — each costs v+π, where π = 2v if condition is true, otherwise π = 0.
PBZ, PBP, PBNN — each costs 3v-π , where π = 2v if condition is true, otherwise π = 0.
POP — costs 3v time units.

Fig. 5. Related timing rules for MMIXAL [2]

Fig. 6. Tree generation sequence (arrows) for PureARDE in MMIXAL

3   Assembly Level Analysis (See Fig. 5, 6 and 7)

Sedgewick [3] showed that assembly level analysis is required to compare pure permutation algorithms **fairly**. In this section, we analyze PureARDE at assembly level with MMIXAL [2] and compare it with the fastest known pure permutation in [4] (we call it Sedgewick02).

The PureARDE MMIXAL code in Fig. 7 refers to the improved PurePerm(int np) in [1]. We use P to represent the number of times that PurePerm procedure (line 40) is called (see Fig. 1, 6 and 7). PurePerm is called when $np \geq 3$, therefore $P = \sum_{j=4}^{N}(nodes\ at\ tree\ level\ p = j) = 1 + N + N(N-1) + N(N-1)(N-2) + ... + N*...*6*5 = N! \sum_{k=4}^{N} 1/k! = \lfloor N!(e-1-1/1!-1/2!-1/3!)\rfloor$. That is, $P = \lfloor 0.051615N!\rfloor$. We use F to represent the number of times that npGt3 (line 93) is executed (see Fig. 6). Line 93 is called by PurePerm procedure when $np > 3$, therefore $F = \sum_{j=5}^{N}(nodes\ at\ tree\ level\ p = j) = 1 + N + N(N - 1) + N(N - 1)(N - 2) + \ldots + N * \ldots * 6 = N! \sum_{k=5}^{N} 1/k! = \lfloor N!(e - 1 - \frac{1}{1!} - \frac{1}{2!} - \frac{1}{3!} - \frac{1}{4!})\rfloor$. That is, $F = \lfloor 0.009949N!\rfloor$.

Similar to [3], we add one column next to the MMIXAL code for the corresponding C code, and one more column for the time spent on the line. As Sedgewick [3] did, we exclude the time for the Output() call since it is the same for all algorithms. Each line cost equals (the MMIXAL command execution time) * (the number of the times the line is executed), based on Fig. 1, 5, and 6.

```
 1 MaxLen   IS    21
 2 N        GREG  0           % N>=4
 3 np       GREG  0
 4 s        GREG  0
 5 d0       GREG  0           % d[0]
 6 d1       GREG  0           % d[1]
 7 d2       GREG  0           % d[2]
 8 d3       GREG  0           % d[3]
 9 a3       GREG  0           % a[3]
10 t        GREG  0           % temp variable
11          LOC   Data_Segment
12          GREG  @
13 prompt   BYTE  "Input a string with length >= 4:",#a,0
14 outHead  BYTE  "Output Permutation line by line:",#a,0
15 d        GREG  @
16 str      LOC   @
17          LOC   @+MaxLen
18 a        GREG  @
19          LOC   @+MaxLen
20 Input    OCTA  str,MaxLen
21
22          LOC   #200
23 Main     PUSHJ 0,getinput   % getinput
24          SUB   np,N,1        % np=N-1;
25          SET   $1,np         % $1=np-N-1;
26          PUSHJ $0,PurePerm   % PurePerm($1):
27 endMain  TRAP  0,Halt,0
28
29 getinput LDA   $266,prompt   % Display prompt
30          TRAP  0,Fputs,StdOut
31          LDA   $266,input    % Read str and d[]
32          TRAP  0,Fgets,StdIn
33          SUB   N,$266,1       % Get N
34          SET   t,@a           % init a[N]
35          STB   t,a,N
36          LDA   $266,outHead   % Output HeadLine
37          TRAP  0,Fputs,StdOut
38          POP
39
40 PurePerm CMP   t,$0,3         % if(np>3)        vP
41          BP    t,npGt3        % goto npGt3;     vP+2vF
42
43 pNP3     SET   s,3            % s=3;            vM/24
44          LDB   a3,d,3         %                 (μ+v)M/24
45          STB   a3,a,3         % a[3]=d[3]       (μ+v)M/24
46          SET   d3,a3          %                 vNI/24
47
48 pOUTS    LDB   d0,d,0         % d0=d[0];        (μ+v)M/6
49          LDB   d1,d,1         % d1=d[1];        (μ+v)M/6
50          LDB   d2,d,2         % d2=d[2];        (μ+v)M/6
51          STB   d0,a,0         % a[0]=d0;        (μ+v)M/6
52          STB   d1,a,1         % a[1]=d1;        (μ+v)M/6
53          STB   d2,a,2         % a[2]=d2;        (μ+v)M/6
54          LDA   $266,a         % Output():

55          TRAP  0,Fputs,StdOut
56          STB   d2,a,1         % a[1]=d2;        (μ+v)M/6
57          STB   d1,a,2         % a[2]=d1:        (μ+v)M/6
58          LDA   $266,a         % Output():
59          TRAP  0,Fputs,StdOut
60          STB   d2,a,0         % a[0]=d2:        (μ+v)M/6
61          STB   d0,a,1         % a[1]=d0:        (μ+v)M/6
62          LDA   $266,a         % Output():
63          TRAP  0,Fputs,StdOut
64          STB   d1,a,0         % a[0]=d1:        (μ+v)M/6
65          STB   d2,a,2         % a[2]=d2:        (μ+v)M/6
66          LDA   $266,a         % Output():
67          TRAP  0,Fputs,StdOut
68          STB   d2,a,1         % a[1]=d2:        (μ+v)M/6
69          STB   d0,a,2         % a[2]=d0:        (μ+v)M/6
70          LDA   $266,a         % Output():
71          TRAP  0,Fputs,StdOut
72          STB   d2,a,0         % a[0]=d2:        (μ+v)M/6
73          STB   d1,a,1         % a[1]=d1:        (μ+v)M/6
74          LDA   $266,a         % Output():
75          TRAP  0,Fputs,StdOut
76          BZ    s,pUP          % if(s==0) goto pUP;   v(6M/24)
77
78          STB   a3,d,s         % d[s]=a[3]:      (μ+v)(3M/24)
79          SUB   s,s,1          % s--:            v(3M/24)
80          LDB   a3,d,s         % a[3]=d[s]:      (μ+v)(3M/24)
81          STB   a3,a,3         %                 (μ+v)(3M/24)
82          STB   d3,a,s         % d[s]=d[3]:      (μ+v)(3M/24)
83          JMP   pOUTS          % goto pOUTS:     v(3M/24)
84 pUP      STB   a3,d,0         % d[0]=a[3]:      (μ+v)M/24
85          POP                  % return:         3vNI/24
86
87 _np      IS    $0             % Local np
88 _s       IS    $1             % Local s
89 _ds      IS    $2             % Temp d[s]
90 _drp     IS    $4             % Temp d[np]
91 next_np  IS    $6             % Para of next PurePerm
92
93 npGt3    GET   $3,rJ          % Recursive keep      vF
94          LDB   _drp,d,_np     % a[np]=d[np]:         (μ+v)F
95          STB   _np,d,_np      %                      (μ+v)F
96          SUB   next_np,_np,1  % $6=np-1:             vF
97          PUSHJ $5,PurePerm    % PurePerm($5):        vF
98 next_s   SUB   _s,_np,1       % s=np-1:              vF
99          LDB   _ds,a,_np      % d[s]=d[np]:          (μ+v)(P-1-F)
100         STB   _np,a,_np      % a[np]=ds:            (μ+v)(P-1-F)
101         STB   _ds,d,_np      % d[s]=d[np]:          (μ+v)(P-1-F)
102         SUB   next_np,_np,1  % $6=np-1:             v(P-1-F)
103         PUSHJ $5,PurePerm    % PurePerm($5):        v(P-1-F)
104         STB   _ds,d,_s       % d[s]=ds:             (μ+v)(P-1-F)
105         SUB   _s,_s,1        % s--:                 v(P-1-F)
106         PBNN  _s,next_s      % if(s>=0) goto next_s: v(P-1+F)
107         PUT   rJ,$3          % Recursive recover    vF
108 endGt3  POP                  % return:              3vF
```

Fig. 7. PureARDE MMIXAL source code (see [1] for both source code and executable)

- Based on the definition of P, line 40 costs vP since each CMP costs v.
- Based on the definition of F, line 93 costs vF since each GET costs v.
- In line 41, condition $np > 3$ happens the same number of times as line 93 does. So the BP, based on MMIXAL rules, costs $vP + 2vF$.
- Lines 43 to 46 happen at tree level $p = 3$. So they happen $N!/24$ times since there are $N!/24$ nodes for $s = 3$ at level $p = 3$.
- The lines from 48 to 73 are executed $N!/6$ times, once for each time we have to directly output the last three elements. Each LDB or STB costs $\mu + v$, so each line costs $(\mu+v)N!/6$. We do not calculate Output() time for any of the algorithms since they are all the same.

- In line 76, condition $s = 0$ at tree level $p = 3$ happens $N!/24$ times, so the BZ line, based on MMIXAL rules, costs $vN!/6+2vN!/24 = v(6N!/24)$.
- Lines 78-83 happen $3N!/24$ times when $s > 0$ at tree level $p = 3$.
- Lines 84-85 happen $N!/24$ times when $s = 0$ at tree level $p = 3$.
- Lines 94-98 happen F times as line 93 does.
- Line 103 happens for each PurePerm call except for the calls from lines 26 and 97. So line 103 happens P-1-F times. Therefore, lines 99 to 105 happen P-1-F times.
- Lines 107 and 108 happen F times as line 93 does.
- Line 106 happens the same number of times as line 105 does. The number of times that $s \geq 0$ happens equals the number of times that line 105 happens minus the number of times that line 107 happens, which is F. Therefore, the PBNN line, based on MMIXAL rules, costs $3v(P$-1-$F)$-$2v(P$-1-F-$F) = v(P$-1+$F)$.

After assessing the time for each line, we obtain the time formula for PureARDE:

PureARDE_Time(N) = {Sum of the executing time of lines from line 40 to 108}
= $(10P+6F+96N!/24-8)v + (4P+79N!/24-2F-4)\mu \approx 4.5758N!v +3.4782N!\mu$.
Based on [1, 3, 4], for $N \geq 4$,
Sedgewick02_Time(N) = {Sum of the executing time of lines with μ or v}
= $(26N!/6+3A_N/6+9B_N/6+5N-1)v + (19N!/6+2B_N/6+N-2)\mu$
$\approx 5.575N!v + 3.406N!\mu$

We assume, with Sedgewick [3], that "instructions which reference data in memory take two time units, while jump instructions and other instructions which do not reference data in memory take one time unit." So, $\mu+v = 2v$. Therefore $\mu = v$ and Sedgewick02_Time(N)/PureARDE_Time(N) = 1.115.

References

1. Chen, R.J., Reschke, K.S., Tong, M.: Code of evaluation and comparison for multiset permutations (2010),
   http://software.xmu.edu.cn/View/ArticleShow.aspx?aid=3108
2. Knuth, D.E.: MMIX: A RISC Computer for the New Millennium. In: The Art of Computer Programming. Fascicle, vol. 1. Addison-Wesley, Co., Inc., Reading (2005)
3. Sedgewick, R.: Permutation generation methods. ACM Computing Surveys (CSUR) 9, 137–164 (1997)
4. Sedgewick, R.: Permutation generation methods. Dagstuhl Workshop on Data Structures, Wadern, Germany (2002),
   http://www.cs.princeton.edu/~rs/talks/perms.pdf (accessed March 22, 2009)

GRP_CH Heuristic for Generating Random Simple Polygon

Sanjib Sadhu, Subhashis Hazarika, Kapil Kumar Jain, Saurav Basu,
and Tanmay De

Department of Computer Science and Engineering,
National Institute of Technology, Durgapur
{sanjibsadhu411,hsubhashis,jainkkapil,
saurabh.basu.cs,tanmoydeynitd}@gmail.com

Abstract. A heuristic 'GRP_CH' has been proposed to generate a random simple polygon from a given set of 'n' points in 2-Dimensional plane. The "2-Opt Move" heuristic with time complexity $\mathcal{O}(n^4)$ is the best known (referred in [1]) among the existing heuristics to generate a simple polygon. The proposed heuristics, 'GRP_CH' first computes the convex hull of the point set and then generates a random simple polygon from that convex hull. The 'GRP_CH' heuristic takes $\mathcal{O}(n^3)$ time which is less than that of "2-opt Move" heuristic. We have compared our results with "2-Opt Move" and it shows that the randomness behaviour of 'GRP_CH' heuristic is better than that of "2-Opt Move" heuristic.

Keywords: Simple polygon, Convex Hull,Visibility of a line segment.

1 Introduction

There are several importances of generating a simple polygon randomly from a given set of points, e.g. verifying time complexity for geometric algorithm and generating the test instances of geometric problem. In this paper, our attempt is to generate randomly a simple polygon from a set $S=\{s_1,s_2,\ldots,s_n\}$ of n points which lie on a 2-dimensional plane.

Although there exists a lot of heuristics to generate a random polygon from n points, still it is an open problem to generate it uniformly at random, i.e. generating a polygon with probability $(\frac{1}{j})$ if there exists j simple polygons on set S in total. We have designed a new heuristic which generate a large number of unique random polygons than the existing heuristic and also with less time complexity. Our heuristic is based on the construction of convex hull of point set and the visibility of a line segment from a point.

A polygon P is said to be **simple**[7] if it does not contain any hole and no edges of it intersect with each other, except that neighbouring edges meet only at their common end point known as vertices of the polygon.

A subset S of the plane is called convex[7] if and only if for any pair of points $p, q \in S$ the line segment $l(p,q)$ is completely contained in S. The **Convex Hull** $CH(S)$ of a point set S is the smallest convex set that contains S.

S. Arumugam and B. Smyth (Eds.): IWOCA 2012, LNCS 7643, pp. 293–302, 2012.

An edge $e(u, v)$ of a simple polygon is said to be **fully visible** from a point r if and only if both the end points u & v of the edge e are visible [6] to the point r.

In this paper, we have designed a new heuristic known as "GRP_CH" for generating a random simple polygon from a set of n points and this runs with time complexity $\mathcal{O}(n^3)$. Since the "2-Opt Move" heuristic[1] is the best known (referred in [1]) among the existing heuristics with time complexity $\mathcal{O}(n^4)$ to generate a simple polygon, we have compared our result with that of "2-Opt Move" heuristic. Our result shows that the randomness behaviour of "GRP_CH" heuristic is much better than that of "2-Opt Move" heuristic.

This paper is organized as follows: In Section 2, the existing heuristics are described. Section 3 deals with detailed description of our proposed heuristic "GRP_CH". In Setion 4 results are discussed. We conclude our paper in Section 5.

2   Literature Survey

From 1992, the generation of geometric objects became an interesting research topic to the researchers. Epstein[3] studied random generation of triangulation. Zhu[5] designed an algorithm to generate an x-monotone polygon on a given set of vertices uniformly at random. A heuristics for generating simple polygons was investigated by ORourke et. al.[2] in 1991. However, the vertices move while creating a polygon in their algorithm. The 2-Opt Moves heuristic was first proposed to solve the traveling salesman problem by J.van Leeuwen et. al.[4] In 1996, Thomas Auer et. al.[1] presented a study of all heuristics present at that time and reported a variety of comparison among them. There implementations are now part of RANDOMPOLYGONGENERATOR, RPG which is publicly available via http://www.cosy.sbg.ac.at/~held/projects/rpg/rpg.html. The existing heuristics for random polygon generating simple polygon are discussed below:

- Permute & Reject
  "Permute & Reject" has been designed by Thomas Auer et. al[1]. The algorithm creates a permutation of S and check whether this permutation corresponds to a simple polygon. If the polygon is simple then it is output; otherwise a new polygon is generated. Obviously, the actual running time of this method mainly depends on how many polygons need to be generated in order to encounter a simple polygon. Clearly, "Permute & Reject" produce all possible polygons with a uniform distribution, but this algorithm is not applicable to anything but extremely small set of points.
- Space partitioning
  To generate a random simple polygon Thomas Auer et. al.[1] designed "Space partitioning" algorithm which recursively partitions a set of point S into subsets and those subsets have disjoint convex hulls. The algorithm has been described in detail in [1]. In the worst case, this algorithm takes $\mathcal{O}(n^2)$ time. Unfortunately, Space Partitioning does not generate every possible polygon on S.

– Steady Growth

"Steady Growth" has also been designed by Thomas Auer et. al. [1]. As initialization, Steady Growth randomly select three points s1, s2, s3 from the set S such that no other point of S lies within convex hull, CH(s1, s2, s3). This convex hull is taken as a start polygon. In each of the following iteration steps a point s is chosen in such a way that by appending s to the polygon, it's convex hull again does not contain any further points of the point set. Then an edge (u,v) of the polygon that is completely visible from s is searched and replaced by the chain (u,s,v). This way the polygon is extended with the point s. By using "Steady Growth", one can compute a simple polygon in at most $\mathcal{O}(n^2)$ time. Unfortunately "Steady Growth" does not generate every possible polygon on S.

– 2-Opt Move

Although Zhu et al.[5] gave the idea of "2-Opt Move" first, it was designed by Thomas Auer et. al.[1] in 1992. This algorithm first generates a random permutation of S, which again is regarded as the initial polygon P. Any self intersections of P are removed by applying so called "2-Opt Move". Every "2-Opt move" replaces a pair of intersecting edges (v_i,v_{i+1}), (v_j,v_{j+1}) with the edges (v_{j+1},v_{i+1}) and (v_j,v_i) as shown in Fig1. In this application, at each iteration of the algorithm one pair of intersecting edge is chosen at random and the intersection is removed. Leeuwen et al.[4] has shown that

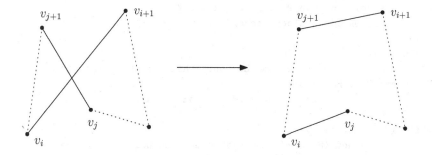

Fig. 1. An example of a 2-opt move

for obtaining a simple polygon, at most $\mathcal{O}(n^3)$ many "2-Opt move" required to be applied. Thus, an overall time complexity of $\mathcal{O}(n^4)$ can be achieved. The "2-Opt Move" heuristic will produce all possible polygons, but not with a uniform distribution[5]. However, this is reported as the best heuristic[1] among all the existing ones in the sense that it produces variety of different simple polygons from the point set S.

3   Proposed Heuristics

The proposed heuristic **GRP_CH** is a randomized algorithm which generates a random simple polygon from a set of n points lying on a 2-dimensional plane.

3.1   Assumption

The point set S lie on a plane in general position that means no three points or vertices are co-linear. The polygon is represented by clockwise orientation of its vertices.

3.2   GRP_CH Algorithm

The convex hull of the point set S is first constructed by using Graham's scan[7] algorithm. Therefore all the remaining points will lie inside the convex hull. This convex hull is taken as polygon P which will be modified later. Now randomly select a point v_i and find out how many edges of the polygon P are fully visible from that point v_i. Select any of those visible edges randomly, say $e(v_1, v_2)$. Therefore, each point as well as each visible edges has equal chance of being selected. Now connect the two ends (v_1, v_2) of that edge with the corresponding point v_i and delete that selected edge $e(v_1, v_2)$ of P. After this, the modified polygon will remain simple one. Repeat the same procedure for the remaining points.

```
Procedure GRP_CH
Input: A set S of points in 2-D Plane.
Output: A simple polygon P with vertex set V
        represented in clockwise order.
 Begin
  Q = CH (S);
      // Q is the vertices on the convex hull of the point set S
  P'= ConvexHull_Edges(S);
           //P' is set of edges of the convex hull
  S = S \ Q ;
  while (S is not empty) do
        v = Random (S);  // randomly select one point  from the set S
        E = FullyVisibleEdges(P', m, v);
               //m is no of vertices of P'
               // E is the set of edges fully visible from the point v
        if (E==NULL) then exit;
           // simple polygon is not possible in this case
        e = Random (E);
           // randomly select one edge from the set E
        P' = P' \ e;
        P' = P' U {v1, v} U {v, v2};
        S = S \ v;
  EndWhile
  P= P'
 End
```

```
FullyVisibleEdges (P, m, v)
Input: Polygon P with m vertices,
       a point v from which visibility is to be checked.
Output: the edge set E, visible from point v.
 Begin
   E=NULL;
   for i=1 to m
           if (the two end points of edge Ei are visible from v)
           // Edge Ei is visible from point v
           E=E U { ei }
           Endif
   Endfor
   Return E
 End
```

The Fig 2 describes the algorithm step by step graphically on a set of 8 points. The GRP_CH algorithm generates all possible polygons that can be generated from a given set of fixed points. There exists always a sequence in which our algorithm can construct any simple polygon(which is really possible to draw using those set of fixed points) from its convex hull.

3.3   Analysis of GRP_CH Algorithm

Computation of the convex hull [Graham's Scan algorithm] takes $\mathcal{O}(n \ log n)$ time. To find out whether the two lines intersect or not can be implemented in constant time. So, a particular edge which is visible or not from a point, can be checked in $\mathcal{O}(k)$ time where k is the convex hull edges. Since k$\leq$ n, this time will be $\mathcal{O}(n)$ and hence, the set of edges which are fully visible from a point can be computed in $\mathcal{O}(n^2)$ time. The while loop runs over $n - k$ times, hence overall GRP_CH algorithm takes $\mathcal{O}(n^3)$ time.

3.4   Limitation of GRP_CH Algorithm

There may arise a particular situation when there is no fully visible edge of polygon P for the randomly selected point p as shown in Fig.3. In such cases, our algorithm cannot generate a simple polygon. We stop and restart our algorithm. It is less likely that again the same points and the same edges will be selected in same order like before.

4   Result and Discussion

The "GRP_CH" algorithm has been implemented in C++ programming language. For random number generation rand() function has been used and this rand() function belong to standard C library. Our code handles input and output

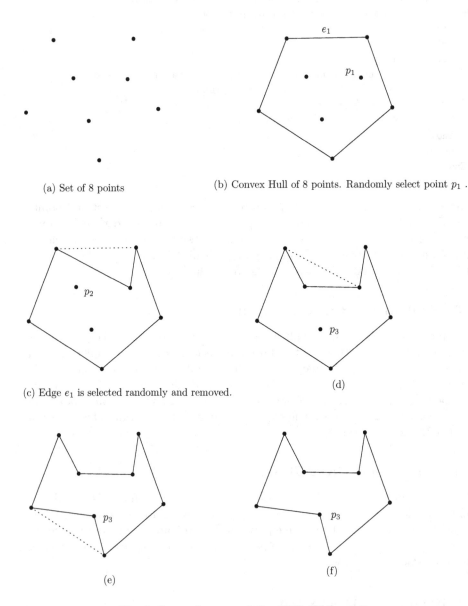

(a) Set of 8 points

(b) Convex Hull of 8 points. Randomly select point p_1 .

(c) Edge e_1 is selected randomly and removed.

(d)

(e)

(f)

Fig. 2. Successive step of the GRP_CH heuristic

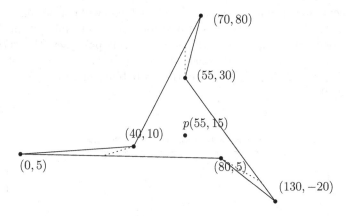

Fig. 3. The point p cannot see any edge of the polygon fully. In this instance, the GRP_CH heuristic fails.

in floating point format. Also, all numerical calculations are based on standard floating point arithmetic. We emphasize that we have not experienced any robustness problems in use of our code. All other software, hardware specification given below

Platform:
Operating System (Ubuntu), Ubuntu Release 10.04 (lucid), Kernel Linux 2.6.32-30-generic,GNOME 2.30.2

Hardware:
Memory : 2.0 GiB, Processor 0 : Intel(R) Core(TM)2 Duo CPU T5550 @ 1.83 GHz, Processor 1 : Intel(R) Core(TM)2 Duo CPU T5550 @ 1.83 GHz

Software:
GNU g++ 4.4.3, CGAL 3.7
Two different series of experiments has been performed:

1. **Probability of Success of GRP_CH Algorithm**
   The "GRP_CH" algorithm takes the random point set as input. The cardinality of point set in each simulation are 5, 10, 15, ..., 195, 200. Each simulation consists of 10,000 runs and the position of the point sets remains unaltered throughout a simulation. However, in next simulation new position of the point sets are taken. For each such simulation, a log file has been generated to count the number of simple polygon generated. It has been observed that the success rate to obtain a simple polygon is 100 percent when cardinalities of point set is not large. However the success rate of generation of random simple polygon drops slightly for large point set. This is due to the limitation of our algorithm (Section 3.4). The Fig 4 shows the probability of success of generating a random simple polygon from the point set S_i.
   The Fig 3 shows a set of seven points(along with their co-ordinates) from which a polygon consisting of six vertices has been generated using "GRP_CH"

heuristic; however in next step the simple polygon will not be generated since the next point to be selected lies at position (55,15) which is within the "non-visible-zone" of the generated polygon. For this point set, the "GRP_CH" heuristic has been executed 10,000 times out of which simple polygons are generated 99,472 times. Therefore, the probability of getting a simple polygon is 0.994.

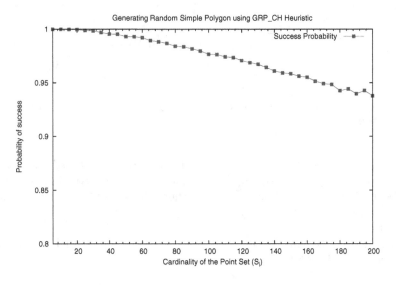

Fig. 4. The Probability of success for generating random simple polygon from a set of points lying on a 2-dimensional plane

2. **Unique polygons generated for given cardinality of the point set**
   We found out the number of unique polygons generated from the set of the all simple polygons generated on each simulation for the point sets of cardinalities 5, 6, 7,..., 12, 13. Here also, for each simulation which consists of 10,000 runs, the position of the point sets remains unaltered. The results of "GRP_CH" algorithm and "2-Opt Move" for the same set of input configuration has been shown in the Table 1. The Fig 5 shows graphically the average of the values of Table 1. For lower cardinality set, our results are comparable to "2-opt Move" algorithm while as the cardinality of the point set increases, "GRP_CH" produces more no. of unique polygons than "2-opt Move" heuristic. Therefore, we can infer from the result that the randomness behaviour of "GRP_CH" heuristic is much better than that of "2-Opt Move" heuristic, although each simulation consists of 10,000 run. However, "2-Opt Move" has probability of success to generate the simple polygon is always one.

Table 1. The number of unique polygons generated by "2-Opt Moves" and "GRP_CH" algorithm

Simulation Number	$|S| \longrightarrow$	5	6	7	8	9	10	11	12	13
1	2-Opt	4	12	17	57	83	135	167	1171	3569
	GRP_CH	4	12	17	57	83	136	179	1792	6125
2	2-Opt	8	5	19	60	251	341	1987	813	2065
	GRP_CH	8	5	19	60	254	359	2696	976	3277
3	2-Opt	4	5	17	57	105	888	484	2611	2022
	GRP_CH	4	5	17	57	105	980	534	4061	3558
4	2-Opt	4	13	19	67	210	440	552	2578	3615
	GRP_CH	4	12	17	57	83	136	179	1792	6125
5	2-Opt	4	13	17	72	34	590	1041	2450	2225
	GRP_CH	4	13	17	72	34	631	1358	3874	3341
6	2-Opt	4	5	19	61	95	357	1227	681	3197
	GRP_CH	4	5	19	61	95	366	1606	939	5911
7	2-Opt	4	12	19	26	30	127	1168	3913	2777
	GRP_CH	4	12	19	26	30	130	1507	6052	4777
8	2-Opt	4	12	6	27	83	312	562	1694	2128
	GRP_CH	4	12	6	27	83	326	611	2424	3353
9	2-Opt	8	13	19	61	216	372	499	2491	4614
	GRP_CH	8	13	19	61	217	405	557	4026	7076
10	2-Opt	8	12	46	163	102	343	2137	760	2042
	GRP_CH	8	12	46	163	102	361	2790	936	3536

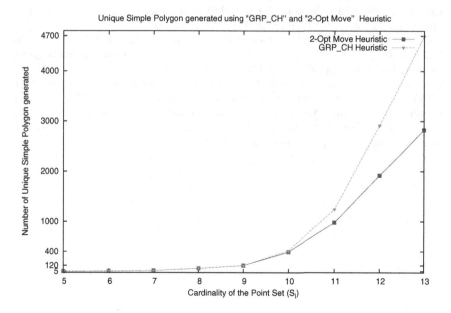

Fig. 5. The number of unique simple polygon generated in "GRP_CH" is larger than that of "2-Opt Move" when more number of point sets are taken as input

5   Conclusion and Future Work

We have presented a heuristic for generating random simple polygons. It has been shown that the probability of successful generation of a random simple polygon by "GRP_CH" is very good for point sets of cardinality below 100. However it tends to decrease as the cardinality of point sets increases. The experiment has been carried out also to find out the number of unique polygons generated on 10,000 runs for the given set of point and compared the results with existing "2-Opt Move" method. We have shown also that the number of unique simple polygons generated by "GRP_CH" algorithm are larger than that of "2-Opt Move" and hence the randomness behaviour of "GRP_CH" heuristic is much better than that of "2-Opt Move" heuristic.

From a theoretical point of view, it remains an open problem to generate polygons on a given set of points uniformly at random. Our heuristic has improved the present existing heuristics although it has a limitation and the removal of that limitation will be a future work.

References

1. Auer, T., Held, M.: Heuristics for the Generation of Random Polygons. In: Proc. 8th Canadian Conference Computational Geometry, Ottawa, Canada, pp. 38–44. Carleton University Press (1996)
2. O'Rouke, J., Virmani, M.: Generating Random Polygons. Technical Report 011, CS Dept., Smith College, Northampton, MA 01063 (1991)
3. Epstein, P., Sack, J.: Generating triangulation at random. ACM Transaction on Modeling and Computer Simulation 4(3), 267–278 (1994)
4. Leeuwen, J.V., Schoone, A.A.: Untangling a travelling salesman tour in the plane. In: Muhlbacher, J.R. (ed.) Proc. 7th Conference Graph-theoretic Concepts in Computer Science (WG 1981), pp. 87–98 (1982)
5. Zhu, C., Sundaram, G., Snoeyink, J., Mitchel, J.S.B.: Generating random polygons with given vertices. Comput. Geom. Theory and Application 6(5), 277–290 (1996)
6. Ghosh, S.K.: Visibility Algorithm in the plane. Cambridge University Press (2007)
7. Kreveld, M.V., Berg, M.D., Schwartskopf, O., Overmars, M.: Computational Geometry, Algorithm and Application. Springer (1996)

Author Index